Democratizing Risk Governance

Monica Gattinger
Editor

Democratizing Risk Governance

Bridging Science, Expertise, Deliberation and Public Values

palgrave
macmillan

Editor
Monica Gattinger
Institute for Science, Society
and Policy
University of Ottawa
Ottawa, ON, Canada

ISBN 978-3-031-24270-0 ISBN 978-3-031-24271-7 (eBook)
https://doi.org/10.1007/978-3-031-24271-7

This Palgrave Macmillan imprint is published by the registered company Springer Nature Switzerland AG
The registered company address is: Gewerbestrasse 11, 6330 Cham, Switzerland

Acknowledgements

This volume is based on the project *@Risk: How to Strengthen Risk Governance in Canada*. *@Risk* comprised a multidisciplinary research team of more than two dozen scholars and graduate students from eleven Canadian and US universities, along with half a dozen senior practitioners from five partner organizations: Canadian Nuclear Laboratories, the Canadian Nuclear Safety Commission, the Canadian Public Health Association, the Genetic Engineering and Society Center at North Carolina State University, and the Institute for Science, Society and Policy (ISSP) at the University of Ottawa. *@Risk* was funded by a Partnership Development Grant (File #890-2016-0035) from the Social Sciences and Humanities Research Council of Canada (co-funded by Genome Canada), along with contributions from the project partners and the University of Ottawa's Faculty of Social Sciences. We are grateful to all of the organizations who supported the project financially.

We are especially grateful to the practitioner members of the research team: Keith Dewar (Canadian Nuclear Safety Commission), Dr. Adrienne Ethier (at the time with Canadian Nuclear Laboratories), Renée Silke (Canadian Nuclear Laboratories), Dr. Bob Walker (retired government senior executive), and Dr. Frank Welsh (at the time with the Canadian Public Health Association). These unique 'pracademics' gave generously of their time, experience, and insights throughout the research study. Their contributions strengthened the framing, analyses, and policy relevance of the volume substantially and helped to ensure the research was grounded in the 'real worlds' of risk governance. Thanks are also owed to Dr. Brenda Kenny (Alberta Innovates), University of Ottawa professors Scott Findlay (Biology) and Dan Paré (Communication), and Professor Éric Montpetit (Université de Montréal), for their early contributions to the project.

As with any book, this volume is the product of many hands beyond the editor's. I have been fortunate to have outstanding research support for this project from beginning to end from staff at the Institute for Science, Society and Policy. I am indebted to Dr. Stewart Fast, ISSP Research Director (2015–2018), for his leadership assembling the research team, developing the project proposal, and shepherding the study in its early years. Without Stewart's support, this study would not have been possible. Likewise, I am immensely grateful to ISSP Research Director Dr. Marisa Beck (2018–2022) for coordinating the study and research team, for her participation in the research and writing of the newborn screening chapter, and, most especially, for her intellectual leadership on the motivated reasoning chapter. I would also like to thank ISSP postdoctoral fellow Dr. Patricia Larkin for stepping in to coordinate the project briefly in 2018, and for her participation in the project as an author of the chapter on carbon capture, utilization, and storage. A heartfelt thank you to Rafael Dezordi, ISSP Communications and Outreach Specialist (2017–2021), for his outstanding communications and logistical support for workshops and events. I will be forever indebted to Katherine Fafard, ISSP Head of Planning and Resource Management, for helping navigate the intricacies of financial administration for this project. Thanks are also owed to Geneviève Dunn, ISSP Communications and Outreach Specialist, for her communications support launching the book, and to Sandy Chung, ISSP Research and Communications Administrative Assistant for her invaluable assistance selecting a cover image for the volume. At Palgrave Macmillan, I would like to extend my appreciation to Rachael Ballard, Global Head and Publisher, Geography and Environment, for believing in the project from the outset, and to Redhu Ruthroyoni, for such able support preparing the manuscript for publication. Many thanks also to the book proposal and manuscript peer reviewers, whose useful comments strengthened the volume immensely. As with any academic publication, responsibility for the final product rests with the authors.

Contents

List of Contributors

Sana Abdelkarim Department of Public Administration & Policy, University at Albany, State University of New York, Albany, NY, USA

Rukhsana Ahmed Department of Communication, University at Albany, State University of New York, Albany, NY, USA

Jeanette Altarriba Department of Psychology, University at Albany, State University of New York, Albany, NY, USA

Mary C. Avery Department of Psychology, University at Albany, State University of New York, Albany, NY, USA

Marisa Beck Ottawa, ON, Canada

Stephen Bird Clarkson University, Potsdam, NY, USA

Duane Bratt Department of Economics, Justice, and Policy Studies, Mount Royal University, Calgary, AB, Canada

Gabriela Capurro Department of Community Health Sciences, University of Manitoba, Winnipeg, MB, Canada

Kim H. Chuong Department of Psychology, College of Social and Applied Human Sciences, University of Guelph, Guelph, ON, Canada

Elizabeth J. Cooper University of Regina, Regina, SK, Canada

Xavier Deschênes-Philion Ottawa, ON, Canada

Heather Douglas Department of Philosophy, Michigan State University, East Lansing, MI, USA

S. Michelle Driedger Department of Community Health Sciences, University of Manitoba, Winnipeg, MB, Canada

Brendan Frank Ottawa, ON, Canada

Monica Gattinger Institute for Science, Society and Policy, University of Ottawa, Ottawa, ON, Canada

Cindy Jardine Faculty of Health Sciences, University of the Fraser Valley, Chilliwack, BC, Canada

Simon J. Kiss Digital Media and Journalism, Wilfrid Laurier University, Brantford, ON, Canada

Jennifer Kuzma Genetic Engineering and Society Center, North Carolina State University, Raleigh, NC, USA

Patricia Larkin Institute for Science, Society and Policy, University of Ottawa, Ottawa, ON, Canada

Ryan Maier Department of Community Health Sciences, University of Manitoba, Winnipeg, MB, Canada

Sara Minaeian University of Oxford, Oxford, United Kingdom

Stuart G. Nicholls Clinical Epidemiology Program, Ottawa Hospital Research Institute, Ottawa, ON, Canada

Laura Nourallah Nepean, ON, Canada

Kieran C. O'Doherty Department of Psychology, College of Social and Applied Human Sciences, University of Guelph, Guelph, ON, Canada

Andrea M. L. Perrella Department of Political Science, Wilfrid Laurier University, Waterloo, ON, Canada

Dina Refki Department of Public Administration & Policy, University at Albany, State University of New York, Albany, NY, USA

Amanda Rotella Department of Psychology, Northumbria University, Newcastle Upon Tyne, UK

Erting Sa Department of Communication, University at Albany, State University of New York, Albany, NY, USA

Ketan Shankardass Department of Health Sciences, Wilfrid Laurier University, Waterloo, ON, Canada

Jordan Tustin School of Occupational and Public Health, Toronto Metropolitan University, Toronto, ON, Canada

Teshanee T. Williams School of Government, University of North Carolina, Chapel Hill, NC, USA

Gregor Wolbring Cummings School of Medicine, University of Calgary, Calgary, AB, Canada

List of Figures

List of Tables

CHAPTER 1

Risk Governance Dilemmas and Democratization: Public Trust, Risk Perception and Public Participation in Risk Decision-Making

Monica Gattinger

The SARS-CoV-2 global health pandemic has underscored that public trust in risk decision-making is crucial. Whether trust in the safety of vaccines, trust in the necessity of lockdown measures, or trust in the very existence of the pandemic itself, successfully addressing the crisis has hinged on public confidence in government decisions. The pandemic has also made visible how perceptions of risk can diverge among and between experts and the public. Public health specialists, epidemiologists, economists, sociologists, political scientists and psychologists have not always agreed on the necessity of things like school closures, travel restrictions or vaccine mandates. Public opinions about risk, for their part,

M. Gattinger (✉)
Institute for Science, Society and Policy,
University of Ottawa, Ottawa, ON, Canada
e-mail: mgatting@uottawa.ca

M. Gattinger (ed.), *Democratizing Risk Governance*,
https://doi.org/10.1007/978-3-031-24271-7_1

have been as varied as the values, ideologies and life experiences of the people holding them, and public perceptions of risk have been forever vulnerable to misinformation and disinformation. In this context, accurate and trusted risk communication has been pivotal. So has listening to citizens, communities and stakeholders, who expect their views to be taken into account in government decision-making. Political leaders who ignore the needs, concerns and interests of their electorates have quickly faced vocal opposition.

The centrality of public trust to effective risk governance, the fragmentation of perceptions of risk and growing expectations for public involvement in risk decision-making, all characterize risk governance in the twenty-first century. Scholars have written extensively about this for decades. And yet, for the most part, the 'real worlds' of risk governance have been failing to successfully address these imperatives. Why? Sometimes because risk practitioners don't see that things have changed. Other times because they view these developments as challenges to—if not outright inimical to—the evidence- and expertise-based underpinnings of risk assessment and management. In yet other cases, risk practitioners are keen to democratize risk decision-making but are uncertain how best to proceed. For others, democratizing risk governance gives them pause because it could reconfigure longstanding power relations in ways those holding power would prefer to avoid.

This chapter frames these challenges to reforming risk decision-making as risk governance dilemmas. Effective risk governance requires confronting differences in expert and public perceptions of risk successfully, engaging the public meaningfully and fostering public trust in decisions. And yet, all three objectives can challenge fundamental epistemological, cultural and ontological underpinnings of risk governance. Understanding the reasons why this is the case (and why not), carefully disentangling causes and effects, and providing case studies of real-world efforts to address the dilemmas, lays the groundwork for informed reform of risk governance arrangements. So does supporting decision-makers to deepen their understanding of contemporary imperatives, to reframe their thinking about public perceptions of risk, to revise their interpretation of what constitutes 'evidence', to identify and leverage opportunities to learn and to deftly navigate new tensions and trade-offs.

This volume aims to contribute to that process. Part provocation, part evaluation, part handbook, part call to action, it aims to advance scholarly and empirical understandings of public participation in risk decision-making, of ways to conceptualize and address differences in public and expert perceptions of risk, and means to foster public trust in risk governance. Risk scholars and practitioners alike are grappling with how best to govern risk in the face of growing calls and rationales for democratization. There is much to be learned about the strengths—and limitations—of opening risk decision-making processes to public participation. This volume shares the fruits of @Risk, a multiyear research project convening some thirty established and emerging scholars and risk practitioners from a wide variety of disciplines and sectors in Canada and the United States. What bound the team together was a shared interest in examining the benefits and drawbacks of public participation in risk governance, in better understanding what drives differences in public and expert perceptions of risk (and what to do about them), and in strengthening risk governance. The book explores risk perception through the lens of 'motivated reasoning' with a focus on government decision-making for emerging and established technologies. As the chapters attest, research team members did not always agree about the desirability or benefits of 'opening up' risk decision-making to citizens, nor did they always agree about what drives differences in public and expert perceptions of risk—nor, crucially, what to do about them. They likewise did not always agree about how best to strengthen risk governance, including whether to democratize it at all. These debates are themselves a contribution: the volume does not intend to be the final word on the issues, but rather, a tool to promote scholarly and practitioner reflection, debate and action, and to inform risk decision-making. On the need for such a tool, team members were united. Failure to reform risk decision-making to proactively and constructively address democratization imperatives will compromise the effectiveness and legitimacy of risk governance now and in the future.

Overview of Volume

The volume comprises two parts. The first focuses on cross-cutting concepts and issues explored throughout the volume. The second presents case studies analysing public participation in a variety of risk decisions in the fields of energy, public health and genomics, and includes a special section dedicated to COVID-19.

The three chapters in Part I examine the concept of motivated reasoning, the interplay and complex relationships among science, values and citizens, and the development of tools to identify inequities in risk decision-making arrangements (see chapter abstracts for descriptions of each text). The chapters problematize the rational, objective, fact-based underpinnings of expert and scientific decision-making for risk, to uncover the values, hierarchies and worldviews that inform all perspectives and decisions. The authors identify theoretical and empirical implications of their analyses and provide readers with concrete recommendations for action. In Chapter 2, Marisa Beck and collaborators highlight the importance of integrating insights from research on motivated reasoning into risk governance. Chapter 3 is a reprint of an open access Descartes Lecture delivered by Heather Douglas in which she explores the complex entanglement of science and values. In Chapter 4, Gregor Wolbring presents the BIAS FREE Framework, a tool for risk practitioners and science and technology educators to help surface inequities and biases in risk decision-making.

Parts II–VI features contributions from a wide variety of authors, disciplines, theoretical perspectives and methodological approaches. What binds them together is their analytical focus on the extent and nature of public participation in risk decision-making (risk assessment, management and communication) and their efforts to identify challenges, opportunities and insights from their cases for risk governance. In conducting the research, authors were asked to explore various dimensions of public participation:

- Transparency: How readily can stakeholders access information about risk-related decision processes and outcomes?
- Inclusiveness and representativeness: Do those who are impacted or concerned by risk issues have formal opportunities to make their voices heard (inclusiveness)? Is a broad range of stakeholders involved, including marginalized social groups, and do participants adequately reflect potentially affected or concerned populations (representativeness)?
- Deliberative quality: How 'deliberative' are participation processes? Do participants have an opportunity to engage in dialogue and exchange with one another or are the processes unilateral or unidirectional? Are voices heard and seriously considered in decision-making?
- Accountability: What are the accountability arrangements for risk decision-makers? Are elected officials accountable to the public

through the legislature? How are unelected officials (e.g., public servants) or institutions (e.g., independent regulatory agencies) held accountable?

Some contributors examined all four dimensions, while others focused on a narrower set of considerations. Chapters also varied in the participation mechanisms under investigation, with some examining public engagement processes undertaken by governments or legislatures, others examining the role of referenda and public deliberations, while still others explored shared decision-making arrangements. Researchers drew on a diversity of theoretical approaches to frame their analyses, including responsible innovation, the Advocacy Coalition Framework, risk governance, motivated reasoning, cultural worldviews and public deliberation. Others anchored their analyses in the applied literature in their field (public health, genomics or energy). The cases represent a mix of federal, provincial and local levels of government, with some chapters examining multiple jurisdictions and others including international comparisons. Almost all cases focus on Canada and examine various phases of risk decision-making, from policy development through to regulation and decision-making within existing policy frameworks. The cases include a mix of established and emerging technologies.

Part II features three case studies in the energy sector. In Chapter 5, Duane Bratt, Patricia Larkin and Xavier Deschênes-Philion examine the Nuclear Waste Management Organization's site selection process for the permanent deep geological repository to store Canada's high-level nuclear waste. The process turns conventional infrastructure siting on its head by committing to the requirement that a site will only proceed with a 'willing host community'. In Chapter 6, Laura Nourallah explores the interplay among risk perception, trust and public engagement for policies respecting hydraulic fracturing for natural gas in New Brunswick, Canada. The analysis highlights the ways in which trust in public institutions shapes citizen learning, perceptions of risk and policy preferences in the context of scientific uncertainty over the environmental and health impacts of an emerging technology. Patricia Larkin, Monica Gattinger and Stephen Bird analyse public confidence in risk decision-making for carbon capture, utilization and storage (CCUS) at large industrial sites in Chapter 7. The research finds that a wide variety of actions are needed to foster confidence in CCUS and the regulatory frameworks guiding its use.

Part III includes two chapters exploring cases in the field of genomics. In Chapter 8, Jennifer Kuzma and Teshanee Williams examine how values and science shape risk governance for food derived from genetically engineered animals. Using the case of parliamentary hearings over genetically engineered salmon, they argue that Canadian risk governance should provide more opportunities for deliberative public engagement and incorporate broader 'science-plus' perspectives into decision-making. Chapter 9 focuses on newborn bloodspot screening. Through an analysis of decision-making processes to add new tests to screening panels in a number of provinces, Marisa Beck, Brendan Frank, Sara Minaeian and Stuart Nicholls find that there are relatively few opportunities for public participation. They identify how greater participation could strengthen decision-making in the future.

Part IV, chapters 10–12, explore cases in public health. In Chapter 10, Michelle Driedger, Elizabeth J. Cooper and Ryan Maier reveal how cancer screening guidelines for mammography and prostate cancer tests can be in tension with models of shared decision-making between doctor and patient. While shared decision-making can foster patient trust in the healthcare system, it may result in calls for unnecessary tests, particularly in the context of shifting guidelines and aggregated evidence that doesn't align with individual doctor and patient values and preferences. Chapter 11 examines how public deliberation processes can help inform childhood vaccination policies. Kim Chuong, Amanda Rotella, Elizabeth J. Cooper and Kieran C. O'Doherty report on the Ontario Vaccine Deliberation, a first of its kind academic-led public deliberation on childhood vaccination, in which citizens deliberated and made recommendations on mandatory vaccination, nonmedical exemptions and compensation for serious adverse events after immunization. The study's findings and recommendations are especially salient in the context of COVID-19. To this end, the chapter includes an appendix that examines the application of democratization processes and public deliberation to the pandemic. In Chapter 12, Andrea M. L. Perrella, Simon J. Kiss and Ketan Shankardass examine controversies over water fluoridation that have led multiple municipalities to discontinue this longstanding and safe public health intervention following local referenda. Using an experimental survey administered in Canada and the United States, they test how narratives based on scientific fact versus those based on normative concerns can influence peoples' views of fluoridation. The findings suggest that it is easier to foster doubt in the safety of fluoridation than it is to foster support for the practice.

Part V, the final two case study chapters, focus on vaccine decision-making for COVID-19. In Chapter 13, Rukhsana Ahmed, Dina Refki, Jeanette Altarriba, Erting Sa, Mary Avery and Sana Abdelkarim study the influence of various information sources on the immunization decisions of members of four culturally and linguistically diverse communities in the United States. The authors find a positive relationship between the ability to find trusted information and the decision to get immunized. The chapter identifies where culturally and linguistically diverse communities turn for trusted information and recommends targeting these sources to foster greater vaccine uptake. Chapter 14 focuses on public participation in decision-making for vaccine priority groups. Michelle Driedger, Gabriela Capurro, Cindy Jardine and Jordan Tustin share insights from a number of focus groups organized to seek input from the general public about who should be prioritized to receive the COVID-19 vaccine. The focus groups underscored that citizens do not always consider the full range of ethical considerations raised by public health decisions, that people who participate in consultation processes are often privileged, and that their views aren't necessarily representative of the broader population.

Before diving into the chapters, the following synthesizes key dilemmas of risk governance that run through the contributions to this volume. This book does not purport to resolve these dilemmas—these tensions will continue to play out in scholarly and practitioner communities—but rather, to shed light on their existence and to help readers, whether students, scholars or practitioners, to further their understanding and ability to meaningfully grapple with and make considered decisions about them in their research, learning and decision-making.

Risk Governance Dilemmas: Public Trust, Risk Perception and Public Participation

The classic definition of 'risk' comprises the existence of some hazard to human or environmental health and the probability of that hazard having a detrimental impact. A convenient example is water, which is both essential to human and environmental health, but also a hazard if ingested in excessive quantities and lethal in the case of drowning. This points to the importance of 'dosing' in assessing risks, along with accurately evaluating the severity of consequences of a hazard for human and environmental health. It also points to the importance of risk mitigation, as risks can be mitigated through various means (e.g., roping off deep water at public beaches to reduce the risk of drowning).

Science is fundamental to risk assessment and management. Understandings of hazards, dosing, probabilities, consequences and mitigation options are anchored mainly in the natural and health sciences and engineering. Crucially, uncertainty is inherent in all risk assessments, whether in terms of the probability of an occurrence, its consequences or the effectiveness of risk mitigation options. Not surprisingly, then, debates over risk assessment and management often centre on uncertainty, especially when the science is incomplete or lacks a strong consensus within the scientific community. Debates also centre on benefit and opportunity as 'risks' are evaluated in pursuit of a benefit of some sort—whether to public health (e.g., risks of adverse health consequences of a vaccine that protects against deadly disease), the economy (e.g., risks of adverse environmental impacts of an energy project that will provide long-term jobs, energy security and economic growth) or the environment (e.g., risks of adverse local environmental impacts of a renewable energy project that will lower global greenhouse gas emissions).

Societal decisions about risk are inherently complex and involve myriad decision-making frameworks, institutions, actors and judgements, often captured by the term 'risk governance' (Renn 2008). The locus of risk governance is frequently centred in specialized policy or regulatory agencies that undertake, oversee or make decisions about multiple aspects of risk. This includes risk assessment, which involves understanding the risk context, estimating the likelihood and severity of the consequences of hazards (determining their risk) and identifying risk mitigation options that could reduce potential adverse outcomes to health, the environment, the economy and the like (Krewski et al. 2007). Risk management, which generally follows risk assessment, involves decision-makers selecting a risk mitigation option. Options can be categorized in a variety of ways, including with the REACT framework, an acronym that refers to regulatory, economic, advisory, community-based and technological options, or some combination thereof (ibid.; this framework is used by a number of authors in this volume). Risk communication, for its part, involves communicating risks and risk decisions (the control options selected) to the public or other stakeholders. Compliance, enforcement and monitoring comprise ongoing assessment to identify intended and unintended outcomes (e.g., the effectiveness of risk mitigation strategies or the occurrence of adverse outcomes) with a view to altering mitigation tools and changing course when necessary. In theory, these aspects of risk governance proceed sequentially, but in practice, they often take place simultaneously (e.g., during a crisis or emergency situation as has been the case with the COVID-19 pandemic).

This volume is concerned predominantly with decision-making involving the first three aspects—risk assessment, risk management and risk communication—although compliance, enforcement and monitoring are touched on in a number of the chapters. Of central concern are public and expert perceptions of risk throughout these phases, along with the level and nature of public engagement in decision-making processes. Crucially, the book's chapters highlight various elements of three risk governance dilemmas with which scholars and risk decision-makers are grappling—and for which risk decision-makers must make informed judgements and decisions.

Dilemma 1—Public Trust in Risk Governance: Holy Grail or Holy Terror?

Public trust in government, industry and the media has declined across Western industrialized democracies in the postwar period, and citizens' deference to authority of various sorts (elite, government, industry, medical, etc.) has declined over the decades (Nevitte 1996, 2011). In an era of "fake news" and social media, the 2017 Edelman Trust Barometer declared "trust is in crisis around the world" (Edelman 2017). In 2018, the media emerged as the least trusted institution globally (Edelman 2018). For more than half of respondents, lower trust in the media "led to an inability to identify the truth" (59%) and an inability to trust "government leaders" (56%) (ibid.). While the 2018 Barometer showed a "revival" of trust in experts, the 2019 Barometer stated that "trust has changed profoundly in the past year", and identified that people had "shifted their trust to the relationships within their control", notably to their employers (Edelman 2019). Importantly, the 2019 survey revealed a "trust gap" between the informed public and the mass population, with the former being more trusting than the latter, where nearly half believed "the system is failing them" (ibid.). This division reflects a broader trend towards a more fractured and polarized political environment both globally and within countries or regions (Nanos 2018). Recent interest in understanding the role of group identity, tribalism, the "politics of resentment" and populism in supporting or undermining liberal democracy are emblematic of these tendencies (Chua 2018; Fukuyama 2018). So is the mounting attention to the capacity for social media to foster the rapid spread of misinformation and disinformation, and to create so-called echo chambers or 'identity bubbles' of people rarely exposed to positions contrary to their own, all of which challenge democracy (see, e.g., Iosifidis and Nicoli 2020; Persily and Tucker 2020) and, concomitantly, the practice of risk governance.[1]

This context raises vexing challenges for risk decision-makers.

On the one hand, it positions trust as the 'holy grail' of effective risk governance. Risk scholars have been beating this drum for decades, pointing to the importance of citizen trust to the effectiveness, legitimacy and support for risk decision-making (see, e.g., Renn and Levine 1991, Stern and Fineberg 1996). This can be seen readily across various risk governance contexts. In public health, for example, researchers contended over a decade ago that the biggest threat to the Canadian healthcare system was a loss of public confidence (Chafe et al. 2011). This has become increasingly apparent when it comes to vaccination (the subject of Chapters 11, 13 and 14). In the context of COVID-19, research on vaccine uptake finds that trust predicts uptake (Latkin et al. 2021). But peeling the onion on trust quickly reveals the complexities of fostering public confidence in politically fragmented societies. Multiple factors shape trust, and, as the Edelman findings above suggest, it can be fickle: who or what people trust, why they trust and with what level of commitment, is in a constant state of flux. Research on vaccine hesitancy, for example, contends that hesitancy should be understood as public mistrust in scientific institutions and experts, including scientists' ties to pharmaceutical companies and the values that underpin scientific consensus (Goldenberg 2016; see Chapter 11). Meanwhile, research on COVID-19 vaccines has found that mistrust is rooted in a variety of issues, including concerns about the safety, effectiveness and need for the vaccine, and the speed with which vaccine trials were conducted (Latkin et al. 2021; see Chapter 14). Layered onto these drivers of trust are tendencies

[1] While the impact of social media, disinformation and misinformation are not the focus of this volume, they figure in a number of the chapters and are areas requiring much further research to advance understanding of their impact on risk governance and risk decision-making.

towards trust and mistrust among different sociodemographic communities: mistrust in COVID-19 vaccines has been found to be higher in ethnocultural communities with an existing distrust in government institutions and the healthcare system (Thomas et al. 2021; see Chapter 13). This complexity leads to uncertainty over what, specifically, will enhance trust in vaccines. As discussed in Chapter 11, the Canadian Medical Association voted down a motion to create a national vaccine injury compensation programme because of concerns that it would amount to admitting the potential harms of vaccines and would reduce public trust and willingness to get vaccinated (Browne 2016). Other experts, meanwhile, have contended that compensation for vaccine injuries would enhance trust and vaccine rates (Keelan and Wilson 2011; Law Reform Commission of Saskatchewan 2009; Manitoba Law Reform Commission 2000; as noted in Chapter 11, Health Canada instituted a no-fault compensation programme for injuries sustained as a result of receiving an authorized vaccine). In these contexts, a comprehensive understanding of what drives trust and perceptions of risk is crucial—both within and between expert and lay communities. So is understanding the best tools and instruments to cultivate trust. Any choice of instrument can have intended and unintended consequences. Compulsory childhood vaccination, for example, may lead to higher vaccination rates, but, as Navin and Largent have noted, could also result in parents taking their children out of public schools or day care, and could drive political polarization (2017; see Chapter 11).

On the other hand, putting public trust at the centre of risk governance can raise concerns that risk decisions become hostage to uninformed public views with a weak or inaccurate grasp of the scientific evidence—or no concern for the science at all. Through this lens, public trust becomes 'holy terror' for risk governance, with governments making risk decisions on the basis of short-term political imperatives rather than a careful analysis of risks and benefits over the long term. In this scenario, people might trust decisions in the short term, but at what cost to society over time? Chapter 12 unpacks this at the municipal level, where the longstanding and safe practice of water fluoridation has been rolled back in multiple communities through effective citizen mobilization against it. Chapter 11 explores a weaker version of these concerns in the case of cancer screening decisions by primary care physicians, where shared decision-making between doctor and patient can lead to unnecessary screening tests. Concerns that people give more weight to evidence from sources they trust, regardless of their knowledge or expertise (close friends, social media campaigns, celebrities or NGOs), are emblematic of public trust as 'holy terror'. So are concerns that people mobilize against science or mobilize doubt or uncertainty in ways that counter scientific evidence and evidence-based decision-making.

Public trust raises multiple questions for risk decision-makers. How to strengthen public trust in decisions when trust itself can shift rapidly and vary markedly across sociocultural groups? How to foster trust in an era of political fragmentation, social media and misinformation, when government institutions as a whole may be mistrusted? How to take decisions informed by robust scientific evidence when short-term political imperatives pull in the opposite direction? How to establish a basis for constructive dialogue and risk decision-making in the face of polarized debates in which perceptions of risk are diametrically opposed? How to build trust in risk decisions when trust in decision-making processes may be just as crucial—even rival—trust in the substance of decisions? Answers to these questions require considered and informed judgement on the part of risk decision-makers. And the questions foreshadow the other two risk dilemmas highlighted by the contributions in this volume: risk perception and public participation.

Dilemma 2—Risk Perception and Motivated Reasoning: Valuing Values or Devaluing Science?

It is no surprise that people can perceive the same risks differently. Risk controversies often revolve around differing perceptions of risk among members of the general public, among experts or between the public and experts. Risk governance scholarship has highlighted and documented this for years (see, e.g., Renn 2008; Jardine et al. 2009; Krewski et al. 2012; Webler and Tuler 2018).

Why do perceptions differ? The go to answer for many scientific and risk practitioners is the 'knowledge deficit model', i.e., that perceptions differ because people lack understanding of the science underlying a given risk. Naturally, this puts the emphasis on 'educating' the public with scientific 'facts' to persuade people of the 'correct' assessment of the risk. But the knowledge deficit model flies in the face of decades of scholarly research illustrating that the reasons risk perceptions differ are multiple, multifaceted and complex. Yes, in some instances, it is lack of understanding. But in many others, differences can be driven by psychology-based factors (Slovic 1987), cultural factors (Douglas and Wildavsky 1982) and social processes (Kasperson et al. 1988; Pidgeon et al. 2003). Recent years have seen a surge in scholarly attention to the concept of 'motivated reasoning' (Druckman and Bolsen 2011, Kahan 2012; Lachapelle et al. 2014), which draws on cultural theories of risk perception to analyse how people process information in their assessment of risk. This strand of research examines the ways in which various social and psychological motivations—group identity and belonging, consistency with individual values, worldviews, prior beliefs or attitudes—shape peoples' assessment of evidence about risk and their corresponding perceptions of it. As discussed in Chapter 2, motivations affect reasoning in a number of ways, including biased memory search, selective use of inferential rules and biased evaluation of scientific evidence (Kunda 1990).

One of the more prevalent approaches to studying motivated reasoning is the cultural cognition thesis, which contends that people process information in ways that are 'identity-protective' and consistent with their cultural worldviews (Kahan 2012; see Chapter 2). The thesis draws on the cultural theory of risk (Douglas and Wildavsky 1982) with a fourfold typology of worldviews—egalitarian, hierarchical, individualistic and fatalistic—based on the extent to which people believe individual behaviour should be regulated and the extent to which society acts as a collective. Through survey research in the United States, Kahan and collaborators have shown that cultural worldviews associate with different opinions on everything from climate change and nuclear power, to abortion, nanotechnology and the Zika virus (Kahan et al. 2007, 2009, 2010, 2011, 2017). Further, survey experiments have shown that differences in cultural orientation are associated with different levels of scepticism over information presented as fact, with biased assimilation of information, and with different assessments of policy positions depending on the cultural identity of the person advancing the position (Kahan et al. 2007). Studies in Canada have shown similar results for risk perception and views of expert credibility (Lachapelle et al. 2014; Perrella and Kiss 2015), but they have revealed that cultural worldview survey questions must be adapted for use in the Canadian context (Kiss et al. 2016).

Research on motivated reasoning underscores that there are important limitations to a narrow science-based understanding of the knowledge deficit model of risk communication (i.e., that more education and knowledge about an issue and the science underpinning it will lead to shared perceptions of risk). Instead, there is evidence to suggest the opposite: those with higher levels of education and more knowledge about an issue may in fact have greater tendencies to reason in a motivated fashion (Hochschild and Sen 2015; Kahan et al. 2012). Instead of the knowledge deficit model, studies in this vein call for more attention to information framing techniques in communicating risk (see Druckman and Bolsen 2011, Hart and Nisbet 2011, and Chapter 12 in this volume), along with sensitivity to things like multicultural differences in trusted information sources and channels (Ahmed 2015, and Chapter 14 in this volume) and empowering people to make informed decisions about personal and public risks (Driedger 2008, Greenberg 2013, and Chapter 10 in this volume).

Importantly, analysing risk perception through the lens of motivated reasoning emphasizes the pivotal role of values in assessing and managing risk, and helps to identify the values and value conflicts implicit in or obscured from view by traditional science-based assessments. It also points to the values anchoring science and scientific research, values that may or may not align with those of the general public (see Chapter 3 in this volume) and lays bare that experts themselves reason in a motivated fashion. Studies of the US nuclear sector have shown tendencies towards motivated reasoning in both scientists and the public whereby expected benefits of the technology shape policy preferences at least as much as risks (Jenkins-Smith et al. 2009). Studies of risk experts in the UK nuclear and genomics sectors have likewise shown that experts advance evidence and analysis in an unconsciously motivated fashion (Wynne 1992, 2006). In contrast to these studies, a survey of soil remediation experts in Québec found that more knowledgeable experts displayed less value-based heuristics and less tendencies towards motivated reasoning (Montpetit and Lachapelle 2016), although motivated reasoning was still in evidence among the experts who participated in the study (ibid.).

The foregoing raises a host of complex questions for risk practitioners, chief among them whether these insights about risk perception and motivated reasoning *open the door* to valuing values in risk governance, or whether they constitute a *trap door* to devaluing science in risk decision-making. On the former, valuing values in risk governance helps to advance understanding of risk controversies and differences in risk perception. These insights can help strengthen risk decisions and public confidence in them. Research on public attitudes towards genetically engineered animals, for example, shows that values and attitudes can shape views, with ethical concerns extending beyond scientific safety to include whether or not the technology should even be developed in the first place (Thompson 1997, see Chapter 8). Risk governance systems designed for and anchored in safe deployment of a technology will be blind to these concerns. But even when safety is the primary concern of risk decision-making, teasing out differences in values can help to explain differences in risk perception and risk mitigation preferences. In primary health care, for example, peoples' political, economic, cultural and personal values all inform their weighting of risks and benefits—and this applies to both patients and specialists (Atkins et al. 2005, see Chapter 10). Diving deeper, motivated reasoning can lead to differences in 'inductive risk reasoning' and thus different standards of proof to judge whether the risks of something (e.g., vaccines, fracking, etc.) are being mitigated appropriately by government and stakeholders (Douglas 2016; Druckman 2012). In this case, motivated reasoning leads not so much to disagreement on the existence of hazards or the desirability of a particular technology, but rather, to differences in the threshold of tolerance for exposure. By way of example, many parents think about the personal needs of their children when it comes to what constitutes a reasonable risk for vaccines, rather than weighing the risks of serious adverse reactions against the benefits of population-level immunization (Goldenberg 2016, see Chapter 11).

On the other hand, emphasizing values, subjectivity and the role of motivated reasoning in risk governance can be perceived as a trap door to removing the scientific ('evidence-based') foundations of risk decision-making—if not the very foundations of science itself. Laying bare that science, scientific evidence and scientists themselves are value-laden can be profoundly destabilizing to the positivist ontological and epistemological foundations of mainstream scientific research, training and practice. The separation of (objective) 'facts' and (subjective) 'values', the independence of the observer from the observed and the existence of universal truths are all tenets challenged by understandings of risk perception that foreground values, subjectivity and motivated reasoning—and not just of the public, but of experts themselves. For many risk practitioners, this raises the spectre that all forms of evidence—from scientific evidence to opinions to religious or other belief systems—will be perceived as equal and deserving of equal weighting in risk decisions. Through this lens, valuing values risks stepping on a trap door to a world that devalues science.

This undesirable prospect is already evident in 'anti-science movements' against vaccination, climate change, fluoridation and most recently in the context of COVID-19, masking (see Chapter 12). In these movements, people are listening to each other—not to science (Camargo and Grant 2015, see Chapter 12). The ideological and partisan connection to some of these tendencies is particularly troubling: scepticism about science and public health measures often comes from those on the right or parties on the right (Kirst et al 2017, see Chapter 12). Scientific literacy can be low among the general public (Durant et al 1989), which leaves people vulnerable to not understanding risk issues. As explored in Chapter 10, this creates dilemmas between following evidence-based recommendations for cancer screening and pursuing shared decision-making between patient and doctor: patient preferences can lead to unnecessary tests. Opening the door to values likewise raises the prospect of reconfiguring power relations in ways that empower marginalized views or groups, something that may or may not sit well with political, public or social power holders. Research in the field of biotechnology, for example, has documented the marginalization of views that are not 'science based' in debates (Thompson 2007; Meghani and Kuzma 2011; see also Chapter 8). Scholarship has likewise documented how the perspectives, values and interests of people with disabilities, women and people of colour have been marginalized in risk decision-making for scientific and technological advancements (see Chapter 4).

One response to concerns about devaluing science is to call for enhancing scientific literacy to strengthen public understanding of the principles of scientific inquiry and methodology, with a particular focus on how science addresses uncertainty. This is akin to the knowledge deficit model, but in this case applied to science as a means of inquiry, rather than to the science underpinning a particular risk context. Doubling down on science may be a necessary but insufficient condition. Research on inequity and social hierarchy unmasks the power relations embedded in prevailing risk governance ontologies and epistemologies (see Chapter 4). It also makes explicit the social, political and philosophical complexities of reforming contemporary risk governance. There are no easy ways to address or navigate this dilemma. Further scholarly research is needed, and risk decision-makers will need to explicitly confront these tensions. Tools like the BIAS FREE Framework presented in Chapter 4 can be a useful place to start.

Dilemma 3—Public Participation in Risk Decision-Making: Stronger or Weaker Decisions?

Since at least the 1990s, governments have become increasingly attentive to involving the general public in decision-making processes. A rise in research on topics like public consultation, citizen engagement, deliberative policymaking, citizen juries and co-development evidences the trend (see, e.g., Fischer 2003; Phillips with Orsini 2002, Fishkin 2011), as does the development of frameworks for categorizing the multiple approaches, tools and techniques for engaging the public (e.g., the public participation spectrum developed by the International Association for Public Participation [IAP2 2022]).

As might be expected, risk governance is no stranger to these tendencies, but given that risk decision-making is anchored in technical, scientific and expert analyses, movements towards public participation raise many questions. Will involving the public lead to stronger or weaker risk decisions? Will it compromise evidence-based decision-making for risk or help to inform it? Will it amplify risk controversies or help to resolve them? Will it politicize risk decision-making or foster democratic accountability? Will it build or erode public trust in risk decisions?

Most scholarship in the field assesses public involvement positively, pointing to numerous aspects of participation that strengthen risk decisions. Since at least the 1990s, scholars have noted the important role of citizens in risk communication and management (Fiorino 1990; Slovic 1993). They have likewise noted that expert perceptions of risk may differ from those of the public and have highlighted the need for public involvement in risk decision-making (Renn 2008; Jardine et al. 2009; Webler and Tuler 2018). The rationales evoked for public involvement are multiple and relate to both the procedural and substantive legitimacy of decisions. Public participation is noted as fundamental to democratic accountability (Nicholls et al. 2016) and for its capacity to strengthen democratic legitimacy for and trust in government decisions about controversial technologies (Renn and Levine 1991). It may also decrease conflict and foster trust in or acceptance of decisions (Stern and Fineberg 1996). This has been documented in the energy sector, for example, where empirical research highlights that when decision-making processes are perceived as open, transparent and unbiased, people are more likely to support decision outcomes—even if they don't agree with them (Cleland et al 2016, Simard 2018, see Chapters 5–7). Engagement is also a means to foster confidence in broader policy systems. As noted above, in the Canadian healthcare system, Chafe, Levinson and Hébert asserted in 2011 that losing public confidence was the largest threat to the system; they argued that greater public engagement would help to restore confidence (see Chapter 11).

Over time, understandings of public participation have evolved from the deficit model of informing people about the science underpinning decisions, to recognizing and valuing the knowledge that citizens bring to the risk decision-making table (Petts and Brooks 2006). This includes forms of knowledge beyond the scientific, including life experiences, and local and cultural knowledge (Coburn 2003; Wynne 2007). In this view, *everyone* learns in public participation processes—citizens ***and*** experts—and decision outcomes are stronger as a result. In health service delivery and health research, for example, public engagement has been found to improve service quality, responsiveness, legitimacy, accountability and transparency (Esmail et al 2015; Kovacs Burns et al. 2014; Manafo et al 2018, see Chapter 11). Moreover, as alluded to in the discussion above on motivated reasoning, engagement processes may help to reveal the value judgements rooted in expert assessments (Kuzma 2016). This opens the door to learning about public values, perspectives and concerns that should be taken into account in risk decision-making. In this view, differences in public and expert perceptions of risk are not a 'problem' to be solved, but rather, an invitation to learn more and better understand.

However, as the chapters in this volume reveal, despite decades of scholarship, risk decision-making does not always include opportunities for public participation, or, if it does, does not always do it well. Why?

As Beck et al. note in Chapter 9, "The inclusion of public(s) in policy decision-making challenges traditional notions about science and politics that underlie models of evidence-based decision-making. First, it problematizes the notion that science and politics—or facts and values—are separate and need to stay separate. Second, it undercuts the position that effective decision-making about risk should rely on scientific and expert knowledge alone". This can be a tough pill for risk practitioners and institutions to swallow, and can lead to government aversion to public participation, or to begrudgingly engaging the public as a box-ticking exercise that lacks authenticity.

But as a number of the chapters in this volume highlight, reservations about public participation go beyond mere heel-dragging and include situations in which involving the public can lead to weaker decisions.

Chapters 10 (cancer screening), 12 (water fluoridation) and 14 (COVID-19 vaccine priority groups) explore these issues. In primary health care, the tool of shared decision-making (SDM) between physician and patient may, on the one hand, foster trust in the healthcare system, but, on the other, may lead to demands for cancer screening tests that are unnecessary. In decision-making for municipal water fluoridation, citizens with an insufficient grasp of the science of fluoridation may be vulnerable to developing inaccurate and false perspectives. Opponents of fluoridation can capitalize on this with information framing techniques that raise doubt and cultivate risk aversion. When voting in a referendum, people may prefer to err on the side of caution and vote against the practice. And if an election is fought on a topic, emotional and psychological factors have greater currency. Peoples' views (and doubts) can be shaped more by images, framing and narratives than by evidence or engagement with technical information. The Internet and social media only heighten these tendencies.

Broad public participation creates a platform for misinformation, disinformation and marginal views. Participation in engagement processes is often self-selective, amplifying the voices of some groups—often those with greater privilege—to the detriment of others (see Chapter 14). Moreover, citizens have a multiplicity of concerns and often hold contradictory opinions simultaneously. The chapter on vaccine priority groups reveals that some people who were unwilling to get vaccinated due to safety concerns found it acceptable to prioritize other groups (vulnerable people, health practitioners) for vaccination. All of this points to the importance of more study to understand the conditions under which greater democratization is warranted and the appropriate tools with which to pursue it. Chapter 12, for example, reveals the important limitations of referenda as a tool. Risk practitioners must make careful judgements about the level of democratization beneficial to a given risk context, and the appropriate participation tools to deploy.

Where to from Here? Resolving Dilemmas or Finding Insights Within Them?

There are no simple answers to the questions raised by the above three dilemmas. As is clear from the contributions to this volume, there is a wide diversity of approaches to risk governance and democratization in different fields and contexts, each with varying outcomes for the process and substance of risk decision-making, and the legitimacy of and public trust in risk governance. There are no simple one-size-fits-all prescriptions. That said, the chapters underscore that public trust in risk decision-making is increasingly a necessary condition for effective risk governance. Whether it is framed as holy grail or holy terror, it is here to stay. But it is not the only condition to effective risk governance. Science and scientific expertise are also necessary conditions—but they are likewise not sufficient. The concept of motivated reasoning helps to explain why, by surfacing the values embedded in both social and scientific processes and by revealing the pivotal role of identity, world-views and values to risk perception and decision-making by experts and non-experts alike. Rather than devaluing science, careful attention to the phenomenon of motivated reasoning helps clarify for decision-makers the nature and underpinnings of divergent perceptions of risk (including their own perceptions), giving them greater purchase over the complexities of contemporary risk decision-making, and, importantly, the potential to take more informed choices. Research in this volume also demonstrates that public participation has the potential to generate more robust decisions, bringing to the fore lived experience, on-the-ground knowledge and broader perspectives than purely scientific assessments. This book also highlights that risk practitioners must undertake a careful assessment of the strengths, limitations and appropriate level and nature of public participation for a given risk decision. Ensuring participation is undertaken on the right things, at the right times and with the right tools is crucial. While this volume offers many insights to that end, more research is needed to advance understanding of the drivers of trust—or mistrust—of risk practitioners, risk decisions and risk governance, to deepen knowledge of the phenomenon and impacts of motivated reasoning and to identify the appropriate levels and mechanisms to democratize risk governance in each context. Contemporary risk decision-making demands it.

References

Ahmed, R. (2015). Intercultural Competence in Healthcare. In J. M. Bennett (Ed.), *Sage Encyclopedia of Intercultural Competence* (pp. 487–490). Thousand Oaks, CA: Sage.

Atkins, D., Siegel, J., & Slutsky, J. (2005). Making Policy When the Evidence Is in Dispute. *Health Affairs*, 24(1), 102–113.

Browne, K. (2016, May 13). Vaccine Injury Compensation and the Common Good (Impact Ethics post). Retrieved from https://impactethics.ca/2016/05/13/vaccine-injury-compensation-and-the-common-good/.

Camargo Jr., K., & Grant, R. (2015). Public Health, Science, and Policy Debate: Being Right Is Not Enough. *American Journal of Public Health*, 105(2), 232–235.

Chafe, R., Levinson, W., & Hébert, P. C. (2011). The Need for Public Engagement in Choosing Health Priorities. *CAMJ*, 183(2), 165. https://doi.org/10.1503/cmaj.101517.

Chua, A. (2018). *Political Tribes: Group Instinct and the Fate of Nations*. New York: Penguin Press.

Cleland, M., Bird, S., Fast, S., Sajid, S., & Simard, L. (2016). A Matter of Trust: The Role of Communities in Energy Decision-Making. Ottawa: Positive Energy, University of Ottawa, and Canada West Foundation. Retrieved 31 October 2019, from https://www.uottawa.ca/positive-energy/sites/www.uottawa.ca.positive-energy/files/mattertrust_report_24nov2016-1_web.pdf.

Coburn, J. (2003). Bringing Local Knowledge into Environmental Decision Making: Improving Urban Planning for Communities at Risk. *Journal of Planning Education and Research*, 22(4), 420–433.

Douglas, H. (2016). Descartes Lecture 3: Science Communication: Beyond the Deficit Model. 5th René Descartes Lectures 2016, Science, Values and Democracy. The Netherlands: Tilberg University and the Tilburg Center for Logic, Ethics and Philosophy of Science.

Douglas, M., & Wildavsky, A. (1982). *Risk and Culture: An Essay on the Selection of Technological and Environmental Dangers*. Berkley, CA: University of California Press.

Driedger, S. M. (2008). Creating Shared Realities Through Communication: Exploring the Agenda-Building Role of the Media and Its Sources in the *E. coli* Contamination of a Canadian Public Drinking Water Supply. *Journal of Risk Research*, 11(1–2), 23–40.

Druckman, J. N. (2012). The Politics of Motivation. *Critical Review*, 24(2), 199–216.

Druckman, J., & Bolsen, T. (2011). Framing, Motivated Reasoning and Opinions About Emerging Technologies. *Journal of Communication*, 61, 659–688.

Durant, J. R., Evans, G. A., & Thomas, G. P. (1989). The Public Understanding of Science. *Nature*, 340(6228), 11–14.

Edelman. (2017). 2017 Edelman Trust Barometer. Accessed February 22, 2019. https://www.edelman.com/research/2017-edelman-trust-barometer.

Edelman. (2018). 2018 Edelman Trust Barometer. Accessed February 22, 2019. https://www.edelman.com/research/2018-edelman-trust-barometer.

Edelman. (2019). 2019 Edelman Trust Barometer. Accessed February 22, 2019. https://www.edelman.com/trust-barometer.

Esmail, L., Moore, E., & Rein, A. (2015). Evaluating Patient and Stakeholder Engagement in Research: Moving from Theory to Practice. *Journal of Comparative Effectiveness Research*, 4(2), 133–145. https://doi.org/10.2217/cer.14.79.

Fiorino, D. J. (1990). Citizen Participation and Environmental Risk: A Survey of Institutional Mechanisms. *Science, Technology, & Human Values*, 15(2), 226–243.

Fischer, F. (2003). *Reframing Public Policy: Discursive Politics and Deliberative Practices: Discursive Politics and Deliberative Practices*. Oxford: Oxford University Press.

Fishkin, J. S. (2011). *When the People Speak: Deliberative Democracy and Public Consultation*. Oxford: Oxford University Press.

Fukuyama, F. (2018). *Identity: The Demand for Dignity and the Politics of Resentment*. New York: Farrar, Straus and Giroux.

Goldenberg, M. J. (2016). Public Misunderstanding of Science? Reframing the Problem of Vaccine Hesitancy. *Perspectives on Science*, 24(5), 552–581. https://doi.org/10.1162/POSC_a_00223.

Greenberg, J. (2013). Risk Communication and the Disclosure Dilemma: The Case of Ottawa's Endoscopy Infection 'Scare'. *Journal of Professional Communication*, 2(1).

Hart, P., & Nisbet, E. (2011). Boomerang Effects in Science Communication: How Motivated Reasoning and Identity Cues Amplify Opinion Polarization About Climate Mitigation Policies. Communication Research. https://doi.org/10.1177/0093650211416646.

Hochschild, J., & Sen, M. (2015). Technology Optimism or Pessimism About Genomic Science Variation Among Experts and Scholarly Disciplines. *The Annals of the American Academy of Political and Social Science*, 658(1), 236–252.

International Association for Public Participation. (2022). Public Participation Spectrum. https://iap2canada.ca/foundations.

Iosifidis, P., & Nicoli, N., eds. (2020). *Digital Democracy, Social Media and Disinformation*. London: Routledge.

Jardine, C., Turtiak, M., & Driedger, S. M. (2009). Public Participation and Risk Governance: Opportunities and Barriers. *International Journal of Risk Assessment and Management*, 13(3/4), 260–275.

Jenkins-Smith, H. C., Silva, C. L., & Murray, C. (2009). Beliefs About Radiation: Scientists, the Public and Public Policy. *Health Physics*, 97(5), 519–527. https://doi.org/10.1097/HP.0b013e3181ad7eec.

Kahan, D. (2012). Cultural Cognition as a Conception of the Cultural Theory of Risk. In S. Roeser (Ed.), *Handbook of Risk Theory*, 724–759. Berlin: Springer.

Kahan, D. M., Braman, D., Cohen, G. L., Gastil, J., & Slovic, P. (2010). Who Fears the HPV Vaccine, Who Doesn't, and Why? An Experimental Study of the Mechanisms of Cultural Cognition. *Law and Human Behavior*, 34, 501–516

Kahan, D., Braman, D., Slovic, P., Gastil, J., & Cohen, G. (2007). The Second National Risk and Culture Study: Making Sense of—And Making Progress In—The American Culture War of Fact. GW Law Faculty Publications & Other Works. 211.

Kahan, D. M., Braman, D., Slovic, P., Gastil, J., & Cohen, G. (2009). Cultural Cognition of the Risks and Benefits of Nanotechnology. *Nature Nanotechnology*, 4(2), 87–90. https://doi.org/10.1038/nnano.2008.341.

Kahan, D. M., Jamieson, K. H., Landrum, A., & Winneg, K. (2017). Culturally Antagonistic Memes and the Zika Virus: An Experimental Test. *Journal of Risk Research*, 20(1), 1–40. https://doi.org/10.1080/13669877.2016.1260631.

Kahan, D. M., Jenkins-Smith, H., & Braman, D. (2011). Cultural Cognition of Scientific Consensus. *Journal of Risk Research*, 14(2), 147–174.

Kahan, D. M., Peters, E., Wittlin, M., Slovic, P., Ouellette, L. L., Braman, D., & Mandel, G. (2012). The Polarizing Impact of Science Literacy and Numeracy on Perceived Climate Change Risks. *Nature Climate Change*, 2(10), 732–735.

Kasperson, R. E., Renn, O., Slovic, P., Brown, H. S., Emel, J., Goble, R., & Kasperson, J. X., & Ratick, S. (1988). The Social Amplification of Risk: A Conceptual Framework. *Risk Analysis*, 8, 177–187.

Keelan, J., & Wilson, K. (2011). Designing a No-Fault Vaccine-Injury Compensation Programme for Canada: Lessons Learned from an International Analysis of Programmes. Toronto, ON: The University of Toronto's Munk School of Global Affairs. Retrieved from https://munkschool.utoronto.ca/research-articles/designing-a-no-fault-vaccine-injury-compensation-programme-for-canada-lessons-learned-from-an-international-analysis-of-programmes/.

Kirst, M., Shankardass, K., Singhal, S., Lofters, A., Muntaner, C., & Quiñonez, C. (2017). Addressing Health Inequities in Ontario, Canada: What Solutions do the Public Support? *BMC Public Health*, 17(1), 1–9.

Kiss, S., Lachapelle, E., & Montpetit, E. (2016). Motivated Reasoning and Cultural Cognition Theory in the Canadian Context. Presentation made at the workshop *@Risk: Risk Management, Democratization and Evidence-Based Decision-Making*. Ottawa.

Kovacs Burns, K., Bellows, M., Eigenseher, C., & Gallivan, J. (2014). 'Practical' Resources to Support Patient and Family Engagement in Healthcare Decisions: A Scoping Review. *BMC Health Services Research*, 14, 175. https://doi.org/10.1186/1472-6963-14-175

Krewski, D., Hogan, V., Turner, M. C., Zeman, P. L., McDowell, I., Edwards, N., et al. (2007). An Integrated Framework for Risk Management and

Population Health. *Human and Ecological Risk Assessment: An International Journal*, 13(6), 1288–1312. https://doi.org/10.1080/10807030701655798.

Krewski, D., Turner, M. C., Lemyre, L., & Lee, J. E. C. (2012). Expert vs. Public Perception of Population Health Risks in Canada. *Journal of Risk Research*, 15(6), 601–625.

Kunda, Z. (1990). The Case for Motivated Reasoning. *Psychological Bulletin*, 108, 480–498.

Kuzma, J. (2016). Policy: Reboot the Debate on Genetic Engineering. *Nature*, 531, 165–167.

Lachapelle, E., Montpetit, É., & Gauvin, J. P. (2014). Public Perceptions of Expert Credibility on Policy Issues: The Role of Expert Framing and Political Worldviews. *Policy Studies Journal*, 42(4), 674–697.

Latkin, C. A., Dayton, L., Yi, G., Konstantopoulos, A., & Boodram, B. (2021). Trust in a COVID-19 Vaccine in the U.S.: A Social-ecological Perspective. *Social Science & Medicine*, 270, 113684. S0277953621000162 https://doi.org/10.1016/j.socscimed.2021.113684.

Law Reform Commission of Saskatchewan. (2009). Vaccination and the Law: Report to the Minister of Justice. Saskatoon, Saskatchewan: Law Reform Commission of Saskatchewan. Retrieved from https://lawreformcommission.sk.ca/vaccinationfinal.pdf.

Manafo, E., Petermann, L., Mason-Lai, P., & Vandall-Walker, V. (2018). Patient Engagement in Canada: A Scoping Review of the 'How' and 'What' of Patient Engagement in Health Research. *Health Research Policy and Systems*, 16, 5. https://doi.org/10.1186/s12961-018-0282-4

Manitoba Law Reform Commission. (2000). Compensation of vaccine-damaged children (report #104). Winnipeg, Manitoba: Manitoba Law Reform Commission. Retrieved from http://www.manitobalawreform.ca/pubs/pdf/archives/104-full_report.pdf.

Meghani, Z., & Kuzma, J. (2011). The "Revolving Door" Between Regulatory Agencies and Industry: A Problem That Requires Reconceptualizing Objectivity. *Journal of Agricultural and Environmental Ethics*, 24(6), 575–599.

Montpetit, É., & Lachapelle, E. (2016). Information Values and Expert Decision-making: The Case of Soil Decontamination. *Policy Sciences*, 49(2), 155–171. https://doi.org/10.1007/s11077-015-9225-x.

Nanos, N. (2018). *The Age of Voter Rage*. London: Eyewear Publishing.

Navin, M. C., & Largent, M. A. (2017). Improving Nonmedical Vaccine Exemption Policies: Three Case Studies. *Public Health Ethics*, 10(3), 225–234. https://doi.org/10.1093/phe/phw047.

Nevitte, N. (1996). *The Decline of Deference: Canadian Value in Change in Cross National Perspective*. Toronto: University of Toronto Press.

Nevitte, N. (2011). The Decline of Deference Revisited: Evidence After 25 Years. Paper presented at Mapping and Tracking Global Value Change: A Festschrift Conference for Ronald Inglehart. March 11, University of California, Irvine.

Nicholls, S. G., Newson, A. J., & Ashcroft, R. E. (2016). The Need for Ethics as Well as Evidence in Evidence-Based Medicine. *Journal of Clinical Epidemiology*, 77, 7–10.

Perrella, A., & Kiss, S. (2015). Risk Perception, Psychological Heuristics and the Water Fluoridation Controversy. *Canadian Journal of Public Health*, 106.

Persily, N., & Tucker, J. A., eds. (2020). *Social Media and Democracy: The State of the Field and Prospects for Reform*. Cambridge: Cambridge University Press.

Petts, J., & Brooks, C. (2006). Expert Conceptualisations of the Role of Lay Knowledge in Environmental Decision-Making: Challenges for Deliberative Democracy. *Environment and Planning A*, 38(6), 1045–1059.

Phillips, S., with Orsini, M. (2002). Mapping the Links: Citizen Involvements in Policy Processes. CPRN Discussion Paper No. F21. Ottawa: Canadian Policy Research Networks.

Pidgeon, N., Kasperson, R. E., & Slovic, P. (2003). *The Social Amplification of Risk*, 447. Cambridge: Cambridge University Press.

Renn, O., & Levine, D. (1991). Credibility and Trust in Risk Communication. In *Communicating Risks to the Public* (pp. 175–217). Dordrecht: Springer.

Renn, O. (2008). White Paper on Risk Governance: Toward an Integrative Framework. In O. Renn & K. D. Walker (Eds.), *Global Risk Governance: Concept and Practice Using the IRGC Framework* (pp. 3–73). Netherlands: Springer. https://doi.org/10.1007/978-1-4020-6799-0_1.

Simard, L. (2018). How to Decide—Engagement: Information and Capacity. Positive Energy, Univeristy of Ottawa and Canada West Foundation. Ottawa. Retrieved 2 July 2017, from https://www.uottawa.ca/positive-energy/sites/www.uottawa.ca.positive-energy/files/pe_louis_simard_final.pdf.

Slovic, P. (1987). Perception of Risk. *Science*, 236, 280–285.

Slovic, P. (1993). Perceived Risk, Trust, and Democracy. *Risk Analysis*, 13(6), 675–682.

Stern, P. C., & Fineberg, H. V. (1996). Understanding Risk: Informing Decisions in a Democratic Society. Committee on Risk Characterization, National Research Council. https://doi.org/10.17226/5138.

Thomas, C. M., Osterholm, M. T., & Stauffer, W. M. (2021). Critical Considerations for COVID-19 Vaccination of Refugees, Immigrants, and Migrants. *The American Journal of Tropical Medicine and Hygiene*, 104(2), 433–435. https://doi.org/10.4269/ajtmh.20-1614.

Thompson, P. B. (1997). Science Policy and Moral Purity: The Case of Animal Biotechnology. *Agriculture and Human Values*, 14(1), 11–27.

Thompson, P. B. (2007). *Food Biotechnology in Ethical Perspective*. Springer.

Webler, T., & Tuler, S. (2018). Four Decades of Public Participation in Risk Decision Making. *Risk Analysis*, 41(3), 503–518.

Wynne, B. (1992). Misunderstood Misunderstanding: Social Identities and Public Uptake of Science. *Public Understanding of Science*, 1(3), 281–304.

Wynne, B. (2006). Public Engagement as a Means of Restoring Public Trust in Science—Hitting the Notes, but Missing the Music? *Public Health Genomics*, 9(3), 211–220.

Wynne, B. (2007). Public Participation in Science and Technology: Performing and Obscuring a Political–conceptual Category Mistake. *East Asian Science Technology and Society: An International Journal*, 1(1), 99–110. https://doi.org/10.1215/s12280-007-9004-7.

PART I

Motivated Reasoning, Science and Values: Rethinking Risk Perception and Public Participation in Risk Governance

CHAPTER 2

Motivated Reasoning and Risk Governance: What Risk Scholars and Practitioners Need to Know

Marisa Beck, Rukhsana Ahmed, Heather Douglas, S. Michelle Driedger, Monica Gattinger, Simon J. Kiss, Jennifer Kuzma, Patricia Larkin, Kieran C. O'Doherty, Andrea M. L. Perrella, Teshanee T. Williams, and Gregor Wolbring

Introduction

A growing body of research investigates how people process information and form opinions and how these reasoning processes can have considerable consequences for risk governance. Specifically, scholarship examines the ways in which individuals' information processing deviates from a

M. Beck (✉)
Ottawa, ON, Canada
e-mail: mbpenslar@gmail.com

R. Ahmed
Department of Communication,
University at Albany,
State University of New York, Albany, NY, USA

M. Gattinger (ed.), *Democratizing Risk Governance*,
https://doi.org/10.1007/978-3-031-24271-7_2

normative model of learning where accuracy is the only goal. The concept of 'motivated reasoning', developed by political psychologists and political scientists, describes and explains these deviations. Theories of motivated reasoning seek to understand how reasoning works when accuracy is not the primary or sole goal directing individuals' reasoning processes. Empirical studies indicate that individuals collect, process, and interpret information in a goal-driven fashion, which enables them to arrive at conclusions that are useful to them in some way—notably because they

---

H. Douglas
Department of Philosophy, Michigan State University, East Lansing, MI, USA

S. M. Driedger
Department of Community Health Sciences, University of Manitoba, Winnipeg, MB, Canada

M. Gattinger
Institute for Science, Society and Policy, University of Ottawa, Ottawa, ON, Canada

S. J. Kiss
Wilfrid Laurier University, Brantford, ON, Canada

J. Kuzma
Genetic Engineering and Society Center, North Carolina State University, Raleigh, NC, USA

P. Larkin
Institute for Science, Society and Policy, University of Ottawa, Ottawa, ON, Canada

K. C. O'Doherty
Department of Psychology, College of Social and Applied Human Sciences, University of Guelph, Guelph, ON, Canada

A. M. L. Perrella
Department of Political Science, Wilfrid Laurier University, Waterloo, ON, Canada

T. T. Williams
School of Government, University of North Carolina, Chapel Hill, NC, USA

G. Wolbring
Cummings School of Medicine, University of Calgary, Calgary, AB, Canada

align with their prior beliefs, worldviews, or the positions of social groups they belong to. Because accuracy is not the sole goal, such reasoning is often perceived as 'wrong' by others and not sufficiently empirically based.

Gaining a better understanding of the sources, mechanisms, and implications of motivated reasoning can help scholars and practitioners of risk governance to anticipate, understand, and address differences in people's risk perceptions, as well as differences in their level of trust in scientific evidence about risk. In practice, differences in perception and processing of risk information can lead to conflicts—even over the factual evidence itself. Prominent examples of risk issues at the centre of long-standing public disputes over the underlying evidence include climate change (e.g., Druckman and McGrath 2019), vaccinations (e.g., Kahan et al. 2010), and new technologies (e.g., Druckman and Bolsen 2011). Motivated reasoning research may help risk practitioners to see that apparent conflicts over 'the evidence' related to risk may be rooted in differences in people's values, identities, and prior beliefs. Importantly, motivated reasoning research also suggests that such conflicts are not easily overcome by simply presenting more evidence to people.

Despite the growing scholarship on motivated reasoning, fundamental conceptual challenges remain. This chapter provides an analysis and discussion of this body of work to increase awareness among risk practitioners and scholars of its insights and contributions. The chapter begins with a review of the main contributions to the literature in psychology, political science, and communication studies. The review finds that prominent theorists in the field use the term 'motivated reasoning' to explain patterns of behavior using different theoretical accounts—and they may even be describing different phenomena altogether.

In addition to identifying and exploring these discrepancies, the chapter also focuses on the normative evaluations inherent in particular uses of the concept of motivated reasoning. These judgments typically include ideas about what it means for individuals to reason in a 'rational' manner and for society to govern risks 'rationally'. We find the use of 'rationality' problematic in assessments of motivated reasoning in the context of risk decision-making in part because some of the theoretical accounts of motivated reasoning suggest that reasoning can be perfectly rational. Given the historical abuses of 'rationality' to dismiss the beliefs of marginalized groups (e.g., women, Indigenous peoples, people of color, or disabled people), we urge caution in assessments of the rationality of the beliefs of marginalized groups and motivated reasoning more generally. While some kinds of motivated reasoning are clearly irrational, others are not.

The chapter proceeds as follows. Section one, "The Theory of Motivated Reasoning", presents basic motivated reasoning concepts and identifies key theoretical models. Section two, "Where Theoretical Models of Motivated Reasoning Diverge", examines the conceptual differences between these models, while the third section, "Is Directional Motivated Reasoning a Problem?", explores the normative implications of motivated reasoning. Section four, "Where to from Here? Theoretical Implications, Empirical Implications, Practical Implications", discusses implications for theoretical and empirical research on motivated reasoning, along with implications for practice and policy. The final section offers conclusions.

The Theory of Motivated Reasoning

Basic Concepts

Generally, motivated reasoning is understood as a psychological description of how people process information and form/update their beliefs and/or attitudes about objects/events/issues. Motivated reasoning generally refers to how people's goals or motivations affect their reasoning and judgments (Kunda 1990). When people pursue accuracy as their sole goal, they strive to reach a correct conclusion; when their goals are directional, they "unconsciously conform assessment of factual information to some goal *collateral* to assessing its truth" (Kahan 2016a, 2, emphasis in original). People may pursue both accuracy and directional goals to different degrees simultaneously (Kunda 1990), with directional goals, whether conscious or unconscious, exhibited as biases in people's search for, interpretation and evaluation of information. In the empirical literature on motivated reasoning, correlations between people's worldviews, goals, or values and their reasoning outcomes are often identified and examined in experimental designs, where study participants holding particular prior beliefs or values are presented with information and asked to assess it in some way (see for example, Lord et al. 1979; Redlawsk 2002; Taber and Lodge 2006).

Some of the key shared conceptual components of the literature addressing motivated reasoning include:

Motivation. The literature often draws on the definition of motivation by Fishbach and Ferguson (2007) as "cognitive representation of a desired endpoint that impacts evaluations, emotions and behaviors" (491). The terms 'motivation' and 'goal' are commonly used interchangeably.

Reasoning. Reasoning is commonly understood to incorporate multiple cognitive processes, including information collection, processing, and evaluation; memory retrieval; attitude formation; judgment and decision-making (Leeper and Mullinix 2018). This chapter focuses in particular on how motivated reasoning affects people's reasoning about risk and their processing of risk information.

System 1 vs. System 2 thinking. Multiple theoretical accounts of motivated reasoning draw on this distinction. In the framework developed by Kahneman (2011), System 1 cognition is immediate and intuitive, while System 2 is deliberate and slow. Traditionally, biases in judgment are attributed to affect-driven System 1 reasoning. However, as we will see below, some argue that it is System 2 cognition that is centrally deployed in motivated reasoning.

Bayesian updating/learning. Motivated reasoning processes are often contrasted with truth-seeking Bayesian learning (Gerber and Green 1999; Redlawsk 2002). According to this model, individuals hold initial estimates of the probability that a hypothesis is true (the prior). The prior is updated when people receive new, relevant evidence. Importantly, a normative accuracy-seeking Bayesian model prescribes that people collect, assess, and adopt new evidence <u>independently</u> of their prior. As a consequence, people with opposite prior views should converge in their opinions when exposed to the same information. Motivated reasoning deviates from this Bayesian ideal because the process of updating is influenced by directional goals (Druckman and McGrath 2019). This means that uptake of new evidence is explicitly <u>dependent</u> on prior beliefs. As discussed below, exposure to the same evidence may then lead to the opposite effect on people with different priors, and increase division or polarization.

Bias. The term 'bias' is ubiquitous in the literature, with motivated reasoning commonly understood as leading to 'bias' in judgment and decision-making. Biased reasoning was first defined as a systematic and measurable deviation from the (known) correct answer (e.g., Tversky and Kahneman 1974). In this early work, biases were not correlated with motivations (ibid., p. 1130). The conception of bias has since expanded to include any correlation between a person's beliefs and motivations other than accuracy goals, even if the true, correct answer remains unknown. Interestingly, none of the theoretical accounts reviewed here provides an explicit definition of bias.

Key Theoretical Models of Motivated Reasoning

First developed by psychologists in the second half of the twentieth century, the concept of motivated reasoning was later picked up by political scientists and communication scholars. Multiple theoretical models of motivated reasoning exist, but a small number of models are the backdrop for numerous research studies. The models do not agree on central theoretical components, but each contributes insights that help to understand the implications of motivated reasoning for risk scholarship and practice.

The model of 'biased assimilation'. In their pivotal study on people's views of the death penalty, Lord et al. (1979) find evidence of what they term 'biased assimilation'. Their results show that people holding strong opinions about the death penalty evaluate and interpret new, ambiguous evidence on the topic in the light of their prior views. Study participants—both supporters and opponents of the death penalty—systematically considered evidence in line with their previous viewpoint as more convincing than incongruent evidence. In fact, the presentation of new evidence made both supporters and opponents become more attached to their initial positions, amplifying divisions between the two groups. Importantly, the model of biased assimilation uses a cognitivist approach to explain information processing, namely the objective of achieving "consistency of [...] evidence with the perceiver's theories and expectations" (ibid., 2099) that shape their "judgments about the validity, reliability, relevance, and sometimes even the meaning of proffered evidence" (ibid.).

The model of motivated skepticism. Taber and Lodge (2006) argue that people's prior attitudes and beliefs about a contentious issue influence how they select and evaluate new information about it. In particular, the authors identify a 'confirmation bias' (seeking out evidence that supports prior attitudes), a 'disconfirmation bias' (discounting non-supportive arguments), and a 'prior attitude effect' (considering arguments supporting prior attitudes to be stronger than those contradicting prior attitudes). The result is what the authors term 'motivated skepticism': exposure to balanced information about a contested issue did not lead to people's opinions converging, but rather, led to further polarization and a strengthening of people's prior attitudes.

The John Q. Public (JQP) model of motivated reasoning. This model defines motivated reasoning as an affect-driven, unconscious judgment process that involves post hoc justification and rationalization (Lodge and Taber 2013; Kraft et al. 2015). Affect is considered the key driver: feelings (positive or negative) arise immediately and spontaneously when people are confronted with new information (the 'hot cognition' hypothesis) and these initial feelings are seen to influence all subsequent processing and reasoning processes. Conscious re-writing of spontaneous responses is not impossible, but it is rare and requires time and effort so that only a strong motivation (e.g., accuracy goals) may make it worthwhile for individuals. However, people often engage in conscious deliberations to vindicate their spontaneous, unconscious judgments after the fact in order to justify their positions to themselves and others.

The Politically Motivated Reasoning Paradigm. When the goal in motivated reasoning is identity protection, Kahan (2016a) refers to this as politically motivated reasoning, which he defines as "the formation of beliefs that maintain a person's status in an affinity group united by shared values" (ibid., 3). Kahan emphasizes that 'identity' can be defined in various ways and along various dimensions, including political affiliation, ideology, values (see below), religion, gender, ethnicity, etc. (Kahan 2016a). No matter the group characteristics, the underlying mechanism that directs information processing is the same: people interpret information in ways that signal their agreement with the position associated with their identity-giving social group.

Cultural Cognition Theory of Risk Perception. While the foregoing models concern human reasoning in general—and may be applied to reasoning about risk—this theory focuses on directionality in risk perception. Based on cultural theory and an individual's 'cultural worldview' or value system (Douglas and Wildavsky 1982), cultural cognition posits that people who belong to different cultural groups[1] systematically differ

[1] Empirical research in social psychology (Kahan et al. 2010) shows that a person's cultural worldview can be reliably approximated and measured based on their relative support for specific societal values: hierarchy vs. egalitarianism and individualism vs. collectivism.

in their perception of risk and risk information through both psychological and social processes (Kahan 2012). Specifically, individuals tend to believe that what they value is not a source of risk and vice versa.

Multiple mechanisms of cultural cognition of risk are identified in the literature (Kahan 2012). A key mechanism is, again, identity protection—here, more specifically, *cultural* identity protection. For example, research has shown that white males systematically perceive risks from environmental hazards to be lower than women or non-white males (the 'white male effect') (Kahan et al. 2007).

Where Theoretical Models of Motivated Reasoning Diverge

The above models of motivated reasoning agree on the general idea that directional 'motivated reasoning' (however it is understood in the various accounts) introduces bias in people's reasoning. However, the models differ in how they explain the source and extent of directionality.

What Is the Motivation in Motivated Reasoning?

We distinguish among three goals: (1) consistency with prior beliefs and attitudes, (2) identity commitments, and (3) value commitments.

Consistency with prior beliefs and attitudes. The model of 'biased assimilation' (Lord et al. 1979) and the model of 'motivated skepticism' (Taber and Lodge 2006) understand motivated reasoning to be directed mainly by people's intrinsic goal to uphold and confirm previously held beliefs and attitudes. Specifically, these models argue that people are motivated to select and evaluate more positively new evidence that supports their previously held beliefs and attitudes. Switching off this kind of inertia takes time and effort.

Empirical studies indicate that people's tendency to process and assess new information about an issue in light of their prior positions can lead to conflict and polarization over scientific evidence. These findings underscore that providing people with more risk information—the 'knowledge deficit' model of risk communication—may not promote shared perceptions of risk. In fact, the opposite may obtain: people may diverge further in their beliefs.

This tendency can be positively correlated with peoples' level of knowledge: the study on motivated skepticism by Taber and Lodge (2006) revealed that more knowledgeable individuals, with stronger initial attitudes and beliefs, were more likely to reflect motivated skepticism in their information processing because their prior beliefs and attitudes were comparably stronger. Crucially, this finding suggests that risk experts may not be less, but rather *more*, likely than the general population to reason in a motivated fashion. Simply put, more knowledge does not necessarily produce reasoning focused solely on truth-seeking.

Identity protection. Kahan's model of politically motivated reasoning focuses on one particular goal in people's reasoning—identity protection. In this model, holding on to familiar beliefs despite being confronted with new, contradicting evidence is not a goal in and of itself. Rather, people's goal when processing new evidence is to align their position with that of a relevant social group to maintain and express their membership in it.

Hence, when belief/disbelief in scientific facts about a risk issue become associated with 'identity-defining affinity groups' (Kahan 2016a), individuals are motivated to reason about information in ways that express their group identity. For example, DeFranza et al. (2020) conducted a study focused on how religiosity (i.e., feelings, thoughts, experiences, and behaviors associated with the sacred) affected adherence to shelter-in-place directives in response to COVID-19. Prior to a shelter-in-place directive, religiosity did not affect people's decisions. However, once there was a shelter-in-place directive, higher religiosity resulted in less adherence to shelter-in-place directives.

Value commitments. Cultural cognition theory identifies worldviews and values as key motivators of directionality in people's reasoning about risk. Cultural cognition specifies that individuals seek consistency with their *values* when forming beliefs about risk, and aim for alignment in their risk perceptions with cultural groups bound by the same *values*. Hence, cultural cognition theory includes both the consistency objective and the goal of identity protection as drivers of directionality in human reasoning, but considers these goals through a value lens.

What Is the Role of Affect and 'Hot Cognition' in Motivated Reasoning?

Some of the models of motivated reasoning above suggest that the phenomenon is primarily a result of immediate, affect-driven judgment; others suggest that motivated reasoning is the outcome of a more deliberate cognitive process. In other words, models differ with regard to whether motivated reasoning is theorized to occur mostly in System 1 or System 2 thinking.[2]

The JQP model of motivated reasoning and the model of motivated skepticism specifically emphasize the influence of affect and 'hot cognition' on the formation/updating of beliefs and attitudes in response to new information. These models situate motivated reasoning firmly in immediate, automatic System 1 thinking, where spontaneous, affect-driven processes drive information processing by triggering selective attention, exposure, and judgment processes. The unconscious, immediate 'hot' response to new information determines the direction and strength of subsequent information processing. While people generally "want to get it straight" (Lodge and Taber 2013, 152), they are unconsciously held hostage by their powerful, affective priors. According to such affect-focused explanations of motivated reasoning, conscious deliberations (System 2) in most instances merely serve to *justify* spontaneous, unconscious judgments (System 1) after the fact.

In contrast, Kahan's model of politically motivated reasoning suggests that deliberate, slow System 2 thinking is required to successfully direct reasoning. For example, Kahan (2013; 2016b) argues that when individuals defeat challenging arguments to ensure their position remains loyal to their identity-giving group, it is a deliberate and often sophisticated intellectual act that requires System 2 thinking.

What Are the Limits of Motivated Reasoning?

The studies reviewed above seem to agree that while "all reasoning is motivated" (Taber and Lodge 2006), individuals do not typically engage in directional motivated reasoning ***in an extreme manner all the time***. For example:

- Accuracy motivations can put a limit on the influence of directional motivations (Kunda 1990; Kahan 2013).
- People with weaker beliefs and attitudes about a certain issue are less likely to engage in motivated reasoning about it (Taber and Lodge 2006).

[2] Not all of the motivated reasoning models reviewed here draw on this framework, but some explicitly do.

- People generally have a desire to appear rational and objective to outside observers, and their need to justify their judgment puts constraints on the judgment's outcome (Kunda 1990, 1999).
- Only a constrained number of risk issues bear so much social meaning that an individual's position on the issue signals belonging to a certain social group (Kahan 2013).

It is not clear from the literature whether and how public authorities might intervene to address directional motivated reasoning on contentious societal issues to facilitate consensus building. Research on motivated reasoning is still fairly new, and as such the main focus has been on understanding the underlying mechanisms rather than investigating how to address the issue. However, all accounts agree that whether driven by consistency goals, value commitments, or identity protection goals, directional motivated reasoning about a societal issue is not easily addressed by more or better evidence. Instead of converging around the evidence, people's opposing positions may harden and diverge further. Models also agree that people with greater expertise about an issue may be particularly prone and better equipped to engage in directional motivated reasoning.

Still, some of the theoretical models above suggest some responses, including information campaigns (Kraft et al. 2015) and preventing positions on important policy issues from becoming associated with certain ideological groups (Kahan 2016a). Cultural cognition theory suggests that risks should be communicated in ways that *affirm* rather than threaten cultural worldviews to elicit greater receptiveness and trust in the information (Kahan 2012). In practice, this may include working with culturally diverse risk communicators who enjoy credibility in the target communities. Others argue that more intrusive measures should be taken to prevent motivated reasoning. In particular, Kahan (2013) argues that individuals' incentive structures should be modified in ways that promote the pursuance of accuracy goals rather than directional goals to link their beliefs more firmly to the truth.

This emerging debate about how to address motivated reasoning assumes that it is indeed a problem requiring intervention. Is motivated reasoning a problem for risk decision-making for the individual and/or society? These normative questions are examined next.

Is Directional Motivated Reasoning a Problem?

All of the theoretical models of motivated reasoning above include more or less explicit evaluations of the benefits of motivated reasoning for individual decision-making. As outlined above, processing information in a way that enables people to uphold their prior beliefs and attitudes allows them to build on their previous experiences and knowledge (Lord et al. 1979; Taber and Lodge 2006). This can be efficient at the individual level because updating beliefs is a time and resource-intensive process. Similarly, engaging in reasoning that protects identity and value commitments affords people an immediate benefit from maintaining loyalty to identity-giving groups, in contrast with the longer term (and often more nebulous) benefit from holding a factually accurate position (Kahan 2013, 2016a).

Examining the impacts of motivated reasoning on risk perception and assessment becomes more controversial when considered from a societal perspective. While it is generally fair to assume that motivated reasoning about risks provides some benefit to individuals, others might judge the risk attitudes and beliefs that they arrive at as simply 'wrong' or harmful to those individuals or to society. Even if people benefit from motivated reasoning, one may argue that collective decision-making about risk can suffer as a consequence. Kahan (2013; 2016a) argues that the benefits to individuals may cost democratic society as a whole since evidence-based decision-making about risks becomes increasingly difficult when new evidence has little impact on people's beliefs.

Judging the effects of motivated reasoning from a societal perspective requires a normative criterion to define 'good reasoning' about risk. The literature often uses 'rationality' as a criterion for evaluation, which is automatically contrasted with any correlations between values, identity, or prior positions and a person's stated beliefs. However, the models reviewed above draw implicitly on different understandings of 'rationality'.

Serving self-interest. Kahan et al. (2012) argue that evidence of identity-protective motivated reasoning shows "how remarkably well-equipped ordinary individuals are to discern which stances towards scientific information secure their personal interests" (733). Rational belief formation is here construed as what is overall in one's self-interest, which Kahan argues is mostly driven by the need to fit in with one's community. As a result, for the individual, Kahan (2013) does not see identity-protective cognition "as a reasoning deficiency but as a reasoning adaptation suited to promoting the interest that individuals have in conveying their membership in and loyalty to affinity groups central to their personal wellbeing" (418).

Based on a similar understanding of 'rationality' as 'alignment with self-interest', Lord et al. (1979) argue that it is rational for individuals to assess new information as more plausible when it aligns with their previous knowledge and experiences: "Willingness to interpret new evidence in the light of past knowledge and experience is essential for any organism to make sense of, and respond adaptively to, its environment" (ibid., 2107). Giving more weight to one's prior attitudes in the collection and processing of new information is therefore seen as generally efficient and sensible (Taber and Lodge 2006).

Publicly defensible. Kunda (1990) draws on this understanding of rationality when she writes that "The biasing role of goals is thus constrained by one's ability to construct a justification for the desired outcome: People will come to believe what they want to believe only to the extent that reason permits" (483). The need for a justification that could pass muster under the scrutiny of others is one sense of rationality that seems to constrain directional motivated reasoning. The contrast to the first understanding of 'rationality' can be sharpened by considering that it is in many circumstances implausible that 'fit with one's peer community' would be accepted as being a defensible public reason to justify a belief.

Truth-seeking. Finally, directional motivated reasoning is generally considered irreconcilable with traditional, enlightenment-era ideas of rationality. Goal-oriented motivated reasoning by definition interferes with accuracy-seeking, dispassionate decision-making as idealized by the norms underlying the accuracy-seeking Bayesian model. The JQP model explicitly considers 'hot cognition' (System 1 thinking) as driving human judgment and therefore suggests that humans process information generally in an 'irrational' manner. While others argue that directional motivated reasoning strongly engages System 2 thinking (traditionally equated with this conception of 'rational' thinking), the general assumption that slow, deliberate thinking necessarily results in accuracy-seeking reasoning does not hold (Kahan 2016c). Importantly, from a risk governance perspective, at the societal level this conception of 'rationality' is typically reflected in calls for basing policymaking and regulation on 'objective' scientific evidence (Sanderson 2006).

'Rationality' as a Contested Concept

What defines rational decision-making is not often explicitly defined by theorists of motivated reasoning. However, their understandings are implicit in the sense that normative evaluations of motivated reasoning phenomena either cast it as 'irrational', in the sense that it leads to assessments and decisions that do not accord with 'the facts' or 'truth' (in line with the third sense of rationality above), or in the sense that the reasoning would not offer a publicly defensible justification for a belief (in line with the second sense of rationality above), or they cast it is 'rational' in the sense that it serves individual purposes, but not those of accuracy (in line with the first sense of rationality above).

We contend, therefore, that a more explicit engagement with what counts as rational in decision-making in the first place is critical to advancing understanding of motivated reasoning phenomena. Specifically, we draw attention to the fact that rationality is a contested concept, as is clear in the different senses of rationality noted above.

An additional line of work important in this regard is that of Gigerenzer and Gaissmaier (2011). A common element in the third sense of rationality noted above is that assessment of the rationality of individuals' decisions relies on whether individuals' reasoning processes followed particular logical or statistical norms (Gigerenzer and Gaissmaier 2011). That is, the assumption is that it is possible to assess the rationality of a decision purely on the basis of universally applied norms, and independent of the particular context in which the decision is made, or of the person making the decision. Todd and Gigerenzer (2012) argue that the assessment of decision-making cannot solely rely on adherence to logical or statistical procedures; it must also take into account the success of decisions in the 'real' world. The authors draw on the notion of ecological rationality to emphasize this particular notion of rationality. Further, as this chapter makes clear, individuals make decisions in the context of particular values, goals, and larger purposes, such that it is rarely possible to identify common ideals about optimal decision outcomes on people in general (e.g., maximizing health, optimizing financial outcomes, etc.).

The use of 'rational' as a desired trait also has societal implications that underscore its contested nature. First, it privileges the views of certain social and demographic groups that have defined what it means to be 'rational', e.g., being accurate, objective, and unemotional. What counts as rational or irrational depends to a large extent, then, on historical, cultural, and political contingencies. Groups and individuals generate diverse narratives of what is considered rational and which meaning prevails depends in part on the power of those putting forward a particular definition. For example, historically, the 'rationality discourse' has been used to disable or discredit groups, including men's power over women, whom they labeled irrational (Wolbring 2008; Buechler 1990; Viola 1986), a tactic still used today (Wolbring 2019; *Daily Star* 2014). The concept of 'irrationality' is also used as a tool to discredit one's opponents in policy or societal debates (see, for example, Wolbring and Diep [2016], Posusney [1993], Van Montagu [2013], Osborne [2014]). Rationality discourse can also be used to question a person's self-perception or self-acceptance. For example, disabled people who perceive their body as a variation that does not need to be fixed—not an aberration—are often told their perspective is not rational because it does not reflect the dominant view (Harris 2001, 2000).

Secondly, the social nature of rationality can be seen when it is used as a standard for making risk decisions. For example, in the governance of emerging technologies, there is always some level of potential risk to consider, but a great deal of uncertainty about its nature, severity, distribution, and probability. In this context, values play a central role in characterizing and mitigating risk based on the evidence. In fact, it is impossible to base societal decisions on scientific information alone (e.g., Kuzma 2018). Yet, regulatory decisions are portrayed as rational and 'science-based', masking the values embedded in decisions that are not made explicit. Those with power and authority have defined what is a rational interpretation of the scientific evidence based on their own values—and often behind closed doors (Meghani and Kuzma 2011). Those outside of the process who hold alternate views are often pegged as irrational Luddites.

In contrast, the idea of 'strong objectivity' challenges the monopoly that powerful actors hold on rationality (Harding 1995, 332). Arising out of feminist standpoint theory, it argues that what we can know is enabled by where we come from socially. Only through the inclusion of diverse standpoints, particularly those from marginalized groups, can we maximize our knowledge and achieve strong objectivity. Strong objectivity redistributes power to groups that have not been at the helm of 'evidence-based' decision-making by defining a more socially robust form of rationality.

Crucially, in contrast to the models of rationality discussed in the previous section, strong objectivity places *importance* on the phenomena driving *motivated reasoning*, such as prior beliefs, values, and identities, in *achieving accuracy* (e.g., through Bayesian updating) (Druckman and McGrath 2019). This is a fruitful insight that deserves more discussion in the literature.

Where to from Here? Theoretical Implications, Empirical Implications, Practical Implications

Our analysis has a number of implications for theory development, empirical studies of motivated reasoning, and the practice of risk governance. The following sections summarize these implications.

Theoretical Implications

As revealed above, the terminology around motivated reasoning is ambiguous. There are discrepancies in key concepts and models, which suggest that not only the theoretical accounts—but indeed the phenomena they describe—vary. We need more theoretical clarity and consistent terminology, tied to empirical practice, to analyze how individuals form beliefs and attitudes.

Further, the normative differences around 'rationality' discussed above were distilled from work within the social sciences literature on motivated reasoning. Additional normative issues arise if one views the issues through the lens of philosophy of science. Work over the past few decades has led philosophers to examine the rational and necessary role of social and ethical values in science, which holds important implications for research on motivated reasoning. There are at least two crucial places where social and ethical values play a legitimate role in scientific reasoning and practice.

The first is in the direction of scientific research effort: deciding what is important to study and how research problems are framed. Public skepticism about scientific claims can arise because some segments of the public view scientific efforts as inappropriately contextualized or directed. For example, if scientists are incentivized to pursue patentable technology solutions to problems of food production but some people are more interested in changing agricultural practices (e.g., shifts to organic farming), those people can view scientific research as fundamentally misdirected and thus results of such scientific work as inadequate for addressing policy issues. Similar concerns have been raised regarding research on the safety of vaccines (Goldenberg 2016).

The second role for values is in the assessment of evidential sufficiency in science. Science is an inherently inductive investigative process and the evidence underpinning scientific claims is never complete. When, then, is the evidence strong enough? Examinations of inductive risk reasoning in science (Douglas 2000; Elliott and Richards 2017) have shown the pervasive need to embed ethical and social values in this judgment. This means members of the public holding different values than scientists might disagree with scientific assessment of evidential sufficiency for value-based reasons—and do so rationally (Douglas 2017).

On the other hand, not all kinds of reasoning can be considered rationally acceptable (in the sense of publicly justifiable). For example, if many segments of the public consider evidence important, there should come a point when the evidence is strong enough for all. If no evidence could convince people, then they would have adopted an unfalsifiable position, which would be irrationally intransigent (as Taber and Lodge [2006] note). This insight can be stated using a Bayesian framework: it is not just priors that diverge among actors, but also likelihood ratios. This can explain why different kinds or levels of evidence might be needed by different actors.

Work by social scientists finding correlations between value-inflected motivations and beliefs or attitudes—including work on cultural cognition theory—tend not to differentiate between rational and irrational influences of values on the assessment of scientific claims. Future work could be geared to do so.

Empirical Implications

Given these theoretical implications, researchers must be more precise in the specific domains or constructs they aim to measure empirically. Take, for example, experimental research that relies on framing effects to evaluate differences in how people process information. Cacciatore et al. (2016) examine how the presentation of information affects people's opinion formation. *Equivalency framing*, drawing largely from psychological literature, examines how otherwise equivalent information can be manipulated to assess if there is an effect on how an individual processes information that is (in)congruent with their beliefs (Druckman 2001).

This makes the approach well-suited to models of motivated reasoning that seek to assess consistency with previously held values or beliefs (e.g., Lord et al. 1979 or Taber and Lodge's model of motivated scepticism). Kahan et al. (2011) found that individuals were more likely to support scientific information congruent with their culturally predisposed position. Equivalency framing studies are most effective when the scientific evidence concerning an issue is fairly well-established, and researchers are seeking to assess which communication strategies may be more effective for a given scenario (Pedersen 2017; Cacciatore et al. 2016).

By contrast, *emphasis framing,* drawing from sociology, examines how presenting specific aspects of an issue unconsciously affects how information is processed. The focus may be on manipulating *what* is received or is salient with different actors, as opposed to ensuring that equivalent content is presented (Cacciatore et al. 2016). Emphasis framing may align with the John Q Public model of political information processing, based on the assumption that unconscious thoughts predict the direction of subsequent reasoning despite conscious deliberation (Taber and Lodge 2016). Emphasis framing may also be a useful strategy in seeking to understand the evolution of a new or emerging risk situation. For example, Driedger et al. (2018) used qualitative thematic analysis to examine how different sets of actors were represented in Canadian news media and on Facebook regarding a controversial hypothesis about a 'promising' new therapy for people suffering from multiple sclerosis. While the need for 'appropriate' and 'standard measures' in following sound science was strongly promoted by scientists and government policy actors, other voices in the debate—patients, advocacy groups, and scientific experts with competing knowledge claims—used oppositional collective action frames to challenge the traditional scientific discourse. By creating a social and political maelstrom, people with multiple sclerosis were able to persuade governments and researchers to respond differently, culminating in the funding of a national clinical trial into a controversial hypothesis that defied all standards of evidentiary support. This type of oppositional collective action might be considered rational skepticism or irrational bias, depending upon the perspective. Nevertheless, while similar studies focused on motivated reasoning have illustrated the presence of bias using similar techniques, it may require more nuanced research approaches to fully understand the causal relationship between stimuli and the value-infused motivations, and to assess the public justifiability of different views.

Further, it may be possible to explore the boundary between rational skepticism and irrational bias by using affective computing and sentiment analysis. Previous studies have used natural language processing techniques to analyze transcripts from interviews with the general public and experts. The research found that people responded positively to information embedded in scientific narrative structures regardless of their stance on the issue (i.e., for or against) (Shanahan et al. 2019). This example may more closely align with emphasis framing. By contrast, it may be more important to examine different types of discourse to understand when individuals respond differently to the same types of information, much like equivalency framing research. One study found that those with different political beliefs often respond to the same types of information positively or negatively in relation to ideology, not facts (Balasubramanyan et al. 2012). Nonetheless, these natural language processing techniques may provide insight to differentiate between rational skepticism as a response to uncertainty and irrational bias.

While an imperfect classification system, equivalence framing is likely more easily assessed with quantitative research and emphasis framing with qualitative studies. That said, looking at motivated reasoning in qualitative research or using non-experimental designs would help researchers to identify and explain motivational biases (Maxwell 2004). Regardless of approach, it is important for researchers to be clear in how they define or use the term 'motivated reasoning', since, as discussed in this chapter, important conceptual differences among models exist. It is also important for them to be explicit about how they define rationality, along with the role and place of values in their research and assessments.

Practical Implications

We identify four key takeaways from this discussion for risk practitioners. First, policymakers and regulators working on risk governance need a better understanding of motivated reasoning and how it affects risk perception. Importantly, research shows that motivated reasoning is a *human* phenomenon—citizens, public authorities, and scientific experts are not exempt from it. In fact, greater expertise on an issue can make individuals more sophisticated in their capacity to reason in a motivated fashion.

Second, the fact that motivated reasoning is inevitably part of any risk decision-making process does not necessarily make decision outcomes flawed or irrational. Rather, the above discussion of rationality deliberately challenges the idea that 'rational', accuracy-oriented, and value-free decision-making processes are superior. Instead, bringing people's values, prior beliefs, and identities *into* public decision-making about risks is crucial to developing and implementing effective solutions and to pursuing democratic legitimacy. Again, rather than chasing an unattainable and ultimately undesirable ideal of solely 'rational' risk governance, greater awareness and better understanding of motivated reasoning (however defined) will better enable policymakers and regulators to detect and address the directional goals, values, and identities that shape people's beliefs and attitudes toward risk issues and to recognize them more effectively in the process—rather than automatically writing them off as irrational and irrelevant distractions.

Third, this discussion also hints at recommendations on what *not* to do in response to motivated reasoning. For instance, simply providing more scientific evidence on a risk issue is not likely to 'cure' people's motivated reasoning by bringing their opinion more in line with science. In fact, research indicates that this strategy may backfire. People may reject messages at odds with their goals and move in the opposite direction of the message (Zhou 2016).

Finally, and perhaps most importantly, since research reveals that *everyone* engages in motivated reasoning, including experts and scientists, the existence of the phenomenon should not be used as an argument *against* efforts to democratize risk governance. In fact, under the guise of 'rationality', doing so makes implicit decisions on *whose* values and objectives matter in risk governance and whose do not, potentially reinforcing the exclusion of marginalized groups.

Conclusion

Research indicates that motivated reasoning is ubiquitous in human thinking and decision-making. But as shown in this chapter, there remain large gaps in our understanding of the phenomenon. We need more clarity around theoretical concepts and models of motivated reasoning, as well as better approaches to studying its effects. Perhaps most importantly, we need to better integrate what we already know about human reasoning into risk governance practice. The normative (if implicit) connotations in research about motivated reasoning should be made transparent and critically discussed. Perceiving motivated reasoning as necessarily harmful to effective risk governance and striving for ‘rationality’ in decision-making about risk ignores the fact that values, identity, and other non-accuracy goals will *always* influence human beliefs and attitudes, and sometimes properly rationally so. Neither experts nor public authorities are immune to directional motivated reasoning. Instead, inclusive and transparent processes that explicitly acknowledge the presence of values and motivations in *all* people’s risk perceptions, assessments, and preferences about risk management open the door to effective and legitimate risk governance.

References

Balasubramanyan, R., Cohen, W.W., Pierce, D., & Redlask, D.P. (2012, June 4–7). Modeling polarizing topics: When do different political communities respond differently to the same news? *Proceedings of the Sixth International Conference on Weblogs and Social Media*. Dublin, Ireland.

Buechler, S.M. (1990). *Women’s movements in the united states: Woman suffrage, equal rights, and beyond*. Rutgers University Press.

Cacciatore, M.A., Scheufele, D.A., & Iyengar, S. (2016). The end of framing as we know it... and the future of media effects. *Mass Communication and Society*, 19(1), 7–23.

Daily Star. (2014, February 7). Japanese women boycott sex with any man who votes for Tokyo’s “menstruating women are irrational” governor. *Daily Star*.

DeFranza, D., Lindow, M., Harrison, K., Mishra, A., & Mishra, H. (2020). Religion and reactance to COVID-19 mitigation guidelines. *American Psychologist*.

Douglas, H. (2000). Inductive risk and values in science. *Philosophy of Science*, 67(4), 559–579.

Douglas, H. (2017). Rejecting the idea of value-free science. In H. Kincaid, J. Dupré, & A. Wylie (Eds.), *Value-free science? Ideals and illusions* (pp. 120–141). Oxford University Press.

Douglas, M., & Wildavsky, A. (1982). How can we know the risks we face? Why risk selection is a social process. *Risk Analysis*, 2(2), 49–58.

Driedger, S.M., Dassah, E., & Marrie, R.A. (2018). Contesting medical miracles: A collection action framing analysis of CCSVI and venous angioplasty (‘liberation therapy’) for people with multiple sclerosis in news and social media. *Science Communication*, 40(4), 469–498.

Druckman, J.N. (2001). The implications of framing effects for citizen competence. *Political Behavior*, 23, 225–256.

Druckman, J.N., & Bolsen, T. (2011). Framing, motivated reasoning, and opinions about emergent technologies. *Journal of communication*, 61(4), 659–688.

Druckman, J.N., & McGrath, M.C. (2019). The evidence for motivated reasoning in climate change preference formation. *Nature Climate Change*, 9, 111–119.

Elliott, K.C., & Richards, T. (2017). *Exploring inductive risk. Case studies of values in science*. Oxford University Press.

Fishbach, A., & Ferguson, M.J. (2007). The goal construct in social psychology. In A.W. Kruglanski, & E.T. Higgins (Eds.), *Social psychology: A handbook of basic principles* (2nd ed., pp. 490–515). Guilford.

Gerber, A., & Green, D. (1999). Misperceptions about perceptual bias. *Annual Review of Political Science*, 2, 189–210.

Gigerenzer, G., & Gaissmaier, W. (2011). Heuristic decision making. *Annual Review of Psychology*, 62, 451–482.

Goldenberg, M.J. (2016). Public misunderstanding of science? Reframing the problem of vaccine hesitancy. *Perspectives on Science*, *24*(5), 552–581.

Harding, S. (1995). 'Strong objectivity': A response to the new objectivity question. *Synthese*, 104(3), 331–349.

Harris, J. (2000). Is there a coherent social conception of disability? *Journal of Medical Ethics*, 26, 95–100.

Harris, J. (2001). One principle and three fallacies of disability studies. *Journal of Medical Ethics*, 27, 383–387.

Kahan, D.M. (2012). Cultural cognition as a conception of the cultural theory of risk. In S. Roeser, R. Hillerbrand, & M. Peterson (Eds.), *Handbook of risk theory: Epistemology, decision theory, ethics, and social implications of risk* (pp. 725–759). Springer

Kahan, D.M. (2013). Ideology, motivated reasoning, and cognitive reflection. *Judgment and Decision Making*, 8(4), 407–424.

Kahan, D.M. (2016a). The politically motivated reasoning paradigm, part 1: What politically motivated reasoning is and how to measure it. In R. Scott, & M. Buchmann (Eds.), *Emerging trends in the social and behavioral sciences*. Wiley.

Kahan, D.M. (2016b). The politically motivated reasoning paradigm, part 2: Unanswered questions. In R. Scott, & M. Buchmann (Eds.), *Emerging trends in the social and behavioral sciences*. Wiley.

Kahan, D.M. (2016c). The expressive rationality of inaccurate perceptions. *Behavioral and Brain Sciences*, 40, 26–28.

Kahan, D.M., Braman, D., Gastil, J., Slovic, P., & Mertz, C.K. (2007). Culture and identity-protective cognition: Explaining the white-male effect in risk perception. *Journal of Empirical Legal Studies*, 4(3), 465–505.

Kahan, D.M., Braman, D., Cohen, G.L., Gastil, J., & Slovic, P. (2010). Who fears the HPV vaccine, who doesn't, and why? An experimental study of the mechanisms of cultural cognition. *Law and Human Behavior*, 34, 501–516.

Kahan, D.M, Jenkins-Smith, H., & Braman, D. (2011). Cultural cognition of scientific consensus. *Journal of Risk Research*, 14(2), 147–174.

Kahan, D.M., Peters, E., Wittlin, M., Slovic, P., Larrimore Ouellette, L., Braman, D., & Mandel, G. (2012). The polarizing impact of science literacy and numeracy on perceived climate change risks. *Nature Climate Change*, 2, 732–735.

Kahneman, D. (2011). *Thinking fast and slow*. Farrar, Straus and Giroux.

Kraft, P.W., Lodge, M., Taber, C.S. (2015). Why people 'don't trust the evidence': Motivated reasoning and scientific beliefs. *Annals of the American Academy of Political and Social Science*, 685, 121–133.

Kunda, Z. (1990). The case for motivated reasoning. *Psychological Bulletin*, 108(3), 480–498.

Kunda, Z. (1999). *Social cognition*. MIT Press.

Kuzma, J. (2018). Regulating gene-edited crops. *Issues in Science and Technology*, 35(1), 80–85.

Leeper, T.J., & Mullinix, K.J. (2018). Motivated reasoning. *Oxford bibliographies political science*. https://doi.org/10.1093/OBO/9780199756223-0237.

Lodge, M., & Taber, C.S. (2013). *The rationalizing voter*. Cambridge University Press.

Lord, C.G., Ross, L., & Lepper, M.R. (1979). Biased assimilation and attitude polarization: The effects of prior theories on subsequently considered evidence. *Journal of Personality and Social Psychology*, 37(11), 2098–2109.

Maxwell, J.A. (2004). Using qualitative methods for causal explanation. *Field Methods*, 16(3), 243–264.

Meghani, Z., & Kuzma, J. (2011). The 'revolving door' between regulatory agencies and industry: A problem that requires reconceptualizing objectivity. *Journal of Agricultural and Environmental Ethics*, 24(6), 575–599.

Osborne, H. (2014, July 1). James Delingpole leads climate change sceptics in trashing IPCC's 'sexed-up' report. *International Business Times*. http://www.ibtimes.co.uk/ipcc-climate-change-report-skeptics-royal-society-509664.

Pedersen, R.T. (2017). Ratio bias and policy preferences: How equivalency framing of numbers can affect attitudes. *Political Psychology*, 38(6), 1103–1120

Posusney, M.P. (1993). Irrational workers: The moral economy of labor protest in Egypt. *World Politics*, 46, 83–120.

Redlawsk, D.P. (2002). Hot cognition or cool consideration? Testing the effects of motivated reasoning on political decision making. *The Journal of Politics*, 64(4), 1021–1044.

Sanderson, I. (2006). Complexity, 'practical rationality' and evidence-based policymaking. *Policy and Politics*, 34(1), 115–132.

Shanahan, E.A., et al. (2019). Characters matter: How narratives shape affective responses to risk communication. *PLOS ONE*, 14(12), e0225968.

Taber, C.S., & Lodge, M. (2006). Motivated skepticism in the evaluation of political beliefs. *American Journal of Political Science*, 50(3), 755–769.

Taber, C.S., & Lodge, M. (2016). The illusion of choice in democratic politics: The unconscious impact of motivated political reasoning. *Political Psychology*, 37(S1), 61–85.

Todd, P.M., & Gigerenzer, G. (Eds.). (2012). *Ecological rationality: Intelligence in the world*. Oxford University Press.

Tversky, A., & Kahneman, D. (1974). Judgment under uncertainty: Heuristics and biases. *Science*, 185(4157), 1124–1131.

Van Montagu, M. (2013). The irrational fear of GM food. *Wall Street Journal*. http://online.wsj.com/news/articles/SB10001424052702303680404579141741399966328.

Viola, L. (1986). Bab'i bunty and peasant women's protest during collectivization. *The Russian Review*, 45(1), 23–42.

Wolbring, G. (2008). Why NBIC? Why human performance enhancement? *Innovation: The European Journal of Social Science Research*, 21, 25–40.

Wolbring, G., & Diep, L. (2016). Cognitive/neuroenhancement through an ability studies lens. In F. Jotterand, & V. Dubljevic (Eds.), *Cognitive enhancement* (pp. 57–75). Oxford University Press.

Wolbring, G. (2019). Ability expectation and ableism governance: An essential aspect of a culture of peace. In Darryl R.J. Macer (Ed.), *Legacies of love, peace and hope: How education can overcome hatred and divide* (pp. 116–126). Eubios Ethics Institute.

Zhou, J. (2016). Boomerangs versus javelins: How polarization constrains communication on climate change. *Environmental Politics*, 25, 788–811.

CHAPTER 3

Science and Values:

The pervasive entanglement

Heather Douglas

Since the mid-twentieth century, insulating science from social and ethical values has been something of an obsession for philosophers of science.[1] Philosophers articulated, and then staunchly defended, a value-free ideal for science. This ideal did not insulate science completely from societal

[1] Whether it was an equal obsession for *scientists* I leave open. There is evidence that asserting the purity of science was central for at least some scientists. See, e.g., Heather Douglas, *Science, Policy, and the Value-Free Ideal* (Pittsburgh, PA: University of Pittsburgh Press, 2009), 64.

This chapter is a reprint of Lecture 1 of the 2016 Descartes Lecture delivered by Heather Douglas (Heather Douglas, *The Rightful Place of Science: Science, Values, and Democracy: The 2016 Descartes Lectures*, Tempe AZ: Consortium for Science, Policy & Outcomes, 2021, 9–36).

H. Douglas (✉)
Department of Philosophy, Michigan State University, East Lansing, MI, USA
e-mail: dougl239@msu.edu

M. Gattinger (ed.), *Democratizing Risk Governance*,
https://doi.org/10.1007/978-3-031-24271-7_3

influence. Philosophers were willing to concede the "context of discovery" to the influence of values (which, in contemporary parlance, includes scientists deciding upon research projects and methodologies), but argued that the "context of justification" had no place for social values. This view was supported by three ideas: (1) that societal values can add no confirmatory weight to empirical claims (and that to think otherwise is to confuse "ought" claims with "is" claims); (2) that values distinctive to scientific theory choice could guide scientists when faced with inferential decisions (i.e., epistemic or cognitive values); and (3) that the authority of science in the public sphere rested on the separation and disentanglement of science from social and ethical values. This final presumption was bound up with hopes for science as a resource in public debates that could transcend divergent societal interests—that science could be a "value-neutral" resource in our democratic discourse.

I will argue that there is something to the first idea—that there is an important conceptual difference between normative and descriptive claims, although in practice they are both used to support each other. Yet because of their logical structures, normative claims cannot provide sole support for descriptive claims, and vice versa. I will argue that the second idea is crucially incomplete—that although there are distinctive epistemic values in science, they cannot decisively guide inference. And, finally, I will argue that the third idea is inadequate as well—that we need a more complex understanding of why we grant science general epistemic authority, with multiple bases supporting that authority.

Descriptive and Normative Claims

It is a standard presumption in philosophy that one cannot derive "is" claims from "ought" claims, nor can one derive "ought" claims from "is" claims. I can make arguments about how one ought to value science or how one ought to value democracy, but that does not mean that the people to whom I am making these arguments *do* value science or democracy. Similarly, I can describe the way the world is in great detail, but someone can always respond, "Yes, but that is not how it ought to be." The difference between "is" and "ought" claims seems crucial for giving us the space to imagine what a better world might be like, even in the face of an (often grim) accurate and detailed description. It also provides space for resisting the automaticity that can follow from a particular "ought" claim. That the world is not like it should be (in some people's eyes) may be a good thing in our view, and we might use that descriptive difference, or the projected costs of change that arise from a detailed description, to resist a normative appeal.

Philosophers debate whether the practical difference between these types of claims is grounded in some metaphysical difference in the nature of normativity. Is the true different from the good? Is the beautiful different from the just? I have no wish to wade into such debates, although it certainly seems plausible that the answer to both these questions is yes. The world does not seem so unified that all the normativities line up. For my purposes here, it suffices to note that making a whole set of descriptive claims (with nothing else) does not make an adequate argument for a normative claim; nor does a whole set of normative claims (with nothing else) make an adequate argument for a descriptive claim. Each kind of claim cannot serve as sole justification for the other. They simply don't interact that way.

They do, however, interact. For example, it is difficult to see how to make an argument about how the world ought to be (or, more pointedly, how we ought to act) without relying upon *some* descriptive claims about the way the world is. We need empirical information about what causes pain, for example, if we are to craft a world with less pain in it. Arguments about what we should do rely upon descriptions of what we can do, what is feasible, what is readily achievable, what comes at higher costs and what those costs are. That we need both kinds of claims in our arguments is indicative that there are in fact two kinds of claims.

Conversely, the question of whether one can make an argument for a descriptive claim without normative claims is a central concern in the current values in science debate, particularly as science is now a major source for our descriptive claims. At issue is whether descriptive arguments rely on (without being built wholly out of) normative claims. The decisions of what to study, how to study it, and when to say the study is completed (when the evidence is sufficient) suffuses normative presuppositions into our descriptive statements. They are not there on the surface (just as descriptive claims are not there on the surface of normative claims) and they do not suffice on their own for arguing for (or supporting) the descriptive claims, but they are part of the overall argument for, and process of, generating the descriptive claims.

But to say that normative claims have a role to play in generating descriptive claims is not to give away the difference altogether. When Carl Hempel argued back in the 1960s that normative claims can provide no confirmatory weight to a descriptive claim, he was right.[2] Saying that the world ought to be a particular way is not a good argument that the world is actually that way. In this form of argument, such normative expressions are more pious hopes than reasons for the accuracy of descriptions. And this is a gap in kind that we want to preserve. The world is often not how we want it to be, and keeping this difference is essential for being able to perceive and to say that.

With this conceptual distinction in place, we can now address the debate over values in science. Acknowledging that there is a conceptual difference between descriptive and normative claims, we can examine more closely how they might (and should) interact in producing science. (For those who are not convinced there is a difference between descriptive and normative claims, the value-free ideal for science doesn't make any sense. In the arguments that follow, I will presume a conceptual difference, but show a practical interdependence, between the two.)

Values in Science

Science is, of course, a human practice. And when we do science, we entangle values, including social and ethical values, in that science. The questions are how values interweave with science, whether it is legitimate and necessary, and ultimately what to do about it.

Critics of the value-free ideal for science initially pointed out how values (particularly social and ethical values) influence the practice and products of science, because science is performed by humans. Feminist philosophers of science showed how sexist values blinded scientists to alternative explanations of phenomena or directed the attention of scientists to some narrow subset of data, a fuller examination of which produced rather different interpretations and results. Examples from archaeology (explanations of how tool use developed), cellular biology (explanations of fertilization processes), and animal biology (explanations of duck genital morphology and mating behavior) demonstrate such influences of values on science in spades.[3] Feminists were quick to point

[2] Carl G. Hempel, "Science and Human Values," in *Aspects of Scientific Explanation and Other Essays in the Philosophy of Science* (New York, NY: The Free Press, 1965), 81.

[3] Alison Wylie, *Thinking from Things: Essays in the Philosophy of Archeology* (Berkeley, CA: University of California Press, 2002); Emily Martin, "The Egg and the Sperm: How Science has Constructed a Romance Based on Stereotypical Male—Female Roles," in

out that problematic science of this sort was not obviously ***bad*** science (scientists were not making up data and not engaged in pseudoscientific practices immune from criticism and revision), but the limitations of it (and the value-influence on it) became obvious once better science was pursued.[4] Looking back on the cases critiqued by feminists, the science looks woefully inadequate and blinkered.

Several feminist philosophers proposed addressing these issues by focusing on the social nature of science.[5] Because science requires communities of scientists, in critical dialogue with each other, feminist scholars looked to the structure of those communities for answers. Improving the diversity of scientists, many argued, would improve the range of explanations pursued and the kinds of phenomena examined, bolstering the epistemic reliability of the sciences. And by having more diversity of participants in science, and more diversity of values through the participants, the value judgments and influences could be more readily spotted by someone in the scientific community rather than disappearing, invisible by virtue of universal acceptance among scientists. In addition, in such an agenda, the virtues of the just and the true could be aligned, as breaking down the barriers to participating in scientific research would be both fairer and produce more accurate science.[6]

Feminism and Science, Evelyn Fox Keller and Helen E. Longino, eds. (New York, NY: Oxford University Press, 1996), 103; Patricia L. R. Brennan, Christopher J. Clark, and Richards O. Prum, "Explosive Eversion and Functional Morphology of the Duck Penis Supports Sexual Conflict in Waterfowl Genitalia," *Proceedings of the Royal Society B* 277, no. 1686 (2010): 1309–1314. See also Donna Haraway, *Primate Visions: Gender, Race, and Nature in the World of Modern Science* (New York, NY: Routledge, 1989); Evelyn Fox Keller and Helen E. Longino (eds.), *Feminism and Science* (New York, NY: Oxford University Press, 1996).

[4] Helen Longino, *Science as Social Knowledge, Values and Objectivity in Scientific Inquiry* (Princeton, NJ: Princeton University Press, 1990); Anne Fausto-Sterling, *Myths of Gender: Biological Theories About Women and Men* (USA: Basic Books, 1985); Sandra Harding, *The Science Question in Feminism* (Ithaca, NY: Cornell University Press, 1986).

[5] Helen Longino, *Science as Social Knowledge, Values and Objectivity in Scientific Inquiry* (Princeton, NJ: Princeton University Press, 1990); Helen Longino, *The Fate of Knowledge* (Princeton, NJ: Princeton University Press, 2001); Lynn Hankinson Nelson, *Who Knows: From Quine to a Feminist Empiricism* (Philadelphia, PA: Temple University Press, 1990); Heidi E. Grasswick and Mark Owen Webb, "Feminist Epistemology as Social Epistemology," *Social Epistemology* 16 (2002): 185–196.

[6] Kristen Intemann, "Why Diversity Matters: Understanding and Applying the Diversity Component of the National Science Foundation's Broader Impacts Criterion," *Social Epistemology* 23, nos. 3–4 (2009): 249–266.

While this is certainly a worthwhile approach to addressing many issues of justice in science and epistemically inadequate science, this approach does not take on the value-free ideal directly. One could argue that the reason for increasing diversity in science is to ferret out those hidden value presuppositions that were distorting the search for truth.[7] Once made clear, one could hope that the values could be removed from the scientific explanations. The called-for diversity in science could be made to serve the ultimate aim of a value-free science. What the feminist critiques showed (for some) is not a problem with the value-free ideal per se but with the past practices of science. The cases of sexist science were weak science, empirically feeble science, and the pursuit of new theories and evidence made science stronger. Stronger science could still aim to be value-free.

Another reason the value-free ideal remained mostly unscathed was that it was narrowly focused on when values need to be kept out. One could still argue that science should be value-free in its justifications, that regardless of how the theories and explanations of empirical phenomena were developed (and feminist critiques showed we needed to improve this process substantially), what mattered when making inferences in science—when deciding what the evidence said—is that scientists try to keep values out of *that* process, and just focus on the evidence at hand (perhaps bolstered by a sense that with diverse participants in science, the evidence at hand is the best available set). The value-free ideal was articulated as being about the moment of inference in science, of being about the practices of justification at one particular point. The idea was that if values were kept out at this point, it could serve as the pure fulcrum for later decisions, that science could be universal and authoritative if and only if values were not part of the justificatory inference. And indeed, the idea that values can offer no confirmatory weight to the pile of evidence, and that if they did we would be blurring the important difference between descriptive and normative claims, added further reason to support the value-free ideal. Scientists needed to make inferences (and justify those

[7] Whether diversity is generally effective at ferreting out value presuppositions is unclear, and probably context dependent. It has certainly helped in some prominent cases examined in the literature, but there is also evidence that implicit biases still pervade scientific practice. Our own tacit value commitments are often opaque even to ourselves. But it does seem that some difference of perspective is needed to make such commitments apparent, whether that comes from within or without, even if the presence of such a difference provides no guarantee of its effectiveness.

inferences) with no regard to social and ethical values, according to the value-free ideal. Maintaining this ideal was crucial to the authority of science, which rested on purity from societal influences at the point of inference.

To upend the value-free ideal, and its presumptions about the aim of purity and autonomy in science, one needs to tackle the ideal qua ideal at the moment of justification. This is the strength of the argument from inductive risk. It points to the inferential gap that can never be filled in an inductive argument, whenever the scientific claim does not follow deductively from the evidence (which in inductive, ampliative sciences it almost never does). A scientist always needs to decide, precisely at the point of inference crucial to the value-free ideal, whether the available evidence is enough for the claim at issue. This is a gap that can never be *filled*, but only stepped across. The scientist must decide whether stepping across the gap is acceptable. The scientist can narrow the gap further with probability statements or error bars to hedge the claim, but the gap is never eliminated.

How is a scientist to decide that the available evidence is enough? That the gap is worth stepping across? That a claim is worth accepting? Some have suggested that epistemic and/or cognitive values can do this. It is time to examine whether there are "canons of inference" that can fulfill this role.

Epistemic and Cognitive Values: What Guidance?

When Isaac Levi suggested in 1960 that there were "canons of inference" that guided decisions of acceptance in science, and that these were sufficient for theory assessment in science, he helped to put in place a crucial piece of the value-free ideal.[8] There has been voluminous work on what became known as "epistemic values" (for some, "cognitive values") in science. Some of the work has focused on particular attributes (e.g., What is the value of simplicity? Does prediction matter more than accommodation? What constitutes a good explanation?),[9] and discussions

[8] Isaac Levi, "Must the Scientist Make Value Judgments?" *The Journal of Philosophy* 57, no. 11 (1960): 345–357.

[9] On simplicity, see Malcolm Forster and Elliott Sober, "How to Tell When Simpler, More Unified, or Less Ad Hoc Theories will Provide More Accurate Predictions," *The British Journal for the Philosophy of Science* 45, no. 1 (1994): 1–35. On explanation,

initially described a collective soup of values that scientists held.[10] More recent work has involved unpacking nuance among the values considered constitutive of science.[11]

It has helped enormously to consider what these values are good for. Instead of merely noting their pervasive importance in science (historically and currently), one could attend to differences in why particular values might be central to science. For example, successful prediction and explanation are values that organize the evidence in relation to theory, and as such help to structure how we assess the strength of the available evidence.[12] Precision in successful explanation and prediction similarly helps assess how strong the evidence is—if precise theories explain or predict precise evidence, we think the evidential support is so much the stronger for the theory. Theories that successfully predict or explain a broad scope of evidence (across a range of phenomena), or theories that successfully predict or explain complex phenomena with simpler theoretical apparatus, also are judged to be supported more strongly by the evidence than competitors without these virtues. These kinds of values are properly epistemic, as they help us judge how good a theory is at this moment, and how strong the currently available evidence is.

Note that while these virtues are very helpful in assessing the strength of the available evidence, they are mute on whether the available evidence is *strong enough* to warrant acceptance by scientists. Such epistemic values do not speak to this question at all.

see Wesley C. Salmon, *Four Decades of Scientific Explanation* (Pittsburgh, PA: University of Pittsburgh Press, 1989), and Heather Douglas, "Reintroducing Prediction to Explanation," *Philosophy of Science* 76, no. 4 (2009): 444–463, for initial orientations. On prediction versus accommodation, a review can be found in Heather Douglas and P. D. Magnus, "State of the Field: Why Novel Prediction Matters," *Studies in History and Philosophy of Science* 44 (2013): 580–589.

10 See, e.g., Thomas Kuhn, "Objectivity, Value Judgment, and Theory Choice," in *The Essential Tension: Selected Studies in Scientific Tradition and Change* (Chicago, IL: University of Chicago Press, 1997), 320–339.

11 Helen Longino, "Cognitive and Non-Cognitive Values in Science: Rethinking the Dichotomy," in *Feminism, Science, and the Philosophy of Science,* Lynn Hankinson Nelson and Jack Nelson, eds. (Dordrecht, Netherlands: Springer, 1996), 39; Heather Douglas, "The Value of Cognitive Values," *Philosophy of Science* 80, no. 5 (2013): 796–806.

12 Heather Douglas, "Reintroducing Prediction to Explanation," *Philosophy of Science* 76, no. 4 (2009): 444–463.

Other traditionally constitutive values in science are more future oriented, and direct our attention to the promise of a theory in the future. These values, such as broad scope over potential (but as yet ungathered) evidence, fecundity in producing predictions (as yet untested), and explanatory power (as yet uninstantiated), are suggestive of the general fruitfulness of a theory. But such future fruitfulness is only a reason to keep working on a theory, to use that potential fecundity to explore the world further, to accept it as a basis for further research, not to accept it generally for other decision-making. I have called these values "cognitive values," because their presence means that a theory will be easier to work with going forward, and thus they have a pragmatic research value for scientists.[13] They are not epistemic, as they do not indicate the general reliability of a theory—they do not tell us that a theory is well supported and likely to produce accurate predictions. They do, however, indicate good research bets. Thus, neither the set of epistemic nor the set of cognitive values can tell us when we have enough evidence. They simply do other jobs.

There are two sources of trouble here for seeing the normative entanglement of science and social values. The first is local: that many of the cognitive values have the same name as the epistemic values, and thus are readily conflated. Predictive power could be a name for past successes (and thus be epistemic) or could be a name for future fecundity (and thus be cognitive). The same goes for explanatory power or scope or even precision and simplicity. That there is a sense of these values that is directed to past success in grappling with and organizing actual evidence (an epistemic sense) and that there is a sense of these values that speaks of the future promise of a theory (a cognitive sense) confuses things. It also makes it seem as though the general list of such values is indeed sufficient for science—for what else do scientists qua scientists need but to assess the strength of evidence and to decide upon the future promise of potential research questions?

But such a conception of scientific practice neglects that we want something else from scientists; that indeed, science is not just pursued for scientists alone. We need to know what to think about the world right now, and not just to know which theories are promising for future research. And we need to know more than how strong the evidence is

[13] Heather Douglas, "The Value of Cognitive Values," *Philosophy of Science* 80, no. 5 (2013): 796–806.

for a particular theory—we need to know whether it is strong enough to use for deciding what to do in the wider world beyond the endeavors of scientific researchers. The inductive gap remains, despite the utility of epistemic and cognitive values, and we have to know what to do about it. Should it be stepped across or not? Even with probability statements or error bars, does the available evidence support the claim enough? Epistemic values can help assess how strong the evidence is; cognitive values can help assess where to place bets for future research. But for the assessment of evidential sufficiency in the moment, we need to look beyond epistemic and cognitive values.

The Necessity of Social and Ethical Values in Science

How do social and ethical values help with this inductive gap? While they can't fill it, they are crucial for deciding when the evidence available (the strength of which is assessed using epistemic values) is strong enough. Strong enough for what? What is this assessment of sufficiency? How does a scientist decide that the inductive gap is acceptably small enough to step across? It is here, at this question, that philosophers and scientists must stop looking at the purely internal practices of science and answer this question with respect to a full understanding of science as it operates *within* societies, rather than *isolated from* societies. When scientists decide the evidence is strong enough, they are deciding not just for themselves, but for anyone who wants to rely upon science for guiding decisions in the broader world. For that, the internal practices and values of science are not sufficient.

Social and ethical values, however, do help with this decision. They help by considering the consequences of getting it wrong, of assessing what happens if it was a mistake to step across the inductive gap (i.e., to accept a claim), or what happens if we fail to step across the inductive gap when we should. In doing so, such values help us to assess whether the gap is small enough to take the chance. If making a mistake means only minor harms, we may be ready to step across it with *some* good evidence. If making a mistake means major harms, particularly to vulnerable populations or crucial resources, we should change our standards accordingly. Social and ethical values weigh these risks and harms, and provide reasons for why the evidence may be sufficient in some cases and not in others.

The difficulty is that there are risks of error in all directions. There are risks of error in prematurely making a claim; there are risks of error in failing to make a claim soon enough; and there are risks of error in saying nothing while we wait for more evidence. There is no perfectly safe space in which to stand. Neither science nor logic can assure us of safety—indeed nothing can. There are no guarantees. What this examination of science, values, and inference can give us is not assurances of success, but assurances that we are doing the best we can—and what that best consists of. Doing our best in science requires the involvement of social and ethical values in the decision that evidence is sufficient.

There are alternatives to involving social and ethical values in evidential sufficiency assessments. We could simply toss a coin when deciding whether to accept or reject a claim. But this would be arbitrary, and thus irresponsible to the authoritative weight that science has in society. And we would still need to decide when the evidence was enough to warrant the coin toss! We could also set standards internal to science: What are the risks to scientific researchers and to the practice of science of accepting or rejecting a claim? But that is also arbitrary—arbitrarily insular: Why should impacts on scientists and research be the only impacts that count? Note that this too would still involve ethical values (some of the impacts on scientists would surely be ethically weighty), but we would be considering only scientists. Why should we do that? With science taking place within a broader society, why should only scientists count in making these decisions? We could ask that scientists never step across inductive gaps, but merely tell us the evidence and how strong they think it is. The practical difficulties of this are insurmountable. As I have argued, the moment of inference is not the only place where inductive risk considerations arise.[14] In addition, we would have to learn how to examine the evidence ourselves, as scientists would no longer be free to tell us what it means (that would be drawing the inference). Finally, we could require that scientists only step across the inductive gap when it is very, very small, and thus be as conservative in their risk-taking as possible. But why is this the right standard? Such a standard presumes that only risks of making a claim incorrectly matter, and ignores the risks of not making a claim when it is true, of waiting too long.

[14] See, e.g., Heather E. Douglas, "Inductive Risk and Values in Science." *Philosophy of Science* 67, no. 4 (2000): 559–579.

To attempt to be value-free in the assessment of evidential sufficiency is to ignore the broader society in which science functions, by being arbitrary, or ignoring the full set of risks, or ignoring the implications of scientific work in the broader society. If science is to be responsible to the broader society in which it functions, if it is to earn its authority, it should not be value-free at all. Instead, it needs to be value-responsive.

Suppose one still wanted to maintain the purity of science from social and ethical values, and that to do so one was willing to institutionally isolate scientists from society. This would involve not only making sure that only risks to scientists and to research were considered in the assessments of evidential sufficiency, but keeping scientists from saying anything publicly about their research. Others would need to maintain and police the border between science and society, deciding what bits of information, which pieces of scientific research, were ready for public consumption and which were not. Communication among scientists would need to fall behind a shroud of secrecy, insulating scientific meetings, publications, and debates from public consumption. Scientists could be free to pursue inquiry indefinitely, and someone else would need to decide when the evidence was enough to instigate other decisions or actions. Scientists would need to eschew the public eye, and would likely need to be physically isolated from the rest of society. We could sever science from society in this way, and thus keep scientists willfully ignorant of the societal implications of their research and from thinking about them. We could have others trained to do this for scientists and have those specialists deciding when evidence was sufficient for a public communication of a claim.

I think we should view such an approach with alarm, and indicative of a misplaced desperation to keep science "pure." Not only would such isolation likely produce questionable science (because the forums for discourse would have to be closed to only professional scientists, who would have to be more strictly credentialed than is currently the case, thus narrowing who was engaged in scientific discourse), but we would need to create and monitor an entirely new social institution. Who would keep track of the boundary policers, and whether they were acting in the public interest or corrupted by a narrower interest? These would be very difficult issues to address. It would also be a very authoritarian institution, as it would require the end of the free exchange of information, and sequestering of the entirety of empirical investigation under confidentiality wraps. The potential for abuse in such an institution is staggering. Despite the complexity we face with the demise of the value-free ideal,

I think addressing the difficulties of relinquishing the value-free ideal is both more manageable and desirable than a truly isolated scientific enterprise.

Nevertheless, the demise of the value-free ideal does leave us with a problem in thinking about science and values: What ideals *should* guide the interaction of science and values?

Searching for New Ideals

That we need some ideals for values in science seems clear. Social and ethical values can have distorting and problematic effects on science, as evidenced by the cases of sexist science uncovered by feminists. Such cases are just one way in which social and ethical values can distort science. Occurrences of manufactured doubt show the influence of social ideologies on scientific research. Because the purveyors of doubt care so much about protecting unfettered capitalism, they are willing to distort the scientific record to forestall unwelcome policies.[15] Social values such as making a profit can lead scientists in the employ of for-profit entities to bend science (e.g., by selectively reporting the results of clinical trials in medical research).[16] And some cases of scientific fraud can be viewed as a pernicious influence of social values, when scientists are so sure of how the world should be, they make up the data to show that it is that way (e.g., the psychologist Cyril Burt and the manufacture of twin data to support his beliefs about the inheritability of intelligence).[17] Social and political values also drove such catastrophic cases as the influence of Trofim Lysenko on Soviet science under Stalin. We should not be sanguine about allowing social and ethical values into science unfettered. Such laissez-faire attitudes about values can make a mess of science.

Philosophers of science have offered several alternative ideals for thinking about how values should operate in science. I will articulate

[15] Naomi Oreskes and Erik Conway, *Merchants of Doubt: How a Handful of Scientists Obscured the Truth on Issues from Tobacco Smoke to Global Warming* (New York, NY: Bloomsbury Press, 2010).

[16] Thomas O. McGarity and Wendy E. Wagner, *Bending Science: How Special Interests Corrupt Public Health Research* (Cambridge, MA: Harvard University Press, 2008).

[17] Stephen Jay Gould, *The Mismeasure of Man* (New York, NY: W.W. Norton, 1981); Raymond E. Fancher, *The Intelligence Men: Makers of the IQ Controversy* (New York, NY: W.W. Norton, 1985).

those ideals here and assess their strengths and weaknesses. We will see that there is no one all-encompassing ideal that can replace the traditional value-free ideal. What relinquishing the value-free ideal requires is that we grapple with a more complex terrain of science-society interactions.[18] Different ideals get at different aspects of scientific practice more or less effectively. Understanding their strengths and weaknesses allows us to see what they are useful for both philosophically and practically.

In the current literature (and I can make no claims to completeness in this fast-moving field), there are at least five different ideals (or norms) for values in science:

1. Placing priority on epistemic values
2. Role restrictions for values in science
3. Getting the right values into science
4. Ensuring proper community functioning
5. Ensuring good institutional structures for scientific practice

Let me describe each, articulating their strengths and weaknesses, and then we can see how they fit together.

Placing Priority on Epistemic Values

Daniel Steel has suggested that the correct ideal for values in science is to make sure they do not hinder the attainment of truth (within the realm of "practically and ethically permissible" science).[19] Ethical values, of course, do restrict our methodologies and the kinds of science we pursue, so Steel does allow those kinds of restrictions on scientific research, even if they do hinder the discovery of new truths. But aside from this restriction, Steel wants no social or ethical values to interfere with the attainment of truth.

This is an interesting ideal, but presents some problems for practical guidance in science. It can be hard in practice to know whether a particular value judgment (whether social, ethical, or cognitive) is helping or

[18] One might say that we are now free to grapple with that more complex terrain, as a purely epistemic approach is no longer remotely adequate.

[19] See, Daniel Steel, "Epistemic Values and the Argument from Inductive Risk," *Philosophy of Science* 77, no. 1 (2010): 14–34; Daniel Steel and Kyle Powys Whyte, "Environmental Justice, Values, and Scientific Expertise," *Kennedy Institute of Ethics Journal* 22, no. 2 (2012): 163–182.

hindering the attainment of truth in the middle of a research project or scientific debate.[20] Part of the excitement of science is not knowing where the truth lies, so whether a value is helping or hindering can be quite unclear without the benefit of hindsight. In addition, one can wonder whether this is the right approach to take even in cases where social and ethical values do hinder the attainment of truth. What counts as ethically permissible science is an ongoing contested arena (as the debate over gain-of-function viral research shows).[21] Sometimes ethical values can inhibit the attainment of truth (because researchers are following their conscience) before the ethical debate is settled, and we might be quite happy about that in retrospect. In short, this ideal works well only when we have settled both what the truth is and what the ethical boundaries of permissibility are, which means guidance *in medias res* is lacking. And we might decide in hindsight that some truths are not worth having, given the ethical costs of getting them. This ideal seems primarily useful for retrospective examinations of scientific debates.

Role Restrictions for Values in Science

In my work, I have emphasized distinct roles for values in science. I have argued that there are two roles for values in science: a direct role (where values serve as a reason to do something, and thus direct the decision) and an indirect role (where values serve to help assess whether the available evidence is sufficient for an inference or choice). I have argued that depending on where one is in the scientific process, different roles are acceptable. For example, a direct role for values is acceptable in deciding which research agenda to pursue (e.g., because the scientist cares about a particular issue) and in deciding which methodologies to employ (e.g., because a particular methodology is ethically preferable). An indirect role is acceptable in these instances as well. But at moments of data characterization and inference (the targeted terrain of the value-free ideal), I have argued that we can maintain scientific integrity while permitting social and ethical values by constraining such values to the indirect role

[20] As noted by Kevin C. Elliott, "Douglas on Values: From Indirect Roles to Multiple Goals," *Studies in History and Philosophy of Science Part A* 44, no. 3 (2013): 375–383.

[21] See Michael J. Selgelid, "Gain-of-Function Research: Ethical Analysis," *Science and Engineering Ethics* 22, no. 4 (2016): 923–964.

only.[22] It is also an ideal that can help guide discourse on contentious scientific issues, as it allows for both the expression of values ("Because of this value, I find the evidence insufficient") and guidance for productive debate ("What evidence would be convincing for you?").

This ideal is a direct counter to the value-free ideal, and targeted as narrowly as the value-free ideal is on these "internal" inferential moments. As such, it has little to say about the direction of research agendas. Further, it cannot help much with methodology selection (or distortion). Finally, it is not much of an ideal in the sense of something to strive for. It is more of a minimum floor, which if one does not meet, one is doing really poor science (such as writing down the data one wishes were accurate or making inferences that one wishes were true). Although I think it is an important norm to hold, it will not suffice for guiding scientific practice.

Getting the Right Values in Science

Several philosophers of science have argued in recent years that the important thing to focus on for values in science is making sure that the right values are influencing scientific research.[23] Such authors have taken an "aims-oriented" approach to the problem of values in science. Janet Kourany, for example, has argued for a "joint satisfaction" ideal for values in science—that only when a decision meets both epistemic and ethical criteria is it a good decision. Kevin Elliott has called attention to the multiple goals of science, including both epistemic aims and social aims.

There are several things to note about this approach. The first is that all the authors that champion this ideal take pains to express concerns for, and support of, the value of inquiry. Both the epistemic aim and

[22] Heather Douglas, *Science, Policy, and the Value-Free Ideal* (Pittsburgh, PA: University of Pittsburgh Press, 2009); Heather Douglas, "Scientific Integrity in a Politicized World," in *Logic, Methodology, and Philosophy of Science: Proceedings of the Fourteenth International Congress*, Peter Schroeder-Heister, Gerhard Heinzmann, Wilfrid Hodges, and Pierre Edouard Bour, eds. (London, UK: College Publications, 2014), 253.

[23] See, e.g., Janet A. Kourany, *Philosophy of Science After Feminism* (New York, NY: Oxford University Press, 2010); Kevin C. Elliott and Daniel J. McKaughan, "Non-Epistemic Values and the Multiple Goals of Science," *Philosophy of Science* 81, no. 1 (2014): 1–21; Daniel Hicks, "A New Direction for Science and Values," *Synthese* 191, no. 14 (2014): 3271–3295.

the ethical/social aim must be met, for example, in Kourany's joint satisfaction ideal. So this ideal is not just about social and ethical values, but about valorizing the general purpose of inquiry and discovery as well. The pursuit of truth matters a great deal to those who argue for this approach.

The second is that this approach successfully addresses concerns about research agenda choices and methodological choices, about which the role-restriction norm has little to say. Because both roles for values are acceptable for these choices, that approach has no normative bite at these stages. Arguing about what the right values are is exactly on target for these choices. For example, in cases where the methodological choices seem to be made to guarantee preselected outcomes, the get-the-right-values-in-science ideal can say that the decisions improperly neglect the value of inquiry, and thus are improper decisions.[24]

Finally, the authors who support this approach tend to want the values utilized to be also the result of good inquiry—not necessarily of the same kind as empirical scientific research, but still informed by good empirical results and robust philosophical debate. Values are not mere contaminants in our process of inquiry with this ideal, but a strong support of it, as they too are open to inquiry.[25]

However, despite its importance, it is doubtful that this ideal is enough. First, what the right values are is often hotly and openly contested. How we know we have the right values can be unclear. So guidance for scientists in practice can be lacking. Second, at the moment of inference (the moment of central concern to the value-free ideal and to the role-restriction ideal), this ideal provides either inaccurate or incomplete guidance. What are we to do when evidence arises that seems to challenge our value commitments? Suppose (and I think this unlikely) that we discover men and women really do have divergent mathematical abilities. Do we reject the evidence because it does not meet the joint satisfaction of ethical and epistemic values? Suppose it is strong evidence (and so meets the epistemic criterion). Do we reject it because it does not fit with our

[24] See also Torsten Wilholt, "Bias and Values in Scientific Research," *Studies in History and Philosophy of Science Part A* 40, no. 1 (2009): 92–101.

[25] This is a pragmatist, Deweyian view on value judgments. See Elizabeth Anderson, "Uses of Value Judgments in Science: A General Argument, with Lessons from a Case Study of Feminist Research on Divorce," *Hypatia* 19, no. 1 (2004): 1–24; and Matthew J. Brown, "Values in Science Beyond Underdetermination and Inductive Risk," *Philosophy of Science* 80, no. 5 (2013): 829–839.

ethical commitments? This seems to conflate the "is" and the "ought," and falls into the trap of wishful thinking and worrisome distortion that the value-free ideal was meant to ward off. It is also a case where the role-restriction ideal serves us well. We can say we want stronger evidence before we are willing to give up on our belief in the general equality of mathematical ability, and we can even say (one would hope) what such evidence should consist of. But rejecting the evidence because we do not like what it says is unacceptable. It is precisely this move that climate deniers often make, and we are rightly frustrated by that.

In short, for guiding scientists in practice, we need both of these ideals—the role-restriction ideal and the get-the right-values-in-science ideal—in operation, although at different levels of granularity. At particular moments of inference, getting roles right is important. And in general, having the right values is important. Indeed, one could justify the roles ideal in terms of the aims ideal—that valuing inquiry properly means, in part, keeping values in the right roles. But as noted above, there is often contention about what the right values are. To address this, we will need a broader communal perspective.

Ensuring Proper Community Functioning

One of the weaknesses of the get-the-right-values-in-science ideal is that it is mute when we don't know what the right values are. What then? Or, what if the right values encompass a plurality of values, all legitimate, with good reasons to support them and reasonable disagreement among them? What kind of ideal can we articulate under these circumstances? Further, the previous ideals generally centre on the impact of values on particular scientific choices. How can we ensure that the conditions that support the requisite critical debate and pluralistic reflection in science are in place?

Philosophers of science (led by feminists) have focused on describing the conditions for proper community functioning to address these concerns. Ensuring that one has a diverse scientific community—with clear forums for debate, expectations for the uptake of criticisms, and effective distribution of research efforts reflecting needed diversity so that alternative theories can be explored—serves to provide essential conditions for the robustness and reliability of science.[26] Such conditions also

[26] See Helen Longino, *Science as Social Knowledge, Values and Objectivity in Scientific Inquiry* (Princeton, NJ: Princeton University Press, 1990); Helen Longino, *The*

provide assurance that value judgments will be elucidated and examined within the scientific community, and that if there are disagreements about which values should be shaping research agendas, those debates can occur in an open and productive way. Having proper community functioning is essential to ensuring that, if there is general agreement on the values, the right values influence science, and, if there is not agreement on the values, some diversity of values will be deployed in making judgments in science.

Some minimum of effective community functioning is needed for producing acceptable science. But we can always do better along the ideals that philosophers like Miriam Solomon and Helen Longino provide for us. This set of ideals, focused as it is on how communities of scholars should work and distribute their efforts, complements ideals 2 and 3, which are more focused on how particular choices should be made in science. The communal functioning ideal calls for proper response and uptake of criticism, for example, but it is from ideal 2, from an articulation of how values can properly play roles in scientific reasoning, that we can see what proper response and uptake consists of. (It is not proper, for example, to say: "I don't accept that empirical claim because it disagrees with my values." It *is* proper to say: "I find that evidence insufficient because of my values and my concern over false positives, so I want stronger evidence before accepting that claim.") That we need ideals both for governing particular choices and for guiding communities should not be surprising. What none of these ideals address, however, is how the scientific community should interact with the broader (democratic) public.

Ensuring Good Institutional Structures for Scientific Practice

While the social epistemological tendencies reflected in ideal 4 are useful for thinking about how we want our scientific communities to work, they do not help inform how the scientific community should think about its role and responsibilities to the broader society or how we want to

Fate of Knowledge (Princeton, NJ: Princeton University Press, 2001); Miriam Solomon, *Social Empiricism* (Cambridge, MA: MIT Press, 2001); Kristen Intemann, "Why Diversity Matters: Understanding and Applying the Diversity Component of the National Science Foundation's Broader Impacts Criterion," *Social Epistemology* 23, nos. 3–4 (2009): 249–266; among others.

structure the science-policy interfaces that so powerfully shape the pursuit and use of science. This area for ideals is the least developed.

It is on this kind of interaction that many ideals articulated by philosophers working on science policy have focused.[27] The trouble is that the science-policy interface is multifaceted, and philosophers have yet to grapple with all the facets in articulating an ideal. What constitutes good institutional structure is very much up for debate.

For now, I hope I have shown that we need some set of nested ideals crafted from those described above. Ideal 2 is the most targeted response to the value-free ideal (both narrowly focused on inferences in science), but once we relinquish this ideal and confront the complexity of science in society, it seems obvious that no one ideal will suffice. Without the value-free ideal narrowing our focus, we have to think about and address all the ways in which values do influence science and consider how that should occur.

The Authority of Science and Ideals for Science

No one ideal for values in science will suffice. We need nested ideals, articulated for individual actors, communal practices, and science-society interfaces, in order to ground the authority of science.

The authority of science rests on the interlocking character of these norms. At the communal level, scientists are expected to continually question and critique each other's work. They are expected to respond to criticisms raised, and to hold no scientific claim above criticism. Such mutual critique is a minimum for granting science prima facie epistemic authority. The more diverse and reflective of the plurality of society the scientific community is, the more taken-for-granted assumptions and

[27] See, e.g., Philip Kitcher, *Science in a Democratic Society* (New York, NY: Prometheus Books, 2011); Justin B. Biddle, "Can Patents Prohibit Research? On the Social Epistemology of Patenting and Licensing in Science," *Studies in History and Philosophy of Science Part A* 45 (2014): 14–23; Heather Douglas, "Inserting the Public into Science," in *Democratization of Expertise? Exploring Novel Forms of Scientific Advice in Political Decision-Making*, Sabine Maasen and Peter Weingart, eds. (Dordrecht, Netherlands: Springer, 2005); Heather Douglas, "Reshaping Science: The Trouble with the Corporate Model in Canadian Government," *Bulletin of the Atomic Scientists* 71 (2015): 88–97; James Robert Brown, "The Community of Science," in *The Challenge of the Social and the Pressure of Practice: Science and Values Revisited*, Martin Carrier, Don Howard, and Janet Kourany, eds. (Pittsburgh, PA: University of Pittsburgh Press, 2008), 189.

unexamined value commitments will (hopefully) be elucidated, the more authority science should have.

But community practices need good individual reasoning practices with which to operate. Maintaining the proper roles for values in science keeps values from acting in place of evidence, which will support the critical interactions needed in science. New evidence should always be able to contest old positions, and this can only happen if values are not used to protect desired positions from unwanted criticism.[28] A scientist can point to their values to argue for why they require more evidence to be convinced, but they can never point to their values to argue for why evidence is irrelevant to the claims they make or protect. Asking for more evidence drives the inquiry dialectic; holding claims above evidential critique does not.

Further, it is not just in individual reasoning integrity (right roles) and communal practices, but in some shared values (operating within proper roles) that science gains its authority in a democratic society. Getting the values right, particularly in the realm of policy-relevant science, strongly supports scientific authority. That scientists are investigating questions we care about, using methodologies that we find morally acceptable and targeted at what we are concerned with, and using values we share for assessing evidential sufficiency, can and should make a big difference for what we think is epistemically authoritative. Thus, elucidating the proper roles and proper values for science is part of what makes science authoritative, rather than undermining the authority of science.

Finally, the authority of science also rests on its raw instrumental success. Relying upon scientific understandings of disease (e.g., in the instance of communicable diseases) has greatly increased lifespans; relying upon scientific understandings of materials has greatly increased the range of what we can manufacture; relying upon scientific understandings of what we can transmit in the air has transformed communication; and so forth. It is this raw instrumental success that is probably at the root of most of the trust that society places in science. But we are running into areas of science where success is not easily measured, especially in the

[28] As Elizabeth Anderson has argued, values in science should not lead to dogmatism. See Elizabeth Anderson, "Uses of Value Judgments in Science: A General Argument, with Lessons from a Case Study of Feminist Research on Divorce," *Hypatia* 19, no. 1 (2004): 1–24.

short term, and the problems we are addressing seem more interrelated than ever. The challenge of science in democracy is still with us.

Implications

There is much work to be done in further fleshing out the ideals for individual, communal, and societal practices in science. We need these levels of norms to mesh together (at least somewhat), so that our communal expectations and societal practices do not place impossible burdens on individual scientists. We need to figure out how these norms align and how to encourage the pursuit of the ideals in real scientific practice.

But we also need to ensure that there is some space between what society might want and what scientists can pursue. While the full autonomy and isolation of science is undesirable, we also don't want a science that only tells us what we want to hear. Some space is crucial for the practice of science. Keeping social values out of a direct role at the moment of inference is part of maintaining this space. Allowing scientists to have a say about research agendas (and to pursue some research for curiosity's sake) is another.

Science cannot be just a mouthpiece for societal interests. If it becomes this, it will not have any claim to distinctive epistemic authority. While we need knowledge to help us pursue our social goals, we also sometimes need to know when such goals are not feasible or desirable (because of what else will come with their successful instantiation). Science needs to be able to tell us when we are running into such issues, to be able to "speak truth to power." This ability is central to its authority in practice.

CHAPTER 4

The BIAS FREE Framework: A Tool for Science/Technology and Society Education to Increase Science and Risk Literacy

Gregor Wolbring

Background

Many academic and non-academic educational efforts are positioned at the intersection between science and/or technology and society with the purpose of increasing the literacy of students and others on the social impact of science and technology and the ability of students to contribute to the academic and non-academic discussions around the social impact of science and technology. To become risk literate about the social risks and other consequences of scientific and technological advancements is a critical aspect of scientific literacy (G. Gardner et al., 2010; G. Gardner and Jones, 2011; Schenk et al., 2019; Ratcliffe and Grace, 2003; Cross, 1993; Lange, 2012). To become risk literate about risk perception is

G. Wolbring (✉)
University of Calgary, Calgary, AB, Canada
e-mail: gwolbrin@ucalgary.ca

M. Gattinger (ed.), *Democratizing Risk Governance*,
https://doi.org/10.1007/978-3-031-24271-7_4

important given the differences in risk perception between groups, the different social risks for different groups and the many groups and their values, circumstances, and motivations one has to take into account in risk governance, as outlined in the introduction to this volume.

The purpose of this chapter is not to engage with risk concepts conceptually or to provide case studies but to introduce the reader to the BIAS FREE Framework (Building an Integrative Analytical System for Recognizing and Eliminating InEquities, BFF), a tool developed for identifying and avoiding biases that derive from social hierarchies by posing 20 analytical questions (Appendix) (Burke and Eichler, 2006; Eichler and Burke, 2006). The premise is that the BFF is a useful tool for educational efforts within and outside academia to enhance risk literacy of scientific and technological advancements and to become literate on topics arising at the intersection of science and society, such as risk perception, science and technology governance, risk governance, public values, science values, trust, biases, and policy development. The BFF benefits risk governance activities as it allows for the unmasking of biases, premises, and positionalities of different actors impacted by science and technology and by revealing differences in risk narratives between different actors and differences in judging risks based on values, ideologies and life experiences of a diversity of people and public perceptions (see Chapter 1).

To have a tool that gives people a set of analytical questions they can use routinely to look for biases, values, premises, and positionalities evident in a given document, whether an academic article, newspaper, social media post or policy document benefits risk communication as it allows one to flag potential problems in risk communication, which, if fixed, could lead to more "accurate and trusted risk communication," seen as "pivotal" to contemporary risk governance (Chapter 1). If this tool is used routinely, it might increase risk literacy of the user, which in turn might also be beneficial for risk governance as "effective risk governance requires successfully confronting differences in expert and public perceptions of risk" (Chapter 1). The BFF could also be useful to engage with motivated reasoning, something seen as an important factor in risk governance endeavours, as discussed in Chapter 2. Moreover, the BFF could help to evaluate at least two of the four dimensions of public participation (inclusiveness and representativeness, and deliberative quality) covered in the chapters of this book. As to this book's focus on risk assessment, risk management, and risk communication, the

BFF could be useful to unravel biases, values, premises, and positionalities in a given discussion or document, which might allow one to see the usefulness/limitation of the analyzed discourse or document. If done well, this could help strengthen trust in risk assessment, management, and communication. The BFF allows one to analyze information framing techniques in communicating risk (the topic of Chapter 12), increases sensitivity to things like multicultural differences in trusted information sources and channels (the subject of Chapter 14), and empowers the person using the BFF by increasing their literacy on the topic they are investigating (an element of Chapter 10). As put forward in the introduction to this volume, "analysing risk perception through the lens of motivated reasoning emphasizes the pivotal role of values in assessing and managing risk, and helps to identify the values and value conflicts implicit in or obscured from view by traditional science-based assessments." The BFF allows the user to think about the values, premises, and positionalities evident in a given document.

The chapter gives first some background on the topics of science and society education and on science and risk literacy. Then it outlines problems identified within the academic literature around science and society education and science and risk literacy. Finally, it introduces the BFF and gives nine examples highlighting the usefulness of the BFF as an educational tool to discuss societal aspects of science and technology and to increase risk and science literacy all of which will result in benefiting risk governance.

Science and Society Education

Many academic educational efforts are positioned at the intersection between science and/or technology and society such as science, environment, society, and technology (SEST); environment education (EE); science, environment, technology, and society (SETS); science technology and literacy (STL); science and education for sustainability; STEM and society; science, technology, and society (STS); socio-scientific issues (SSI); science for citizenship; science and technology in society (SATIS); science through science, technology, and society (S-STS); Science in Social Context (SISCON), social studies of science, public understanding of science, science/technology and human values; science and literacy; environmental education; and science, mathematics, and technology education (SMTE). All these discourses cover educational aspects. For example, science, technology, and society (STS) education is grounded in the belief that science education should include historical, philosophical, cultural, sociological, political, and ethical perspectives and the ability to move from individual cases to system thinking (Pedretti, 1999; Waks, 1989). It is argued that science education and SSI education enable students to partake in socio-scientific decisions (Fensham, 2007; Chowdhury, 2016). Teaching STEM and society courses to teachers has a positive impact on teachers ability to teach SSI within the stem curriculum (Macalalag et al., 2020).

Increasing Science Literacy

Science literacy is one goal of science and society discourses (Birdsall, 2013; Chowdhury, 2016) and a focus of science education reforms (G. Gardner and Jones, 2011). Science literacy is also taught outside teaching institutions. Science museums are, for example, involved in increasing science literacy so that visitors can participate in decision-making, understand social and environmental responsibilities, and engage in sociopolitical action (Iannini, 2019).

Hodson proposes what he calls a "radical form of multicultural science education for sociopolitical action" that includes as goals "raising participation and attainment levels in science for students from ethnic minority groups; and sensitizing all students to racism, and other forms of discrimination and oppression, in science and technology, science education, and contemporary society" (Hodson, 1999: 785). He argues

> Education for critical scientific literacy is inextricably linked with education for political literacy and with the ideology of education as social reconstruction. The kind of social reconstruction envisaged includes, of course, the confrontation and elimination of racism, sexism, classism, and other forms of discrimination, scapegoating, and injustice; it includes a substantial shift away from rampant consumerism and toward a more environmentally sustainable lifestyle that promotes appropriate technology. Adopting appropriate technology entails rejecting technologies that violate our moral–ethical principles, exploit or disadvantage minority groups, or have adverse environmental impact. The curriculum proposals outlined here are intended to produce activists: people who will fight for what is right, good, and just; people who will work to refashion society along more socially-just lines; people who will work vigorously in the best interests of the biosphere. (Hodson, 1999: 789)

According to Hallman, science literacy includes scientific knowledge, knowledge on the impact of a given scientific product or process, and the ability to contribute to science and technology governance discussions within and outside academia (Hallman, 2017).

Science literacy also includes knowledge and the ability to critically think about the interactions among science, technology, society, and environment (Yore and Treagust, 2006), expects an understanding of the impact of societal values, believes in scientific and technological developments (Chowdhury, 2016) and the ability for informed decision-making; it also includes the ability to analyze, synthesize, and evaluate information, to deal sensibly with moral reasoning and ethical issues, and to understand connections inherent among socio-scientific issues (SSI) (Bencze et al., 2012).

STS education "aims to promote scientific and technological literacy in order to empower citizen participation in democratic decision-making and citizen action processes for resolving these issues" (Waks, 1989: 429). Social and civic literacy is one goal of science literacy, and citizenship values are expected to have a role in science education (Alghamdi, 2020).

Increasing Risk Literacy

Risk management includes evidence and normative claims as to what is good, acceptable, and tolerable (Renn and Walker, 2008: 333). "Understanding vulnerability of a target – whether it is a system, an individual, a community or a nation – is an important part of estimating risk" (Renn and Walker, 2008: 338). "Public values, concerns, perceptions of risk are seen as equally important for identifying, understanding, and managing risks" alongside scientific and technological aspects and they have to be included (Renn and Walker, 2008: 333). Risk literacy is argued to be essential for a safe everyday life (Nara and Sata, 2016) and influences risk perception (Kusumi et al., 2017). It includes an understanding of risk perception (Nara and Sata, 2016), and how different groups can perceive risks differently (Renn and Walker, 2008). As such risk literacy is essential for risk governance, which "pertains to the various ways in which many actors, individuals, and institutions, public and private, deal with risks surrounded by uncertainty, complexity, and/or ambiguity" (Van Asselt and Renn, 2011: 431). Risk literacy about scientific and technological advancements is a critical aspect of scientific literacy (G. Gardner and Jones, 2011; G. Gardner et al., 2010; Schenk et al., 2019; Ratcliffe and

Grace, 2003; Cross, 1993; Lange, 2012). However, risk literacy levels of students are often limited (Carmi and Alkaher, 2019). Schenk, drawing from (Hansen and Hammann, 2017), stated the following three aspects of how risk should be taught:

> (1) Scientific knowledge and statistics/probability, including knowledge about the risk issue, basic understanding of statistics and experts' risk judgements as well as reliability of data. (2) Knowledge about science (uncertainty, science in society and science as social practice), moving towards the issue of science in the making and stakeholders' and scientists' role in this. (3) Risk assessment, including risk-benefit analysis, ethical deliberations and decision-making. (Schenk et al., 2019: 1283)

Problems

There are various problems with how science and/or technology and society are taught. Science curricula in Australia are noted to be disconnected from sociopolitical issues (Gough, 2019) and the teaching workforce is unprepared to teach societal aspects of science (Gough, 2019). Schenk argues that risk literacy is not furthered enough in science education (Schenk et al., 2019). Then there is the issue that scientific literacy is influenced by one's views on social and ethical issues and one's values, beliefs, and perceptions (Zeidler et al., 2002; Kolstø, 2001; Solomon, 1987; Kelly and Erduran, 2019; Drummond and Fischhoff, 2017; Sinatra et al., 2014). Furthermore, drawing from the motivated reasoning and cultural cognition theory literature, it is argued that "individual risk perceptions—and the acknowledgment of expert consensus—are shaped by their values in ways to maintain their group identities" (Stecula and Merkley, 2019: 3, see also Kahan et al., 2007; Wolbring and Djebrouni, 2018) and "selectively learn only what is supported by their Worldview" (Webler and Tuler, 2021). Motivated reasoning is argued to hinder science understanding (Sinatra et al., 2014; Lobato and Zimmerman, 2019; Levy, 2019) and academic knowledge and evidence generation (Wolbring and Djebrouni, 2018). "When one's cognition is motivated at arriving at a particular conclusion, then the greater one's ABILITY, the better one is at making one's construal of the evidence yield the desired conclusion" (Hallsson, 2019: 2191).

In the remainder of the chapter, the author introduces the reader to the BIAS FREE Framework as an educational tool to increase science and society literacy, including risk literacy.

The BIAS FREE Framework

The *BIAS FREE* Framework was a tool developed for identifying and avoiding biases that derive from social hierarchies. BIAS FREE stands for "Building an Integrative Analytic System for Recognizing and Eliminating inEquities" (Burke and Eichler, 2006; Eichler and Burke, 2006). The theoretical framework behind the BIAS FREE Framework "draws on a long history of research related to the social dimensions of scientific knowledge, power and access to resources." Burke and Eichler argue that there are no such things as "value-free" or "emotionally detached"(Burke and Eichler, 2006). Social inequities and the biases surrounding them can be divided into three distinct sets of problems. They name the three problems as follows "Type of Bias H—Maintaining an Existing **H**ierarchy, Is dominance of one group over the other in any way justified?", "Type of Bias F—**F**ailing to Examine Differences, Is membership in a non-dominant/dominant group examined as socially relevant and accommodated?" And "Type of Bias D—Using **D**ouble Standards, Are non-dominant/dominant groups dealt with differently?" (Burke and Eichler, 2006; Eichler and Burke, 2006)". To make visible the Bias abbreviated with the letter **H** (Maintaining an Existing **H**ierarchy) Burke and Eichler pose eight analytical questions, H1-H8: accepting hierarchy as natural; denying hierarchy exists; adopting the perspective of the dominant group; applying the norms of the dominant groups to non-dominant groups; objectification of specific persons or groups; pathologization (labeled as deficient if one differs from the norm); victim blaming and appropriation (for the actual wording of all the diagnostic question for H, F, D see Appendix below). To make visible the Bias abbreviated with the letter **F** (**F**ailing to Examine Differences) Burke and Eichler pose four analytical questions, F1–F4: insensitivity to differences, decontextualization, over-generalization, or universalization and assumed homogeneity. To make visible the Bias abbreviated with the letter **D** (Using **D**ouble Standards) Burke and Eichler pose eight analytical questions, D1–D8: overt double standards, underrepresentation or exclusion, exceptional underrepresentation or exclusion, denying agency, treating dominant opinions as facts, stereotyping, exaggerating differences, and hiding double standards (Burke and Eichler, 2006; Eichler and Burke, 2006). Burke and Eichler applied the BFF in various projects such as in an

African setting (Eichler and Burke, 2006), building an inclusive national strategy for disabled children in Kyrgyzstan (Burke and Pupulin, 2009) or gender bias and beyond (Eichler and Burke, 2010).

Making Use of the BIAS FREE Framework

Risk literacy. Qualitative and quantitative tools are needed to help students with forming opinions regarding potential consequences of a risk issue (Schenk et al., 2019). The BFF could be such a tool. Risk perception, risk narratives, and whether one uses the term risk is influenced by many factors (Schmidt, 2004; Mañez et al., 2016). Different actors use the very term risk differently (Noga and Wolbring, 2014). Many of the 20 analytical categories of the BFF could unmask risk perception biases that shape risk narratives. Risk is a term that is instrumentalized by many (Noga and Wolbring, 2014). The term is often used as a descriptor without giving it a second thought due to people accepting dominant narratives. How the term is used often reflects various problems indicated by the 20 BFF analytical questions. To give one example, the phrase "risk of having a child with Down Syndrome" is often used in the context of increasing age of women having pregnancies (Wolbring, 2004, 2017a). However, risk is a judgment not a fact in this case. One has a higher probability of having a child with Down Syndrome at higher age of becoming pregnant. However, the term probability or similar factual terms are much less used (Wolbring, 2004, 2017a). Risk is the mainstream use because it gives voice to the mainstreamed pathologization of Down Syndrome, itself a value judgment about people with Down Syndrome. It is one example of risk perception bias and an example for the BFF analytical questions H3, H4, and H5.

Motivated Reasoning. It is argued that motivated reasoning is impacting individual risk perceptions (Stecula and Merkley, 2019; Kahan et al., 2007; Webler and Tuler, 2021), hinders science understanding (Sinatra et al., 2014; Lobato and Zimmerman, 2019; Levy, 2019), and influences academic knowledge and evidence generation (Wolbring and Djebrouni, 2018). The BFF could be used as a tool to identify premises, biases, and positionalities of documents available to students and others on a given topic. Equity, diversity, and inclusion and similar phrases and frameworks are used as concepts to discuss improvements within the workplace, including universities (Wolbring and Lillywhite, 2021) and places that engage with environmental issues (Salvatore and

Wolbring, 2021, 2022). The BFF could be used to enhance equity, diversity, and inclusion discussions, teaching and research in relation to risk narratives and communication around emergencies and disasters, by identifying the dominant views present and the views missing, such as views representative of EDI-linked groups (Lillywhite and Wolbring, accepted; under review). This increase in awareness could raise the quality of risk education, literacy, narratives, governance, and communication.

Of note, the BFF is not stopping MR behavior as such but could influence MR behavior if people read every document automatically with having in mind the 20 analytical questions of the BFF. Indeed, when the author has used the BFF in the classroom, many of the students become aware of premises, biases, and positionalities they never thought about.

Biases in science textbooks and science education. Various studies have revealed numerous biases in science textbooks (Poredi, 2017; Ragusa, 2013; Ndura, 2004) that could be unearthed by the BFF. Sadker and Zittleman outlined seven biases: invisibility, linguistic bias, stereotyping, imbalance and selectivity, unreality, fragmentation, and cosmetic bias (Sadker and Zittleman, 2010), all of which fit within the BFF Framework. Sadker and Zittleman applied their biases to gender but stated that their work could also be used for "race, ethnicity, the elderly, people with disabilities, gays and lesbians, and limited-English speakers" (Sadker and Zittleman, 2010: 83). They argue that analyzing text through the lens of such biases is an important critical reading skill. Poredi in her 2017 Master theses argued that "science textbooks can be analyzed as operative on a number of levels. With specific reference to gender, one can examine representation from the following vantage points: (a) exclusion/invisibility and the underrepresentation of women; (b) stereotyping and assigning gender roles, undermining women's contribution and those of different ethnic backgrounds to the field of science; (c) linguistic constructions of language that rely on masculine metaphors, undermining the feminine aspects of science; and (d) isolation of women from the realm of science" (Poredi, 2017: 17). Potter and Rosser (1992) argue that "a textbook is a major factor that influences the teaching of science, it stands as an important element that may aid in attracting girls to or deterring them from studying science" (p. 669). Poredi (2017) outlines many studies that found biases related to gender, race, Indigenous people, and disability in science-related textbooks. Many of the biases found fit within the BFF categories. (Ragusa, 2013) looked at visual and written content bias, author's perspective bias, and omission as bias. The BFF can be applied to visual and written content and author's perspectives, and omission is present in the BFF under D2 Underrepresentation or exclusion: Are non-dominant groups under-represented or excluded? Hodson (1999) argued that teachers often have a bias toward supporting dominant views exhibited in their teaching material. "Commonly, these teachers perceive ideas and sources of information such as textbooks and movies that draw on dominant perspectives as 'normal,' while regarding those that deploy minority position perspectives as biased and 'political.'" In other words, the setup of their teaching will include many of the BFF biases. To unmask the BFF biases in textbooks might increase the education related to topics such as environmental issues and emergency and disaster management, common areas for risk education, narratives, governance, and communication.

Unmasking hierarchies. Numerous hierarchies are identified in the STEM and science education literature. Science curricula exhibit a hierarchy in which "values and implicit social messages" are taught (Hodson, 1999: 791). Curricula display a hierarchy that prioritizes scientific theory and technological applications over social and political questions (Hughes, 2000; Breunig, 2005). Other forms of hierarchy identified are: presence of gender binaries (Hughes, 2000; Breunig, 2005); hierarchy of knowledge sources (Breunig, 2005) such as Indigenous science knowledge versus other science knowledge (Higgins, 2019); civic hierarchies based on race, class, gender, and other categories of identified social difference (Burton, 2019), hierarchy of needs (Dillon and Teamey, 2002); hierarchy of credibility (Oh, 2017); hierarchy of values (Gresch et al., 2017; Maarschalk, 1988); and power hierarchies that shape what is taught in science education for what end (Gough, 2019; Henderson and Zipin, 2010) and what counts as scientific literacy (Orpwood, 2007). It is argued that educators are often unaware of how they propagate hierarchies, such as hierarchies of knowledge and civic hierarchies, e.g., gender binaries, and that this something that has to change (Hughes, 2000). Many of these hierarchies can be supported by actions that would be unmasked by using the 20 BFF analytical questions.

Lemke argued that not surprisingly "those who succeed in science tend to be like those who define the "appropriate" way to talk science: male rather than female; white rather than black; middle and upper-middle class, native English-speakers, and standard dialect speakers" with the linked North-European middle-class culture ability expectations of "emotional control, orderliness, rationalism, achievement, punctuality, social hierarchy, etc." (Lemke, 1990: 138). The BFF is a useful tool to analyze documents, such as reports, proposals, academic articles, and education material used in STEM/science/STSE/STS education for all the hierarchies and biases mentioned by Lemke. The BFF has been developed to unmask many of the hierarchies identified in the STEM and science education literature, such as civic hierarchies based on race, class, and gender (Burton, 2019), but also others in relation to disabled people.

Breunig argues that educational and other institutional assumptions including environmental issues, gender issues, issues related to institutional accessibility, and issues of hierarchy and power have to be revealed (Breunig, 2005). The BFF is useful in unearthing the assumptions mentioned by Breunig. Hierarchies of visibility are also present in emergency and disaster management and environmental education and activism (Salvatore and Wolbring, 2021, 2022, Lillywhite and Wolbring, accepted), including risk narratives (Lillywhite and Wolbring, under review). BFF can help to reveal these hierarchies.

Civic hierarchy based on abilities: the case of disabled people. The disabled people's rights movement coined the term ableism (The Editors of Encyclopaedia Britannica, 2013) to provide a term for the cultural reality of an existing hierarchy of body/mind abilities and underscore that a dominant group has the power to set ability norms and to define others as falling short of the ability norms and labelling the other as impaired (disabled people). The disabled people's rights movement also coined the term disablism to make visible the often-negative disabling use of ability-based judgments and norms against disabled people (Wolbring, 2008, 2014; Miller et al., 2004). The disabling use of irrelevant ability norms against disabled people is an example of BFF H4 normalization. Ability expectation in relation to disabled people is one civic hierarchy mentioned in the BIAS FREE Framework in the same way gender, race, and other social hierarchies are mentioned (Burke and Eichler, 2006; Eichler and Burke, 2006). The BFF is being used by the author in disability studies undergraduate classes where students are asked to find examples for the 20 questions of the BFF in relation to disabled people. They have no problem finding examples for all the 20 questions of the BFF and the BFF has been applied to disabled children (Burke and Pupulin, 2009). The medical view of disabled people is seen especially to negatively impact their presence in the discussions around the "social" dimensions of environmental issues (Salvatore and Wolbring, 2021, 2022), including sustainability issues (Wolbring et al.,

2013; participants of the UN Department of Economic and Social Affairs (UNDESA) and UNICEF organized Online Consultation, 8 March to 5 April, Disability inclusive development agenda toward 2015 & beyond, 2013; Lieve Sabbe UNICEF and Vladimir Cuk International Disability Alliance, 2013), including emergency and disaster management in general (Lillywhite and Wolbring under review, paper 1) and risk narratives (Lillywhite and Wolbring under review, paper 2). The BFF could unmask these biases and improve the education, research and policy activities around risk narratives, governance, education, and literacy.

Civic hierarchy based on abilities: the case beyond disabled people. Ability-based judgments and norms impact not only disabled people but the relationship between humans in general, humans and animals, humans and nature and humans and machines (Wolbring, 2012, 2013, 2014, 2019, 2020; Wolbring and Lisitza, 2017; Bentley et al., 2017; Nocella II, 2017; Wolbring et al., 2020). Ability judgments and conflicts between groups are evident and influence how we deal and engage with emergencies and disasters. Many ability-based judgments by and conflicts between people also exist in relation to scientific and technological advancements whereby which abilities are pursued depends on who is the dominant group shaping a discussion. Being stereotyped by the dominant group BFF D6 stereotyping, is one strategy of the powerful (Wolbring, 2020; Perez et al., 2013). Ability-based judgments influence risk perceptions, governance, narratives, and communication. As such, being literate on ability-based judgments, conflicts and power realities is an important aspect of science, society, and risk literacy. The BFF can be used in conjunction with ability expectation exercises to make visible ability-based judgments and conflicts (Wolbring, 2017b; Wolbring et al., 2019) and to increase ability-based literacy, including risk literacy, on the interface between the public and topics covered in risk communications.

Enriching science education. The BFF is a useful educational tool for science education. Sleeter in her book *Multicultural Education as Social Activism,* which covers multicultural science education, outlines numerous problems that justify multicultural education where the BFF can be used to make visible the problems she mentions, namely, to identify (a) who shapes the discourse, how and why (centering); (b) that theories are socially constructed and often generated by dominant groups to provide justification for their dominance but that theories can also be generated by minorities (social construction of theory); (c) that the dominant ones write history to their liking and use this to set the stage for the future (subjugation and liberation); (d) that dominant groups set social, political, and other boundaries, including by categorizing groups to suit themselves (critique and redefinition of social collectives); (e) that dominant groups define identities of groups but can be undone by the minority groups (sense of identity); and (f) that dominant groups use ability expectations in a disabling way to justify their dominance, such as saying that the dominant group is dominant because it exhibits certain desirable abilities others don't, which can be countered by an alternative narrative by minority groups (oppressed groups are durable, strong and active) (Sleeter, 1996, see also Hodson describing Sleeter's work, Hodson, 1999: 792–793).

The BFF can be used to strengthen all levels of learning, namely, the skills of being able to remember, understand, apply, analyze, evaluate, and create (Krathwohl and Anderson, 2009). It can be used to achieve all four curriculum levels of issue-based teaching Hodson outlines for STS and STSE education, which include an engagement with the societal impact of science and technology, an understanding of the forces that influence science and technology advancements, a self-reflection of one's views on the topic and the ability to take action (Hodson, 2010, see also Rubba and Wiesenmayer, 1985).

It is argued that "students see social issues related to science within a narrow individual lived experience framework and not from a societal, dimension" (Hughes, 2000: 433). By using the BFF tool, students can be taught to recognize various social dimensions. Breunig argues that experiential education and critical pedagogy see the development of a socially just world as one purpose of education (Breunig, 2005). The BFF aligns with this vision.

Likewise, the BFF could be used to enhance decision-making in STSE education by inducing reflection and self-regulated learning (Gresch et al., 2017). It can be used for "interrogating the politics of science, technology, engineering, and mathematics (STEM) learning at a global level and for STEM literacies that include robust sociopolitical analysis" (de Roock and Baildon, 2019: 1). Sociopolitical analyses are a recognized part of shaping risk governance and communication (Renn, 2017).

Enriching media literacy and science education. Media literacy and the ability to critically evaluate science-based news to identify fake news and to evaluate scientific literature are part of being scientifically literate and science education (Jarman and McClune, 2010, 2007; Scheufele and Krause, 2019; McDougall et al., 2018; Dani et al., 2010; Magnuson,

2018; Storksdieck, 2016; Jurecki and Wander, 2012; Calado and Bogner, 2013). The BFF is a simple tool students can use to increase their media literacy by marking certain biases, positionalities, and assumptions. Being media literate is essential for risk communication and trust in risk governance.

Teaching science education. Teachers are a catalyst, a change agent, a facilitator, a teacher of action research (Pedretti and Hodson, 1995). The BFF is a useful tool for teachers to fulfill these roles. The BFF is a simple tool that allows students to become aware of problems and realities whereby the BFF could be a catalyst for change. Teaching students to analyze any given text using the BFF as a 20-point checklist might make them less prone to confirmation bias that one reads only what relates to one's view (Knobloch-Westerwick et al., 2015; Reed et al., 2019), something seen as a problem for science literacy.

If free of bias is a criterion of objectivity and if judging the objectivity of documents is one goal of teaching students to evaluate scientific literature and other materials critically (Jurecki and Wander, 2012), then the BFF can help with this goal. Furthermore, a course designed "to teach a mixed group of journalism and science, technology, engineering, and mathematics (STEM) majors to evaluate expertise and bias, examine data and misrepresentation, and develop evidence-based communication on complex scientific topics" (Reed et al., 2019: 212) would benefit from the BFF.

According to Burton in his thesis *Ecopedagogy, the Earth charter, and Manitoba curriculum,* the new curriculum for senior years encourages educators to be more critical, to flag biases in their teaching material, to provide a multitude of perspectives and to increase the literacy of students to identify biases in teaching (Burton, 2019). The BFF can be used for all the points raised by Burton and would enrich teaching risk governance and communication, as well as increase risk literacy.

Conclusion

The BFF is a useful tool to make visible biases, premises, and positionalities in science and society discourses, including risk narratives. It also helps to increase the sensitivity of people to look for such biases, premises, and positionalities by using the BFF as a checklist. The BFF is likewise a useful tool in increasing science and society literacy, including risk literacy, of students and other people engaged with science and society topics. It enables the science and society mandate of STEM and science education, which should include risk education. Although the BFF cannot directly change the behavior of motivated reasoning, by giving people a tool to unmask biases, premises, and positionalities present in documents, it adds to the democratization of science and technology governance and risk governance.

Appendix: The BIAS FREE Framework (BFF) Questions

Type of Bias H—Maintaining an Existing **H**ierarchy, Is dominance of one group over the other in any way justified?

H1 Denying hierarchy:	Is the existence of a hierarchy denied in spite of widespread evidence to the contrary?
H2 Maintaining hierarchy:	Are practices or views that are based on a hierarchy presented as normal or unproblematic?
H3 Dominant perspective:	Is the perspective or standpoint of the dominant group adopted?
H4 Normalization:	Are norms derived from the dominant group and then applied to the non-dominant group without questioning their relevance?
H5 Pathologization:	Is the non-dominant group defined as deficient when it differs from the norms derived from the dominant group?
H6 Objectification:	Is stripping people of their intrinsic dignity and personhood presented as normal or unproblematic?
H7 Victim blaming:	Are victims of personal or societal/systemic violence blamed and held accountable?
H8 Appropriation:	Is ownership claimed by the dominant group for entities that originate(d) in or belong to the non-dominant group?

Type of Bias F—**F**ailing to Examine Differences, Is membership in a non-dominant/dominant group examined as socially relevant and accommodated?

F1 Insensitivity to difference:	Has the relevance of membership in dominant/non-dominant group been ignored?
F2 Decontextualization:	Has the different social reality of dominant and non-dominant groups explicitly been considered?
F3 Over-generalization or universalization:	Is information derived from dominant groups generalized to non-dominant groups without examining if it is applicable to the non-dominant groups?
F4 Assumed homogeneity:	Is the dominant or non-dominant group treated as a uniform group?

Type of Bias D—Using **D**ouble Standards, Are non-dominant/dominant groups dealt with differently?

D1 Overt double standard:	Are non-dominant and dominant groups treated unequally?
D2 Underrepresentation or exclusion:	Are non-dominant groups under-represented or excluded?
D3 Exceptional underrepresentation or exclusion:	In contexts normally associated with non-dominant groups, but pertinent to all groups, is the dominant group under-represented or excluded?
D4 Denying agency:	Is there a failure to consider non-dominant/dominant groups as both actors and acted upon?
D5 Treating dominant opinions as facts:	Are opinions expressed by a dominant group about a non-dominant group treated as opinion or fact?
D6 Stereotyping:	Are stereotypes of non-dominant/dominant groups treated as essential aspects of group membership?

D7 Exaggerating differences:	Are overlapping traits treated as if they were characteristic of only non-dominant/dominant groups?
D8 Hidden double standard:	Are different criteria used to define comparable facts with the effect of hiding their comparability?

References

Alghamdi, A. K. H. (2020). Citizenship education in science curricula: Exploring the Saudi Arabia case. *International Journal of Science and Mathematics Education, 18*, 669–689, https://doi.org/10.1007/s10763-019-09991-2.

Bencze, L., Sperling, E., & Carter, L. (2012). Students' research-informed socio-scientific activism: Re/visions for a sustainable future. *Research in Science Education, 42*(1), 129–148.

Bentley, J. K., Conrad, S., Hurley, S., Lisitza, A., Lupinacci, J., Lupinacci, M. W., et al. (2017). *The intersectionality of critical animal, disability, and environmental studies: Toward eco-ability, justice, and liberation*. Washington, DC: Lexington Books.

Birdsall, S. (2013). Reconstructing the relationship between science and education for sustainability: A proposed framework of learning. *International Journal of Environmental and Science Education, 8*(3), 451–478.

Breunig, M. (2005). Turning experiential education and critical pedagogy theory into praxis. *Journal of Experiential Education, 28*(2), 106–122.

Burke, M. A., & Eichler, M. (2006). Building an integrative analytical system for recognising and eliminating in equities (BIAS FREE framework). https://ethz.ch/content/specialinterest/gess/cis/center-for-securities-studies/en/services/digital-library/publications/publication.html/128225. Accessed 9 October 2022.

Burke, M. A., & Pupulin, A. (2009). Building an inclusive national strategy for disabled children in Kyrgyzstan through an application of the BIAS FREE Framework. In *Disability & international development* (pp. 51–68). Springer.

Burton, W. (2019). Ecopedagogy, The Earth Charter, and Manitoba curriculum: A critical content analysis (Theses). https://mspace.lib.umanitoba.ca/handle/1993/33780. Accessed 9 October 2022.

Calado, F. M., & Bogner, F. X. (2013). A reflection on distorted views of science and technology in science textbooks as obstacles to the improvement of students' scientific literacy. *European Journal of Educational Research, 2*(2), 51–68.

Carmi, N., & Alkaher, I. (2019). Risk literacy and environmental education: Does exposure to academic environmental education make a difference in how students perceive ecological risks and evaluate their risk severity? *Sustainability, 11*(22), 6350, https://doi.org/10.3390/su11226350.

Chowdhury, M. A. (2016). The integration of science-technology-society/science-technology-society-environment and socio-scientific-issues for effective science education and science teaching. *Electronic Journal of Science Education, 20*(5), 19–38.

Cross, R. (1993). The risk of risks: A challenge and a dilemma for science and technological education. *Research in Science & Technological Education, 11*(2), 171–183.

Dani, D., Wan, G., & Henning, J. E. (2010). A case for media literacy in the context of socioscientific issues. *New Horizons in Education, 58*(3), 85–98.

de Roock, R. S., & Baildon, M. (2019). MySkillsFuture for students, STEM learning, and the design of neoliberal citizenship in Singapore. *Cognition and Instruction, 37*(3), 285–305, https://doi.org/10.1080/07370008.2019.1624545.

Dillon, J., & Teamey, K. (2002). Reconceptualizing environmental education: Taking account of reality. *Canadian Journal of Math, Science & Technology Education, 2*(4), 467–483.

Drummond, C., & Fischhoff, B. (2017). Individuals with greater science literacy and education have more polarized beliefs on controversial science topics. *Proceedings of the National Academy of Sciences, 114*(36), 9587–9592.

Eichler, M., & Burke, M. A. (2006). The BIAS FREE framework. *Canadian Journal of Public Health, 97*(1), 63–68.

Eichler, M., & Burke, M. A. (2010). Beyond gender bias: A methodological approach for analyzing inequities. In V. Lasch, U. Sonntag, & U. Maschewsky-Schneider (Eds.), *Equity in access to health promotion, treatment and care for all European women* (pp. 31–45).

Fensham, P. J. (2007). Competences, from within and without: New challenges and possibilities for scientific literacy. https://eprints.qut.edu.au/12883/1/Linnaeus_Tercentenary_Symposium.pdf. Accessed 9 October 2022.

Gardner, G., & Jones, G. (2011). Science instructors' perceptions of the risks of biotechnology: Implications for science education. *Research in Science Education, 41*(5), 711–738.

Gardner, G., Jones, G., Taylor, A., Forrester, J., & Robertson, L. (2010). Students' risk perceptions of nanotechnology applications: Implications for science education. *International Journal of Science Education, 32*(14), 1951–1969.

Gough, A. (2019). Woman being disruptive: Challenging (E) quality in science education. In *Critical voices in science education research* (pp. 257–267). Springer.

Gresch, H., Hasselhorn, M., & Bögeholz, S. (2017). Enhancing decision-making in STSE education by inducing reflection and self-regulated learning. *Research in Science Education, 47*(1), 95–118.

Hallman, W. K. (2017). What the public thinks and knows about science and why it matters. In *The Oxford handbook of the science of science communication* (pp. 61–72): Oxford University Press Oxford.

Hallsson, B. G. (2019). The epistemic significance of political disagreement. *Philosophical Studies, 176*(8), 2187–2202.

Hansen, J., & Hammann, M. (2017). Risk in science instruction. The realist and constructivist paradigms of risk. *Science & Education, 26*(7–9), 749–775.

Henderson, D., & Zipin, L. (2010). Bringing clay to life: Developing student literacy through clay animation artwork to tell life-based stories. In B. Prosser, B. Lucas, & A. Reid (Eds.), *Connecting lives and learning: Renewing pedagogy in the middle years* (pp. 20–41). Kent Town, South Australia: Wakefield Press.

Higgins, M. (2019). Pursuing response-ability in de/colonizing science education. In *Critical voices in science education research* (pp. 223–233): Springer.

Hodson, D. (1999). Going beyond cultural pluralism: Science education for sociopolitical action. *Science Education, 83*(6), 775–796.

Hodson, D. (2010). Science education as a call to action. *Canadian Journal of Science, Mathematics and Technology Education, 10*(3), 197–206.

Hughes, G. (2000). Marginalization of socioscientific material in science–technology–society science curricula: Some implications for gender inclusivity and curriculum reform. *Journal of Research in Science Teaching: The Official Journal of the National Association for Research in Science Teaching, 37*(5), 426–440.

Iannini, A. M. N. (2019). Science museums: Reflections from an Autobiographical Journey. In *Critical voices in science education research* (pp. 73–81). Springer.

Jarman, R., & McClune, B. (2007). *Developing scientific literacy: Using news media in the classroom.* New York, USA: Open University Press.

Jarman, R., & McClune, B. (2010). Developing students' ability to engage critically with science in the news: Identifying elements of the 'media awareness' dimension. *The Curriculum Journal, 21*(1), 47–64, https://doi.org/10.1080/09585170903558380.

Jurecki, K., & Wander, M. C. (2012). Science literacy, critical thinking, and scientific literature: Guidelines for evaluating scientific literature in the classroom. *Journal of Geoscience Education, 60*(2), 100–105.

Kahan, D. M., Braman, D., Gastil, J., Slovic, P., & Mertz, C. (2007). Culture and identity-protective cognition: Explaining the white-male effect in risk perception. *Journal of Empirical Legal Studies, 4*(3), 465–505.

Kelly, R., & Erduran, S. (2019). Understanding aims and values of science: Developments in the junior cycle specifications on nature of science and pre-service science teachers' views in Ireland. *Irish Educational Studies, 38*(1), 43–70.

Knobloch-Westerwick, S., Johnson, B. K., Silver, N. A., & Westerwick, A. (2015). Science exemplars in the eye of the beholder: How exposure to online science information affects attitudes. *Science Communication, 37*(5), 575–601, https://doi.org/10.1177/1075547015596367.

Kolstø, S. D. (2001). Scientific literacy for citizenship: Tools for dealing with the science dimension of controversial socioscientific issues. *Science Education, 85*(3), 291–310.

Krathwohl, D. R., & Anderson, L. W. (2009). *A taxonomy for learning, teaching, and assessing: A revision of Bloom's taxonomy of educational objectives*: Longman.

Kusumi, T., Hirayama, R., & Kashima, Y. (2017). Risk perception and risk talk: The case of the Fukushima Daiichi nuclear radiation risk. *Risk Analysis, 37*(12), 2305–2320.

Lange, J. M. (2012). Education in sustainable development: How can science education contribute to the vulnerability perception? *Research in Science Education, 42*(1), 109–127.

Lemke, J. L. (1990). *Talking science: Language, learning, and values*. Norwood, NJ: Ablex Publishing Corporation.

Levy, N. (2019). Due deference to denialism: Explaining ordinary people's rejection of established scientific findings. *Synthese, 196*(1), 313–327.

Lieve Sabbe UNICEF and Vladimir Cuk International Disability Alliance. (2013). Global thematic consultation on addressing inequalities persons with disabilities online discussion synopsis. http://www.worldwewant2015.org/file/307833/download/334462. Accessed 9 October 2022.

Lobato, E. J., & Zimmerman, C. (2019). Examining how people reason about controversial scientific topics. *Thinking & Reasoning, 25*(2), 231–255.

Maarschalk, J. (1988). Scientific literacy and informal science teaching. *Journal of Research in Science Teaching, 25*(2), 135–146.

Macalalag, A. Z., Johnson, J., & Lai, M. (2020). How do we do this: Learning how to teach socioscientific issues. *Cultural Studies of Science Education, 15*, 389–413, https://doi.org/10.1007/s11422-019-09944-9.

Magnuson, K. (2018). Improving students' scientific literacy through critical evaluation of scientific texts and media. http://csusm-dspace.calstate.edu/handle/10211.3/202466. Accessed 9 October 2022.

Mañez, M., Carmona, M., Haro, D., & Hange, S. (2016). Risk perception. http://pure.iiasa.ac.at/id/eprint/13903/1/Chapter3-ENHANCE.pdf. Accessed 9 October 2022.

McDougall, J., Zezulkova, M., van Driel, B., & Sternadel, D. (2018). Teaching media literacy in Europe: evidence of effective school practices in primary and secondary education, NESET II report. http://eprints.bournemouth.ac.uk/31574/1/AR2_Teaching%20Media%20Literacy_NESET.pdf. Accessed 9 October 2022.

Miller, P., Parker, S., & Gillinson, S. (2004). Disablism how to tackle the last prejudice. http://www.demos.co.uk/files/disablism.pdf. Accessed 9 October 2022.

Nara, Y., & Sata, T. (2016). Construction of the practical model and learning program for risk literacy of everyday life: Based on students' awareness. *Procedia Computer Science, 96*, 1258–1266.

Ndura, E. (2004). ESL and cultural bias: An analysis of elementary through high school textbooks in the Western United States of America. *Language, Culture and Curriculum, 17*(2), 143–153, https://doi.org/10.1080/07908310408666689.

Nocella II, A. J. (2017). Defining eco-ability. In S. J. Ray, & J. Sibara (Eds.), *Disability environmental humanities* (pp. 141–168). Lincoln, NE, USA: University of Nebraska Press.

Noga, J., & Wolbring, G. (2014). The oil and gas discourse from the perspective of the Canadian and Albertan governments, non-governmental organizations and the oil and gas industry. *Energies, 7*(1), 314–333.

Oh, J.-Y. (2017). Suggesting a NOS Map for nature of science for science education instruction. *Eurasia Journal of Mathematics, Science and Technology Education, 13*(5), 1461–1483.

Orpwood, G. (2007). Assessing scientific literacy: Threats and opportunities in promoting scientific literacy: Science education research in transaction. http://uu.diva-portal.org/smash/get/diva2:40649/FULLTEXT01.pdf. Accessed 9 October 2022.

Participants of the UN Department of Economic and Social Affairs (UNDESA) and UNICEF organized Online Consultation - 8 March - 5 April Disability inclusive development agenda towards 2015 & beyond. (2013). Disability inclusive development agenda towards 2015 & beyond. http://www.un.org/en/development/desa/news/social/disability-inclusive-development.html. Accessed 9 October 2022.

Pedretti, E. (1999). Decision making and STS education: Exploring scientific knowledge and social responsibility in schools and science centers through an issues-based approach. *School Science and Mathematics, 99*(4), 174–181.

Pedretti, E., & Hodson, D. (1995). From rhetoric to action: Implementing STS education through action research. *Journal of Research in Science Teaching, 32*(5), 463–485.

Perez, E. R., Schanding Jr, G. T., & Dao, T. K. (2013). Educators' perceptions in addressing bullying of LGBTQ/gender nonconforming youth. *Journal of School Violence, 12*(1), 64–79.

Poredi, S. (2017). The impact of bias present in high school science textbooks. https://yorkspace.library.yorku.ca/xmlui/bitstream/handle/10315/34193/MED00004.pdf?sequence=1. Accessed 9 October 2022.

Potter, E. F., & Rosser, S. V. (1992). Factors in life science textbooks that may deter girls' interest in science. *Journal of Research in Science Teaching, 29*(7), 669–686.

Ragusa, G. (2013). Science literacy and text book biases. https://www.asee.org/file_server/papers/attachment/file/0003/3493/revised-Science_Literacy_and_Text_Book_Biases_ASEE_2013-FINAL.pdf. Accessed 9 October 2022.

Ratcliffe, M., & Grace, M. (2003). *Science education for citizenship. Teaching socio-scientific issues* (Vol. null, null). Maidenhead, Philadelphi: Open University Press.

Reed, K., Hiles, S. S., & Tipton, P. (2019). Sense and nonsense: Teaching journalism and science students to be advocates for science and information literacy. *Journalism & Mass Communication Educator, 74*(2), 212–226, https://doi.org/10.1177/1077695819834415.

Renn, O. (2017). *Risk governance: Coping with uncertainty in a complex world*. Routledge.

Renn, O., & Walker, K. (2008). Lessons learned: A re-assessment of the IRGC framework on risk governance. In O. Renn, & K. D. Walker (Eds.), *Global risk governance: Concept and practice using the IRGC framework* (pp. 331–367). Dordrecht: Springer Netherlands.

Rubba, P. A., & Wiesenmayer, R. L. (1985). A goal structure for precollege STS education: A proposal based upon recent literature in environmental education. *Bulletin of Science, Technology & Society, 5*(6), 573–580.

Sadker, D., & Zittleman, K. (2010). Gender bias: From colonial America to today's classrooms. In J. A. Banks, & C. McGee Banks (Eds.), *Multicultural education: Issues and perspectives* (7th. ed., pp. 135–169). New York: Wiley.

Salvatore, C., & Wolbring, G. (2021). Children and youth environmental action: The case of children and youth with disabilities. *Sustainability, 13*(17), 1–27, https://doi.org/10.3390/su13179950.

Salvatore, C., & Wolbring, G. (2022). Coverage of disabled people in environmental-education-focused academic literature. *Sustainability, 14*(3), 1211, https://doi.org/10.3390/su14031211.

Schenk, L., Hamza, K. M., Enghag, M., Lundegård, I., Arvanitis, L., Haglund, K., et al. (2019). Teaching and discussing about risk: Seven elements of potential significance for science education. *International Journal of Science Education, 41*(9), 1271–1286.

Scheufele, D. A., & Krause, N. M. (2019). Science audiences, misinformation, and fake news. *Proceedings of the National Academy of Sciences, 116*(16), 7662–7669, https://doi.org/10.1073/pnas.1805871115.

Schmidt, M. (2004). Investigating risk perception: A short introduction http://www.markusschmidt.eu/pdf/Intro_risk_perception_Schmidt.pdf. Accessed 9 October 2022.

Sinatra, G. M., Kienhues, D., & Hofer, B. K. (2014). Addressing challenges to public understanding of science: Epistemic cognition, motivated reasoning, and conceptual change. [Article]. *Educational Psychologist, 49*(2), 123–138, https://doi.org/10.1080/00461520.2014.916216.

Sleeter, C. E. (1996). *Multicultural education as social activism*. SUNY Press.

Solomon, J. (1987). Social influences on the construction of pupils' understanding of science. *Studies in Science Education, 14*(1), 63–83, https://doi.org/10.1080/03057268708559939.

Stecula, D. A., & Merkley, E. (2019). Framing climate change: Economics, ideology, and uncertainty in American news media content from 1988 to 2014. *Frontiers in Communication, 4*, 6, https://doi.org/10.3389/fcomm.2019.00006.

Storksdieck, M. (2016). Critical information literacy as core skill for lifelong STEM learning in the 21st century: Reflections on the desirability and feasibility for widespread science media education. [journal article]. *Cultural Studies of Science Education, 11*(1), 167–182, https://doi.org/10.1007/s11422-015-9714-4.

The Editors of Encyclopaedia Britannica. (2013). Ableism. https://www.britannica.com/topic/ableism/additional-info#history. Accessed 9 October 2022.

Van Asselt, M. B., & Renn, O. (2011). Risk governance. *Journal of Risk Research, 14*(4), 431–449.

Waks, L. (1989). New challenges for science, technology, and society education. *Technology in Society, 11*(4), 427–432.

Webler, T., & Tuler, S. (2021). Four decades of public participation in risk decision making. *Risk Analysis, 41*(3), 503–518, https://doi.org/10.1111/risa.13250.

Wolbring, G. (2004). Parents without prejudice. In S. Klein, & J. Kemp (Eds.), *Reflections from a different journey* (pp. 18–22). McGraw Hill Professional.

Wolbring, G. (2008). Why NBIC? Why human performance enhancement? *Innovation; The European Journal of Social Science Research, 21*(1), 25–40, https://doi.org/10.1080/13511610802002189.

Wolbring, G. (2012). Eco-ableism. https://svara98.typepad.com/blog/2012/09/eco-ableism-anthropology-news.html. Accessed 9 October 2022.

Wolbring, G. (2013). Ecohealth through an ability studies and disability studies lens. In M. K. Gislason (Ed.), *Ecological health: Society, ecology and health* (Vol. 15, pp. 91–107, Advances in Medical Sociology). London, UK: Emerald.

Wolbring, G. (2014). Ability Privilege: A needed addition to privilege studies. *Journal for Critical Animal Studies, 12*(2), 118–141.
Wolbring, G. (2017a). The gift of belonging: From parents to society. In K. Scorgie, & D. Sobsey (Eds.), *Working with families for inclusive education: Navigating identity, opportunity and belonging* (pp. 63–70). Bingley, UK: Emerald Publishing Limited.
Wolbring, G. (2017b). Teaching about ability expectation and its governance: The issue of STEM. In K. Sheehy, & A. Holliman (Eds.), *Education and new technologies perils and promises for learners* (pp. 122–140). London, UK: Routledge.
Wolbring, G. (2019). Ability expectation and ableism governance: An essential aspect of a culture of peace. https://eubios.info/yahoo_site_admin/assets/docs/Legaciesbook1October2019.273183519.pdf#page=124. Accessed 9 October 2022.
Wolbring, G. (2020). Violence and abuse through an ability studies lens. *Indian Journal of Critical Disability Studies 1*(1), 41–67.
Wolbring, G., Deloria, R., Lillywhite, A., & Villamil, V. (2019). Ability expectation and ableism peace. *Peace Review (Palo Alto, Calif.), 31*(4), 449–458, https://doi.org/10.1080/10402659.2019.1800929.
Wolbring, G., Deloria, R., Lillywhite, A., & Villamil, V. (2020). Ability expectation and ableism peace. *Peace Review, 31*(4), 449–458, https://doi.org/10.1080/10402659.2019.1800929.
Wolbring, G., & Djebrouni, M. (2018). Motivated reasoning and disabled people. *Interdisciplinary Perspectives on Equality and Diversity 4*(2), no page number.
Wolbring, G., & Lillywhite, A. (2021). Equity/Equality, Diversity, and Inclusion (EDI) in universities: The case of disabled people. *Societies, 11*(2), article number 49, https://doi.org/10.3390/soc11020049.
Wolbring, G., & Lisitza, A. (2017). Justice among humans, animals and the environment: Investigated through an ability studies, eco-ableism, and eco-ability lens. In A. J. Nocella II, A. E. George, & J. Schatz (Eds.), *Weaving nature, animals and disability for eco-ability: The intersectionality of critical animal, disability and environmental studies* (pp. 41–62). Lanham, MD, USA: Lexington books.
Wolbring, G., Mackay, R., Rybchinski, T., & Noga, J. (2013). Disabled people and the post-2015 development goal agenda through a disability studies lens. *Sustainability, 5*(10), 4152–4182.
Yore, L. D., & Treagust, D. F. (2006). Current realities and future possibilities: Language and science literacy—Empowering research and informing instruction. *International Journal of Science Education, 28*(2–3), 291–314.

Zeidler, D. L., Walker, K. A., Ackett, W. A., & Simmons, M. L. (2002). Tangled up in views: Beliefs in the nature of science and responses to socioscientific dilemmas. *Science Education, 86*(3), 343–367.

PART II

Public Trust, Risk Perception and Public Participation: Lessons From the Real Worlds of Risk Governance

Energy

CHAPTER 5

Getting It Right? The Site Selection Process for Canada's High-level Nuclear Waste

Duane Bratt, Patricia Larkin, and Xavier Deschênes-Philion

Introduction and Context

This chapter examines the role of assessment and risk management in the decision-making processes for selecting a permanent site for the long-term storage of Canada's high-level nuclear waste. Geographically, it focuses on two areas: the potential storage sites and, once a site is selected, the transportation routes used to access the site. Temporally, it focuses on the planned site and route selection processes up until the facility is licensed to operate.

Nuclear waste management policy and regulatory frameworks are under federal authority. Canada's radioactive waste management policy

D. Bratt (✉)
Department of Economics, Justice, and Policy Studies, Mount Royal University, Calgary, AB, Canada
e-mail: dbratt@mtroyal.ca

P. Larkin · X. Deschênes-Philion
University of Ottawa, Ottawa, ON, Canada

M. Gattinger (ed.), *Democratizing Risk Governance*,
https://doi.org/10.1007/978-3-031-24271-7_5

framework is defined by Natural Resources Canada (NRCan) while the *Nuclear Safety and Control Act* established the Canadian Nuclear Safety Commission (CNSC) to act as an independent regulator responsible for the use of nuclear material. Under the *Nuclear Fuel Waste Act* (NFWA), the Nuclear Waste Management Organization (NWMO) was established in 2002 as a not-for-profit corporation responsible for the site selection process for a deep geological repository (DGR) to store Canada's used nuclear fuel. The NWMO will make a recommendation to the CNSC in due course.

There were 12,718 cubic meters of high-level nuclear waste at the end of 2019, with a predicted 21,015 cubic meters projected by 2050 (Natural Resources Canada 2021). Temporary storage solutions include wet and dry licensed facilities. A permanent solution must be developed in order to deal with the current and future nuclear waste.

The NWMO's Adaptive Phased Management (APM) plan "requires used fuel to be contained and isolated in a deep geological repository. It also calls for a comprehensive process to select an informed and willing host for the project" (NWMO 2017a). The multi-step process involves finding a willing host community in a geologically suitable region and conducting a vast array of tests, with the support and input, of local citizens and surrounding communities, including Indigenous groups.

The APM framework contains nine steps. The first two steps, Process Initiation and Initial Screening, are completed. The third step, Preliminary Assessments of Suitability, is ongoing. Additional steps are:

- Detailed site evaluations (Step 4)
- Confirmation of acceptance by willing host community (Step 5)
- Formal agreement to host the repository (Step 6)
- Step 6 is concurrent with the CNSC regulatory review and approval process (Step 7)
- Construction of the repository (Step 8)
- Operation of the repository (Step 9).

The NWMO has four key guiding questions: Is there the potential to find a safe site? Is there the potential to foster the well-being of the community? Is there the potential for citizens in the community to continue to be interested in the process through subsequent steps? Is there the potential to foster the well-being of surrounding communities? (NWMO 2017i).

During the Initial Screening phase (Step 2), the NWMO evaluated the potential suitability of the interested communities based on a list of screening criteria.[1] 21 of 22 communities passed initial screening and requested a more formal "assessment of suitability." By the end of 2014, all communities had entered Step 3, with the preliminary assessment, conducted through a series of activities: desktop studies and fieldwork to complete both scientific and technical studies, as well as community well-being assessments. Resources to support communities are available.

The most common reasons to exclude a community from the process are (i) the limited geological potential and (ii) the NWMO's assessment that the community no longer represents a potential host (NWMO 2020). On the last point, however, the NWMO does not provide further explanation.

The NWMO has progressively narrowed the number of communities from 22 to five and then to two: Ignace and South Bruce (NWMO 2020a). The best possible site and willing host could be chosen by 2023, but no time frame was put in place to allow flexible design and iterative planning. As of September 2022, borehole drilling has restarted after a pause due to the COVID-19 pandemic. "Testing and analysis of core samples from both areas is now well underway. These are significant steps that will enhance our understanding of whether the siting areas can meet robust regulatory requirements" (NWMO 2022: 7).

Once identified, the NWMO will apply for a license to build a DGR to contain and isolate used nuclear fuel. In brief, the proposed technology consists of a "multiple-barrier system," with nuclear fuel bundles introduced into copper-coated canisters and encased in bentonite clay boxes, placed 500-meters-deep underground.

[1] The NWMO's list includes five screening criteria: "The site must have available land of sufficient size to accommodate the surface and underground facilities. This available land must be outside protected areas, heritage sites, provincial parks, and national parks. This available land must not contain known groundwater resources at the repository depth that could be used for drinking, agriculture or industrial uses, so that the repository site is unlikely to be disturbed by future generations. This available land must not contain economically exploitable natural resources as known today, so that the repository site is unlikely to be disturbed by future generations. This available land must not be located in areas with known geological and hydrogeological characteristics that would prevent the site from being safe" (NWMO 2017h).

Once the site is selected, and before the repository is built, conceptual routes may be selected to transport the waste from current on-site temporary storage facilities in Manitoba, New Brunswick, Quebec, and Ontario to the DGR. The selection of transportation routes represents another important element of the overall risk management framework, with a draft planning document currently under a public consultation process (NWMO 2020b).

This chapter has five parts. Part one is a methodology statement. Part two describes the risks and uncertainties that are related to nuclear waste management. Part three evaluates people's level of affectedness by the project. Part four analyzes the types of risk management tools involved in nuclear waste siting. Part five provides a preliminary assessment of the level of democratization characterizing Canada's site selection decision-making process.

Methodology

This study relies on official NWMO documents, supplemented by secondary material from other organizations, books, academic articles, and news pieces. Importantly, this study includes eight semi-structured interviews from participants in two potential host communities: Hornepayne (removed from consideration) and Ignace (still involved in the process). Individuals were initially contacted through their involvement with the Community Liaison Committees (CLCs), but we used snowballing techniques to increase the number of respondents. All respondents had closely followed the site selection process over many years. There was a range of opinion within the two separate communities: some wanted their community to host the DGR, others did not, and some were undecided. Respondents were promised anonymity such that remarks are indicated as 'Hornepayne resident' or 'Ignace resident'.

There were no interviews with Indigenous communities that were adjacent to potential sites. The authors had initially arranged for interviews with members of the Constance Lake First Nation, but when Hornepayne was removed from consideration, the band members canceled them. In addition, there were no interviews conducted of community members along transportation routes because these will not be identified until after a host site is selected.

Risks and Uncertainties in Nuclear Waste Management

There are six types of risk (technological, environmental, human health, political, security, and financial) and three categories of uncertainties (epistemic, semantic, and normative) applicable to the nuclear waste management case study. Technological risks deal with the safe construction, operation, and closure of a geological repository, with the storage casks capable of safely containing the nuclear waste for thousands of years, but also accessible should reprocessing solutions become feasible in the future. Environmental risks include geological processes, tectonic movement, and climate change, which could damage the DGR. Risks to the environment also include the potential for contamination of the soil and water around the repository, as well as possible contamination due to accidents along the transportation routes. Human health risks include exposure to radiation through the DGR or contamination of soil or water. These risks to human and environmental health are magnified by having a large quantity of high-level nuclear waste at one localized site and the long-term radioactivity associated with the waste. Accidents along the transportation routes could also impact human health, a frequently noted concern of the interview participants.

In addition, the existence and perception of risks to the environment may have social effects with political, security, and financial implications. Political risks may arise from the NWMO's requirement to consult with the affected communities. These affected communities not only include those considered for the final site of the DGR, but also those communities on or near the transportation routes. There may even be political risks from communities that are nowhere near the DGR or the transportation routes; broader public opinion, at the provincial or national level, may simply be opposed to a permanent nuclear waste site. The government's final decision to approve the DGR construction is also highly political. Will a federal cabinet be willing to support such a project if public opinion becomes highly mobilized against it, and will a prime minister be willing to make such a decision part of his or her political legacy? Security risks include accidents or terrorist attacks during transportation or at the facility itself. Financial risks include the costs of building and maintaining the site and maintaining or improving the methods of transportation to the site (e.g., roads, rail, airport, etc.).

Uncertainty is reflected in the key common questions: How do we know the nuclear waste storage is going to be safe, and the technology durable, for a million years? More broadly, can we keep safely producing nuclear energy, and waste, in Canada in the future? While nuclear advocates believe the long-term waste storage technology is safe and reliable, opponents argue that a series of epistemic modeling risks and uncertainties are left unanswered; corrosion, pressurization, earthquakes, future ice ages, or human interference could result in dramatic consequences for environmental and human health. NWMO replies to those concerns stating that the DGR rock formation will be highly stable and the multiple-barrier system is a proven technology. Semantic uncertainty is also high: How do we warn future populations to stay away from the nuclear repository? The key concern here is how to ensure no one will excavate nuclear waste in the future. To do this, a universal sign that will be understood thousands of years from now to communicate: "Don't dig here!" must be found. The NWMO has not yet addressed this semantic issue. This risk ranked high in the minds of some residents: We "can't forget that we have repositories in our country."

Although not constructive toward dealing with existing nuclear fuel waste, normative uncertainty also remains about whether Canada should continue to produce nuclear energy. Opponents believe that the entire nuclear supply chain should be assessed, rather than a single waste management project, and a phase out from nuclear energy must occur.

It is also important to mention that *not building* a repository for used nuclear fuel is not risk free. An isolated DGR is a benefit with limited surface access (including inadvertently). All Canadian nuclear fuel waste is currently stored in temporary surface facilities that require management and surveillance in perpetuity: nuclear power reactor sites (Bruce, Darlington, and Pickering in Ontario; Gentilly in Quebec; and Point Lepreau in New Brunswick) and Canadian Nuclear Laboratories sites (Whiteshell Laboratories in Manitoba and Chalk River Laboratories in Ontario).

These facilities are licensed by the CNSC under the NFWA, with the regulated "wet/dry" approach designed to last at least 50 years (CNSC 2014b). Life span extensions or repackaging are identified as additional risk management options. Monitoring and regular inspections help to ensure the safety of the interim storage facilities.

While interim storage facilities are considered a safe and reliable option in the short and medium term, these are not a practical solution for thousands of years into the future. According to the NWMO, "Although the used fuel's radioactivity decreases with time, chemical toxicity persists. The used fuel will remain a potential health risk for many hundreds of thousands of years. For this reason, it requires careful management" (NWMO 2018c). Simply put, the "do nothing" option is not risk free.

Level of Affectedness

To assess individuals' level of affectedness and agency in risk exposure, three questions guide the analysis: Do individuals have a voice? Do they have a choice? And are they able to mitigate risks? Nuances may emerge, in part based on the stage of the decision-making process.

This case finds both individual-affected and individual-dependent assessments of risks. Canada's DGR will be located in a willing host community (individual-dependent), but the diffusion of risks may have some effects at a much larger scale (individual-affected). Under the APM approach, early stages of the site selection process are more prone to integrate public inputs, i.e., give Canadians a voice and local communities a choice. But once the site has been selected, and technical work has commenced, the individuals may become more affected and people's agency progressively reduced.

Primary Risk Management Intervention Tools

The REACT risk management framework identifies options in five categories, namely regulatory, economic, advisory, community-based, and technological (Krewski et al. 2007). All five tools are important in the DGR context: a strict federal regulatory framework, economic incentives for host communities to participate in the site selection process, advisory interventions involving knowledge transfer, community involvement in the site selection process, and technological developments and assessments of site suitability, combined with the multi-barrier system. Each tool is reviewed separately here, but it is also important to note that they remain highly intertwined in practice.

Regulatory Interventions

The regulatory framework is central to the project's completion as it determines the boundaries of action for the NWMO, guides the consultation and information process, determines the financial compensations for participation, and determines the benefits to the host community.

The Radioactive Waste Policy Framework is the overarching policy implemented by NRCan, stating that the "federal government is responsible for ensuring that long-term radioactive waste management is carried out in a safe, environmentally sound, comprehensive, cost-effective and integrated manner" (Natural Resources Canada 2022).

Within the policy, nuclear energy, from cradle to grave, is a highly regulated field. The sole Canadian federal regulator is the CNSC, established in 2000 by the *Nuclear Safety and Control Act*, to replace the previous Atomic Energy Control Board. The CNSC's mandate is to regulate "the use of nuclear energy and materials to protect health, safety, security and the environment; to implement Canada's international commitments on the peaceful use of nuclear energy; and to disseminate objective scientific, technical and regulatory information to the public" (CNSC 2018a).

The quasi-judicial CNSC reports to Parliament through the Minister of Natural Resources, although decision-making is at arm's length from the government. The Federal Court of Canada is the only body that can review CNSC decisions, and while it cannot reverse a decision, it can make a referral back to the CNSC for reconsideration if it believes the CNSC acted outside its mandate. The Canadian Parliament can also temporarily bypass CNSC decisions, although this has only occurred once (Bratt 2012, 42–45; CNSC 2014a).

The regulatory framework used by the CNSC to fulfill its mandate consists of the laws passed by Parliament, and the regulations, licenses, and documents that the CNSC develops (CNSC 2017). The four key federal laws are: the *Nuclear Energy Act* (1985) regarding the research and development of nuclear energy in Canada; the *Nuclear Safety and Control Act* (2000) for the use of nuclear energy in Canada; the *Nuclear Liability and Compensation Act* (2017) that defines responsibilities of liability in case of accident; and the *Nuclear Fuel Waste Act* (2002), that, as previously mentioned, provides a framework for decision-making on the management of nuclear fuel waste. In addition to creating the NWMO, the NFWA established a trust fund into which every major owner of used nuclear fuel in Canada must deposit to finance the long-term management of used nuclear fuel (NWMO 2018d).

The NWMO is also subject to obligations under the *Impact Assessment Act*, 2019, that stipulates that an environmental assessment must be conducted by the Impact Assessment Agency for designated nuclear projects and the CNSC supports the agency in the conduct of environmental assessments. Thus, even after completing the public engagement process associated with finding a willing host community, additional public engagement and technical processes will likely be required under the environmental assessment or impact assessment legislation in place, with obligations under the CNSC regulatory process. The latter currently include CNSC Regulatory documents 2.11.1, *Waste Management, Volume I: Management of Radioactive Waste* and *Waste Management, Volume III: Safety Case for Long-Term Radioactive Waste Management* (CNSC 2019c).The CNSC has also drafted guidance relevant for DGRs, REGDOC 1.2.1: *Guidance on Deep Geological Site Characterization* that may be used in subsequent license applications (CNSC 2018b).

While the CNSC is the licensing authority, it works closely with other federal and provincial bodies to ensure that regulatory requirements are met, and it is ultimately the responsibility of the license applicant to meet all applicable requirements. Provinces are also responsible for protecting public health and safety, property and the environment within their borders, and regulating resource exploitation and extraction. Therefore, the provinces will likely require their own studies of the environmental impacts of this project. Both provinces and municipalities also have their own permits, licenses, approvals, and/or bylaws that must be upheld (NWMO 2017k). All of these factors are taken into consideration for each specific project overseen by the CNSC.

The responsibility for regulating the transportation of nuclear waste is shared between the CNSC (*Packaging and Transport of Nuclear Substances Regulations*, 2015—SOR/2015-145) and Transport Canada (*Transportation of Dangerous Goods Regulations*). As a result, the NWMO "will need to demonstrate to these authorities the safety and security of its transportation system" (NWMO 2018a), including the use of a CNSC-certified transport package that meets all the license requirements and regulations and implementation of a Transportation Security Plan and an Emergency Response Plan. Periodic reviews and audits are also part of the plan. A discussion document for "Planning Transportation for Adaptive Phased Management" (NWMO 2016a) and a questionnaire (NWMO 2016b) are available online for those who want to share their thoughts on this issue. According to the NWMO, transportation of nuclear waste to the DGR should not begin before 2040.

It is important to emphasize that citizens can influence regulatory interventions in a number of ways. For example, there are established processes for citizen consultation on all regulatory documents and throughout the licensing and in-service phases of the project.

Residents of Ignace and Hornepayne recognized and in fact were quite knowledgeable about the role of the NWMO and CNSC in the site selection process and approval. They knew that the NWMO's job was finding a technical solution to the nuclear waste problem and solving the political problem by locating a suitable community, with the CNSC's job to evaluate whether or not the project can demonstrate safety. If it becomes licensed, periodic safety evaluations would form part of its licensing basis. Those interviewed believed that the NWMO and CNSC played a "vital" role in the site selection process. Even those interviewees that were openly hostile toward hosting a DGR and very critical of the NWMO mentioned that the CNSC was much more effective and would always answer their questions.

Economic Interventions

There are two types of economic intervention tools in Canada's nuclear waste management sector. The first one is industry-related: the *Nuclear Liability and Compensation Act* (Canada 2015) establishes the compensation and liability regime in case of a nuclear accident. This Act specifies that the nuclear operator is liable for damage caused within Canada at the facility or during transportation of nuclear material (including but not limited to bodily and property damage, economic loss, and environmental damage) of up to $1 billion.

The second type of economic intervention tool is community-related and refers to financial compensation as a core incentive for communities to host the DGR. Beyond the prospect of job creation within the willing host community, interested communities have already received money for participating in the site selection process. Members of the communities of Hornepayne and Ignace who were interviewed noted the benefits of NWMO support for community projects while participating in the site selection process. Each community (including First Nation and Métis communities) that participated in preliminary assessment and engagement processes was allowed to receive between $250,000 and $600,000, depending on how far along they were in the ongoing assessment process. Some Indigenous organizations have also received up to $150,000 for their participation in the process (NWMO 2017c, d, e).

The chosen host community should receive substantial financial compensation for hosting the nuclear waste facility, but the numbers remain unknown. Such financial compensations represent a key economic intervention tool in the process of finding Canada's permanent site for high-level nuclear waste. In this case, the interview participants believed this is the core reason why their communities were/are willing to host the repository. Concerning both the ongoing funding and the DGR compensation, critics said, however, that it was "bribe money."

Advisory Interventions

Advisory intervention tools refer to the knowledge transfer from experts on nuclear energy and waste management to the communities. During the site selection process, the NWMO has sought inputs from a variety of specialists. Two background papers have been written: *Developing a Siting Strategy for a Nuclear Fuel Waste Management Facility* by Richard Kuhn and Brenda Murphy (2006) and *Learning from the Experience of Others: A Selection of Case Studies about Siting Processes* prepared by Stratos Ltd. (2006). Two papers on economic benefits were also published: the *Summary of Economic Benefits Linked to Adaptive Phased Management at an Economic Region Level* (Aecom Canada Ltd. 2009) and *A Preliminary Assessment of Illustrative Generic Community Economic Benefits from Hosting the APM Project* (Aecom Canada Ltd. 2010). Professor William Leiss has also prepared three expert papers exploring the concept of risk and risk communication: *Thinking About Risk and Safety* (Leiss 2009a); *How might communities organize their discussions about hosting a site for used nuclear fuel?* (Leiss 2009b); and *What is happening in other countries?* (Leiss 2009c).

Two other individuals seem to have developed considerable knowledge on the topic over the past few years and have become advisory experts communicating with the communities: Gordon Edwards, from the explicitly anti-nuclear Canadian Coalition for Nuclear Responsibility, and Jason Donev, a physicist from the University of Calgary. Both Edwards and Donev have given presentations to CLCs in special meetings organized by the NWMO. These experts, as one Hornepayne resident put it, "slowly chipped away and increased people's technical, scientific based knowledge."

Community-Based Interventions

Community-based interventions are central to the nuclear waste disposal facility site selection process. The NWMO has decided to use a bottom-up approach by initiating meetings and information sessions only with communities interested in welcoming the DGR. In fall 2019, the NWMO conducted surveys of the residents from the, at the time, remaining five potential host communities. The surveys measured respondents' knowledge and awareness of the NWMO, questions about the site selection process (i.e., transportation, infrastructure, security, and safety), sources of information, preferred methods of NWMO communication, and project areas to learn more about (NWMO 2019b). The eventual local host community, and to some extent the neighboring area, must demonstrate its willingness to welcome the DGR on its territory. As a result, the community interventions (be it asking for more information, risk assessments, expressing questions and concerns about survey work, etc.) are at the heart of the NWMO's approach. However, what type of approval will be required and how the approval from the community will be achieved and sustained remains uncertain. For example, what happens if a community gives approval to the DGR, and then decades later, after the facility has been built and operating, a community changes its mind and wants it closed down? So far, the NWMO argues that it is "committed that the project will only be located in an area with an informed and willing host" and that the "project will only proceed with the involvement of the community, First Nation and Métis communities in the area, and surrounding communities working together to implement it." Community approval, the NWMO suggests, "will need to be supported by a compelling demonstration of willingness" (NWMO 2017j). What constitutes compelling community willingness to host the site has not been substantiated by the NWMO. Questions then arise as to whether this will be done through a referendum, town hall meetings, local political support, or simply through the absence of local opposition?

The NWMO also recognizes the importance of Indigenous Knowledge as a different epistemology to consider in the site selection process:

"Indigenous Knowledge is a complex and sophisticated system of knowledge drawing on millennia of wisdom and experience that constantly grows and expands with the experience of each generation. As we continue to move through the site selection process and engage with communities, there is an opportunity to learn from local Indigenous Knowledge and apply that learning to planning and decision-making processes" (NWMO 2017g). The NWMO also adds that it will "look to Indigenous communities and local Indigenous Knowledge holders in the areas surrounding interested communities to find ways to apply Indigenous Knowledge to the site selection process and protect it in its application" (NWMO 2017g). In order to do so, the NWMO has developed an Indigenous Knowledge policy (NWMO 2016c).

Some criticism has been raised with regard to the actual integration of Indigenous Knowledge into the NWMO's site selection process. Meagan Sarah Weatherdon argued that the NWMO has "interpreted Indigenous spiritual beliefs and philosophies in ways compatible with their own agenda" and has limited First Nations' engagement (2017: 94). Weatherdon also suggested that the NWMO's scientific epistemology seeks to co-opt Indigenous traditional knowledge rather than truly listening to it: "the NWMO interprets indigenous spirituality within its own cosmological and commercial framework, which seems to grant transcendental power to technology and science" (2017: 97). Genevieve Fuji Johnson also raised the flag over First Nations' involvement in the NWMO's consultation process arguing, "there is evidence that participants' perspectives were not weighted equally. There were many accounts of how views were dismissed and excluded from the NWMO's assessment framework and recommendation" (2015: 79). This was especially true for the Assembly of First Nations and the Congress of Aboriginal Peoples who claimed, in the earlier stages of the process, that they did not have sufficient time and funding to assess the project and express their concerns (AFN 2005).

The Aboriginal[2] and treaty rights of Aboriginal peoples in Canada are recognized and affirmed in Section 35 of the *Constitution Act, 1982*. The Supreme Court of Canada has found that the Crown (the federal and provincial governments) has a duty to consult Aboriginal peoples whenever it contemplates conduct that could adversely impact potential

[2] This chapter generally uses the term Indigenous, but some documents continue to use the term Aboriginal.

or established Aboriginal and/or treaty rights. The courts have struck down infrastructure projects that do not meet the standard of the duty to consult. Since all of the potential sites, and transportation routes to them, are on or pass through treaty areas, asserted or established traditional territories, or near First Nations communities, the duty to consult is an essential component of the nuclear waste site selection process.

The CLC, as the residents of Ignace and Hornepayne recognized, was obviously the key mechanism for community-based input. There was a separate CLC for each potential site community and each had between 8 and 12 members. Members went through an open selection process, but were appointed by the town council. Some positions were set aside for specific demographic groups, i.e., Indigenous representative, youth representative, and adjacent communities representative. The town's mayor and 1–2 councillors were also on the CLC. CLC members from Ignace and Hornepayne believed that the CLCs were inclusive and provided a good reflection of the community. A minority, however, felt that the CLCs were stacked to the pro-side. One Hornepayne resident noted that "what we had [with] the nuclear waste Community Liaison Committee [was] that the majority of that committee was all pro-repository members. And that's a fact. So we had no voices at the table asking or digging into tougher questions. Adding, I call it tension or friction. When you add that to the table, you're not all necessarily going to agree. But that tension and friction bring those better answers in the end of how to move forward. So do I have to say no, I don't think all voices individually got heard." It was a new process, so "the NWMO took a lot of lead on setting up community, the committees, the liaison committees. And there was a miscommunication of exactly whose committee that is.... Hindsight [is] 20/20, [and] it would have been better to have a third party group that helps set the community up and not be attached to the NWMO."

Technological Interventions

Techno-scientific expertise has been, and will be, highly solicited during the DGR's site selection and construction process. Additional work considers site maintenance and closure as well as waste transport parameters.

The DGR consists of a "multiple-barrier system." While the first barrier is the site itself, this also includes the technology to encase the waste. Nuclear fuel pellets "made from uranium dioxide powder, baked in a furnace to produce a hard, high-density ceramic," are then inserted in a sealed tube made of Zircaloy, a corrosion-resistant metal, and then bundled with other such tubes to create a log-shaped package. This fuel bundle is introduced in a fuel container made with carbon steel pipe and copper. Finally, each of these containers "will be encased in a highly compacted bentonite clay buffer box" 500-meters-deep underground (NWMO 2017b). The NWMO asserts that placing nuclear fuel deep underground is a proven method with minimal risks. In Finland, where the world's first civilian spent fuel DGR is currently under construction, a very similar technology is being used (Ialenti 2020).

To better familiarize the communities with the technology, the NWMO brought some of the residents to the technology, and in other cases, they brought some of the technology to the community. For example, many of the interviewed residents mentioned the trips that they took to the NWMO's test proof facility in Oakville, Ontario, to examine the canisters and other technological innovations. In addition, as a further example of the advisory-based approach, the NWMO developed a mobile Learn More Centre, containing "3D models, interactive technology, hands-on learning opportunities, videos, and photos" (NWMO 2019a). This mobile unit was sent to the potential site communities. Participants noted that times were set aside for different age groups to attend information sessions, such as for school groups and seniors.

Level of Democratization

This section presents a preliminary assessment of the level of democratization associated with the siting of the DGR. The four criteria, based on democratic principles, include: transparency, inclusiveness, deliberative quality, and accountability. Further assessment could evaluate the level of democratization more systematically.

Interviews with members of the affected communities give a preliminary assessment of these parameters: How did the NWMO interact with local communities? Did local communities feel that their concerns were addressed and their voice heard? Do Indigenous communities perceive that they were properly involved in the decision-making process? These are among the key questions that need to be answered through direct enquiries. By doing so, divergences may be observed between the NWMO's perception of democratic practices and communities' vision of that aspect of the interaction with the project proponent.

Table 5.1 presents the four criteria based on an evaluation of the NWMO siting process. It includes a preliminary assessment done independently by the authors, based on NWMO documentation and their scholarly knowledge of democratization, and the opinions of the residents of Hornepayne and Ignace, who personally experienced the process. Each criterion has been marked out of 10 (with 0 being a total lack of demonstration of the principle and 10 exemplifying a high degree of implementation of the principle). The average for each democratic principle is then calculated, and the total scores for the combined four criteria are aggregated (rounded to one decimal place). The aggregated mark determines the overall level of democratization: 0–10 minimal; 11–30 moderate; and 31–40 high.

Based on this preliminary assessment of the level of democratization, it appears that the nuclear waste management case displays a moderate level of democratization, with an average total score of 26.4. However, there is a wide discrepancy in scores at an individual level. While Bratt and the co-authors assessed a 7–9 for each principle, in most cases the interviewee score was either between 7 and 10 or it was zero. Those who supported the project gave the NWMO very high scores, but those opposed to the project gave them zeros across the board.

The following section describes how the NWMO integrates the four different democratic principles into the process of finding a permanent site for Canada's high-level nuclear waste.

Transparency

Average Score: 6.6

Providing publicly available information about nuclear energy and waste management safety is a central part of both the CNSC's and the NWMO's mandates. The CNSC's REGDOC-3.2.1: *Public Information and Disclosure* (CNSC 2017) clarifies how the licensed nuclear facilities, licensees, and license applicants must "develop and implement a public information program that includes a disclosure protocol. Through an effective public information program, a licensee or licence applicant establishes an atmosphere of openness, transparency and trust."

The CNSC has organized outreach activities for communities and Indigenous peoples who have expressed interest in learning more about the CNSC's regulatory role and the licensing process for any application for a DGR for Canada's used nuclear fuel. For example, in 2018–2019

Table 5.1 Level of democratization in nuclear waste management

Democratic principle	*Criteria*	*Individual assessment*			
		Bratt	*Co-Authors*	*Interviewees Avg*	*Overall Avg*
Transparency	Transparency of risk management decisions: Is the rationale for decisions publicly available? Are the uncertainties involved clearly stated?	8	7	6.3	6.6
Inclusiveness & Representativeness	Presence and scope of formal public participation opportunities: Is there public notification of risk management process? Is input solicited? Are a broad range of stakeholders involved?	7	7.5	6.4	6.7

(continued)

Table 5.1 (continued)

Democratic principle	*Criteria*	*Individual assessment*			
		Bratt	*Co-Authors*	*Interviewees Avg*	*Overall Avg*
Deliberative quality	Non-experts' public involvement in risk management process; length of decision-making process	7	7	5.9	6.2
Accountability	Accountability of those conducting risk management to Parliament, legislatures or other publicly elected representatives	9	8	6.4	6.9
Combined Level of Democratization	Combining the score for each criterion, marked out of 10. Total scores for combined four criteria of 0–10 ("minimal"); 11–30 ("moderate"); and 31–40 ("high")	31	29.5	25.0	26.4

Source for democratic principles: Chapter 1. Authors' own calculations of individual assessment (the last full year before COVID-19 arrived in Canada), the CNSC held 22 "meet the regulator" sessions in communities across Canada as well as a further 30 meetings with Indigenous communities and organizations (CNSC 2019a).Through these meetings, the CNSC informs communities about how they regulate nuclear energy and waste and how communities can participate in the public hearing process. CNSC staff are also interested in hearing about the most effective ways to involve communities and Indigenous groups and how best to provide information to those who want to know more about the CNSC and other relevant matters within the scope of its mandate.

The CLCs, for their part, seek to engage with the local community, provide information and education, and listen to the citizens. They provided advice to the municipal council regarding the NWMO's site selection process and APM framework. Information is centralized on one website[3] in which each municipal government has its own webpages describing their mission statement, committee membership, proceedings (meetings, minutes and agendas, public information sessions, news, open houses, etc.), and a Question and Answers page. The NWMO is often invited to those meetings to make presentations and inform the population about the ongoing site selection process. The CNSC may also be invited to explain their role in regulating the nuclear sector. The CLCs do not claim to advocate for the repository site, but rather seek to gather the best possible information to help inform the decision-making process. For instance, Ignace's CLC was created in November 2011 with the objective "to provide a mechanism for the sharing of information between: the community and the local accountable authority; and the community and the NWMO" (Ignace CLC 2012). Similarly, the Hornepayne CLC was "established by the Township of Hornepayne Municipal Council in October 2011. Our objective is to help Hornepayne learn about Canada's plan for used nuclear fuel and involve Hornepayne residents in these learning activities" (Hornepayne CLC 2019).

Public information and disclosure is key to the very existence of the CLCs. Their educational role is well understood. What seems less clear, however, is how public input is taken into account by the local authorities and the NWMO, and for what purpose. For example, minutes may not detail discussion items and points of view. One other area where questions

[3] See: http://clcinfo.ca/.

about the transparency of the process remain is: Under what criteria is a community deemed unfit to become the deep geological repository's willing host and thus removed from the process?

The residents of Ignace and Hornepayne had mixed views about the NWMO's transparency. A majority of them felt the NWMO did an "excellent job." One Ignace resident said that they were involved in many projects over the years and the NWMO was the "most transparent process that I have ever come across." As another resident put it, "how do you judge transparency if people are apathetic about learning the information in the first place. You know, your window can be as clean as possible, but if people aren't peaking into the window. It's transparent, but nobody is taking advantage of the transparency." However, a minority were adamant that the NWMO was "not very transparent at all." One Hornepayne resident complained "I went to open houses and the meetings and was told, more or less, to keep my mouth shut."

Inclusiveness and Representativeness

Average Score: 6.7

A wide range of inclusion and representation measures exist as part of Canada's nuclear waste site selection process, including First Nations' consultation (e.g., integration of Indigenous Traditional Knowledge), the involvement of local communities and surrounding area through the CLCs, including the participation of local elected officials, and consultation with experts in various fields.

However, critics pointed out that the CLCs were not representative. One Hornepayne resident stated that "the majority" of the CLC were "pro-repository members." This meant that "we had no voices at the table asking or digging into tougher questions." Those that were on the CLC were heard, but those that were not on the CLC had no voice. Opponents of the DGR felt that they were deliberately excluded from the CLCs. However, other residents said that there was plenty of opportunities, but some people wanted to remain outside of the process.

With respect to Indigenous populations, the Council of Elders and Youth "is an advisory body to NWMO management. It provides counsel on the application of Indigenous Knowledge in the implementation of Adaptive Phased Management. In addition, the Council of Elders and Youth provides advice on issues that could enhance the development and maintenance of good relations with Aboriginal communities" (NWMO

2017f). Interviewees suggested that there was ample consultation with respect to Indigenous engagement.

Looking ahead to the implementation phase, the CNSC's REGDOC-3.2.2 *Indigenous Engagement* sets out the requirements and guidance for licensees on Indigenous engagement, in addition to procedural direction in support of the whole-of-government approach to Aboriginal consultation (CNSC 2019b).

Deliberative Quality

Average Score: 6.2
Non-expert public in the interested communities are involved through the different steps of the APM process and invited to collaborate (express concerns, ask questions, participate in local meetings) with the NWMO and the CNSC.

The deliberative qualities of the NWMO's site selection process will have to be assessed continually as the steps unfold over the upcoming months and years. Consultation, surveys, or interviews with directly affected communities would help assess their perception of the quality of the process. Two issues were discussed frequently by interviewees. Regarding the range of expertise the CLCs invited to the meeting, most individuals acknowledged that the "process demanded that NWMO" bring alternative perspectives to the communities. However, one resident noted that this meant that they put on an equal platform people with "years of science and scholarship" and people who had been "environmental activists for 30 or 40 years, but do not have a scientific background." With respect to the timeline, those members of Ignace and Hornepayne who supported the project complained that the NWMO was being too deliberative. "It took too long," was a common refrain.

Accountability

Average Score: 6.9
The concept of public accountability is widely used in democratic systems to describe the fundamental relationship of trust between the citizens (or the public), on one side, and the governing authorities, agencies, and public enterprises who are trusted to make good use of public resources, on the other (Bovens 2005).

The NWMO is at arms-length from the government and is accountable to both the public and the federal government. Oversight of the transportation of nuclear substances is shared between two federal departments, the CNSC and Transport Canada. If the Commission grants a license for a DGR in the future, the CNSC will oversee the NWMO's activities and ensure its compliance. It is also important to note that while the CNSC is the licensing authority, it works closely with other federal and provincial bodies to ensure that all regulatory requirements are met.

According to the NWMO, accountability is one of the organization's key values, along with safety, integrity, excellence, collaboration, and transparency. However, the only mention with regard to accountability is that: "We take responsibility for our actions, including wise, prudent and efficient management of resources" (NWMO 2018b).

A critical issue that will be key in the future is the overall perception of the NWMO in the public's view. Is there a perception gap between public officials and communities with regard to the NWMO's accountability? If so, how does it affect the site selection process? Critics of the NWMO pointed out that it was funded by the nuclear industry and they felt that is whom the NWMO was ultimately accountable to. On the other hand, those that supported the DGR felt that the NWMO was accountable to the potential host community.

Conclusion

This chapter has provided preliminary information to summarize the risk management framework for the site selection process for a DGR for Canada's high-level nuclear waste. It has also described the level and type of affectedness, evaluated the applicable risk management intervention tools through the REACT framework, and assessed the level of democratization in the case. These three assessments will help in understanding, and perhaps strengthening, risk management capacity in Canada.

Acknowledgements The authors would like to thank Bob Walker, Adrienne Ethier, Keith Dewar, Julie Brown, and Renée Silke for their thoughtful comments on earlier drafts. As always, the final version is the sole responsibility of the authors.

References

Aecom Canada Ltd. 2009. *The Summary of Economic Benefits Linked to Adaptive Phased Management at an Economic Region Level.*

Aecom Canada Ltd. 2010. "A Preliminary Assessment of Illustrative Generic Community Economic Benefits from Hosting the APM Project." NWMO SR-2010-09. May 2010.

Assembly of First Nations (AFN). 2005. *Nuclear Fuel Waste Dialogue. Recommendations to the Nuclear Waste Management Organization.* September 30, 2005.

Bourassa, Maureen, Kelton Doraty, Loleen Berdahl, Jana Fried, Scott Bell. 2016. "Support, Opposition, Emotion and Contentious Issue Risk Perception." *International Journal of Public Sector Management*, Vol. 29, No. 2, pp. 201–216.

Bovens, Mark. 2005. "Public Accountability." in Ferlie, Ewan, Flynn, Laurence E. Jr., and Pollitt, Christophe (ed.) *The Oxford Handbook of Public Management*. Oxford University Press: 182–208.

Bratt, Duane. 2012. *Canada, the Provinces, and the Global Nuclear Revival: Advocacy Coalitions in Action.* McGill-Queen's University Press.

Canada. 2015. Nuclear Liability and Compensation Act S.C. 2015, c. 4, s. 120.

Canada. 2018. Packaging and Transport of Nuclear Substances Regulations, 2015. SOR/2015-145. http://laws-lois.justice.gc.ca/PDF/SOR-2015-145.pdf. Regulations are current to November 2, 2020.

Canadian Nuclear Safety Commission (CNSC). 2014a. "Atomic Energy of Canada Limited National Research Universal Reactor Safety System Upgrades and the Canadian Nuclear Safety Commission's Licensing and Oversight Process: IX. Reasons for the Extended Outage following the NRU Reactor Shutdown in November 2007." https://nuclearsafety.gc.ca/eng/resources/publications/reports/lessonslearned/ix.cfm.

Canadian Nuclear Safety Commission (CNSC). 2014b. "Canadian National Report for the Joint Convention on the Safety of Spent Fuel Management and on the Safety of Radioactive Waste Management" https://nuclearsafety.gc.ca/pubs_catalogue/uploads/joint-convention-fifth-national-report-2014b-eng.pdf.

Canadian Nuclear Safety Commission (CSNC). 2017. "REGDOC-3.2.1: Public Information and Disclosure." http://nuclearsafety.gc.ca/eng/acts-and-regulations/consultation/comment/regdoc3-2-1.cfm.

Canadian Nuclear Safety Commission (CNSC). 2018a. "The Commission." https://nuclearsafety.gc.ca/eng/the-commission/index.cfm?pedisable=true.

Canadian Nuclear Safety Commission (CNSC). 2018b. "REGDOC-1.2.1: Guidance on Deep Geological Site Characterization." http://www.nuclearsafety.gc.ca/eng/acts-and-regulations/consultation/comment/regdoc1-2-1.cfm.

Canadian Nuclear Safety Commission (CNSC). 2019a. "2018–2019 Departmental Results Report." http://www.nuclearsafety.gc.ca/eng/resources/publications/reports/departmental/drr-2018-2019a/index.cfm.

Canadian Nuclear Safety Commission (CNSC). 2019b. "REGDOC-3.2.2: Indigenous Engagement." http://nuclearsafety.gc.ca/eng/acts-and-regulations/regulatory-documents/published/html/regdoc3-2-2-ver1.1/index.cfm.

Canadian Nuclear Safety Commission (CNSC). 2019c. "REGDOC-2.11.1: Waste Management, Volume III: Safety Case for Long-Term Radioactive Waste Management, Version 2." https://nuclearsafety.gc.ca/eng/acts-and-regulations/consultation/comment/regdoc2-11-1-vol3-ver2.cfm.

Hornepayne Community Nuclear Liaison Committee. 2019. "About." http://www.clcinfo.ca/hornepayne/about-us.

Ialenti, Vincent. 2020. *Deep Time Reckoning: How Future Thinking Can Help Earth Now*. Cambridge, MA: MIT Press.

Ignace Community Nuclear Liaison Committee. 2012. "February 22, 2012 Meeting Agenda." https://clcinfo.ca/ignace/files/2012/10/ICNLC-Agenda-February-22-2012.pdf.

Johnson, Genevieve Fuji. 2015. "National Consultation and the Nuclear Waste Management Organization." in Johnson, Genevieve Fuji (ed.). *Democratic Illusion: Deliberative Democracy in Canadian Public Policy*. Toronto: University of Toronto Press.

Krewski, D., Hogan, V., Turner, M. C., Zeman, P. L., McDowell, I., Edwards, N. and Losos, J. 2007. "An Integrated Framework for Risk Management and Population Health."*Human and Ecological Risk Assessment: An International Journal*, Vol. 13, No. 6, pp. 1288–1312. https://doi.org/10.1080/10807030701655798.

Kuhn, Richard G. and Brenda L. Murphy. 2006. "Developing a Siting Strategy for a Nuclear Fuel Waste Management Facility." NWMO SR-2006-01. June 2006.

Leiss, William. 2009a. "Paper#1: Thinking About Risk and Safety." NWMO SR-2010-10. November 2009.

Leiss, William. 2009b. "Paper#2: How Might Communities ORGANIZE Their Discussions About Hosting a Site for Used Nuclear Fuel?" NWMO SR-2010-11. November 2009.

Leiss, William. 2009c. "Paper #3: What Is Happening in Other Countries?" NWMO SR-2010-12. November 2009.

Natural Resources Canada (NRCan). 2021. "Inventory of Radioactive Waste in Canada." https://www.nrcan.gc.ca/sites/nrcan/files/energy/pdf/uranium-nuclear/RadWaste%20Inventory%20Report%20-%20ENG.pdf.

Natural Resources Canada (NRCan). 2022. "Canada's Radioactive Waste." https://www.nrcan.gc.ca/our-natural-resources/energy-sources-distribution/nuclear-energy-uranium/radioactive-waste/7719#a.

Nuclear Waste Management Organization (NWMO). 2016a. "Planning Transportation for Adaptive Phased Management." Discussion Document. https://www.nwmo.ca/~/media/Site/Files/PDFs/2016a/08/25/11/44/TransportationDiscussion_EN.ashx?la=en.

Nuclear Waste Management Organization (NWMO). 2016b. "Planning Transportation for Adaptive Phased Management." Discussion Document Questionnaire. https://www.nwmo.ca/~/media/Site/Files/PDFs/2016b/08/25/11/46/transportation_questionnaire_EN.ashx?la=en.

Nuclear Waste Management Organization (NWMO). 2016c. "Indigenous Knowledge Policy." https://www.nwmo.ca/~/media/Site/Reports/2016/10/20/11/02/English_IndigenousKnowledgePolicy_2016.ashx?la=en.

Nuclear Waste Management Organization (NWMO). 2017a. "Who We Are." https://www.nwmo.ca/en/ABOUT-US/Who-We-Are.

Nuclear Waste Management Organization (NWMO). 2017b. "Multiple-Barrier System." https://www.nwmo.ca/en/A-Safe-Approach/Facilities/Deep-Geological-Repository/Multiple-Barrier-System.

Nuclear Waste Management Organization (NWMO). 2017c. "Recognizing Community Leadership." https://www.nwmo.ca/en/Site-selection/Steps-in-the-Process/Step-3-Preliminary-Assessments-of-Suitability/Step-3-Phase-1-Desktop-Studies-and-Engagement/Recognizing-Community-Leadership.

Nuclear Waste Management Organization (NWMO). 2017d. "Acknowledging Aboriginal Contributions." https://www.nwmo.ca/en/Site-selection/Steps-in-the-Process/Step-3-Preliminary-Assessments-of-Suitability/Step-3-Phase-1-Desktop-Studies-and-Engagement/Acknowledging-Aboriginal-Contributions.

Nuclear Waste Management Organization (NWMO). 2017e. "Recognizing Communities." https://www.nwmo.ca/en/Site-selection/Steps-in-the-Process/Step-3-Preliminary-Assessments-of-Suitability/Step-3-Phase-2--Field-Studies-and-Engagement/Recognizing-Communities.

Nuclear Waste Management Organization (NWMO). 2017f. "Council of Elders and Youth." https://www.nwmo.ca/en/ABOUT-US/How-We-re-Governed/Council-of-Elders.

Nuclear Waste Management Organization (NWMO). 2017g. "Indigenous Knowledge." https://www.nwmo.ca/en/Site-selection/About-the-Process/Indigenous-Knowledge.

Nuclear Waste Management Organization (NWMO). 2017h. "Step 2: Initial Screening." https://www.nwmo.ca/en/Site-selection/Steps-in-the-Process/Step-2-Initial-Screening.

Nuclear Waste Management Organization (NWMO). 2017i. "Step 3: Preliminary Assessment of Suitability." https://www.nwmo.ca/en/Site-selection/Steps-in-the-Process/Step-3-Preliminary-Assessments-of-Suitability.

Nuclear Waste Management Organization (NWMO). 2017j. "Steps 4–9: Site Confirmation Construction and Operation." https://www.nwmo.ca/en/Site-selection/Steps-in-the-Process/Steps-4-to-9-Site-Confirmation-Construction-and-Operations.

Nuclear Waste Management Organization (NWMO). 2017k. "Regulatory Oversight." https://www.nwmo.ca/en/Canadas-Plan/Canadas-Used-Nuclear-Fuel/Regulatory-Oversight.

Nuclear Waste Management Organization (NWMO). 2018a. "Transportation Regulations and Oversight." https://www.nwmo.ca/en/A-Safe-Approach/Transportation/Transportation-Regulations-and-Oversight.

Nuclear Waste Management Organization (NWMO). 2018b. "Vision, Mission, Values." https://www.nwmo.ca/en/ABOUT-US/Who-We-Are/Our-Commitment/Vision-Mission-Values.

Nuclear Waste Management Organization (NWMO). 2018c. "How Is It Stored Today?" https://www.nwmo.ca/en/Canadas-Plan/Canadas-Used-Nuclear-Fuel/How-Is-It-Stored-Today.

Nuclear Waste Management Organization (NWMO). 2018d. "Funding." https://www.nwmo.ca/en/ABOUT-US/Who-We-Are/Funding.

Nuclear Waste Management Organization (NWMO). 2019a. "The NWMO's New Mobile Learn More Centre launches with Visit to Manitouwadge, Constance Lake First Nation and Hornepayne." https://www.nwmo.ca/en/More-information/News-and-Activities/2019/07/31/14/11/NWMOs-new-Mobile-Learn-More-Centre-launches-NoS.

Nuclear Waste Management Organization (NWMO). 2019b. "Findings from Community Surveys—Fall 2019." https://www.nwmo.ca/~/media/Site/Reports/2020/02/13/16/42/Community-Surveys-Integrated-Summary-Report--EN.ashx?la=en.

Nuclear Waste Management Organization (NWMO). 2020a. "Areas No Longer Being Studied." https://www.nwmo.ca/en/Site-selection/Study-Areas/Areas-No-Longer-Being-Studied.

Nuclear Waste Management Organization (NWMO). 2020b. "Moving Forward Together: An Invitation to Review a Draft Planning Framework for the Transportation of Used Nuclear Fuel." https://www.nwmo.ca/~/media/Site/Files/PDFs/2020b/08/12/00/38/Draft-Transportation-Planning-Framework.ashx?la=en.

Nuclear Waste Management Organization (NWMO). 2022. "Guided by Science, Grounded in Knowledge, Committed to Partnership: Annual Report 2021." https://www.nwmo.ca/~/media/Site/Reports/2022/03/22/19/59/NWMO-2021-Annual-Report.ashx?la=en.

Stratos Ltd. 2006. "Learning from the Experience of Others: A Selection of Case Studies About Siting Processes." NWMO SR-2006-03. August 2006.

Transport Canada. 2017. "Transportation of Dangerous Goods Regulations Including Amendment SOR/2017-253." Government of Canada. http://www.tc.gc.ca/media/documents/tdg-eng/Consolidated_Marine_Provisions_English.pdf.

Weatherdon, Meagan Sarah. 2017. "Indigenous Knowledge and Contested Spirituality in Canadian Nuclear Waste Management." *Journal for the Study of Religion, Nature and Culture*, Vol. 11, No. 1, pp. 86–108.

CHAPTER 6

Hydraulic Fracturing in New Brunswick: Trust, Deliberation, and Risk Decision-Making

Laura Nourallah

Introduction and Context

This chapter examines the role of trust in energy decision-making for shale gas development in the province of New Brunswick (NB), Canada. Building on theoretical frameworks in public policy studies, the study assesses the role of risk perception in shaping policy actors' preferences in the context of public engagement processes between 2007 and 2017. As with many other types of energy development in Canada, shale exploration and production are under provincial jurisdiction and therefore regulated by the province. With the potential for large-scale shale development in its jurisdiction, the provincial government set out to engage the public in 2011, as the assessment of NB's geology by energy corporations was taking place to test the viability of production. The case of

L. Nourallah (✉)
Nepean, ON, Canada
e-mail: lnourallah9@gmail.com

M. Gattinger (ed.), *Democratizing Risk Governance*,
https://doi.org/10.1007/978-3-031-24271-7_6

NB is therefore interesting to understand from a democratization standpoint—with the potential for shale gas development on one-fifth of the province's land mass, the government set out to put citizens at the heart of decision-making (GNB, 2010) and employed a series of engagement activities in the context of risk and epistemic uncertainty surrounding the practice associated with producing shale gas. However, in 2014, a moratorium was imposed on hydraulic fracturing and it was extended indefinitely in 2016 after a commission established by the government toured the province.

Shale gas development involves the deployment of a technological practice of hydraulic fracturing, a practice that fractures rock horizontally and employs chemical compounds in large volumes of water to release gas from the rock that is then pumped back to the surface (NRCan, 2020), at which point the gas is separated from the water for transportation and use. A relatively recent breakthrough in the technology that enabled multi-stage hydraulic fracturing has led to the boom in production in North America. More commonly referred to as fracking, the practice has raised considerable fears about its impacts on the environment and human health. Potential risks include groundwater contamination, earthquakes, increased greenhouse gas emissions, and risks to health from endocrine disruption to impacts on newborns (Council of Canadian Academies, 2014). Shale development is not new to Canada; however, large-scale production has mainly taken place in Canada's Western provinces (with a few small-scale activities taking place in NB).

Throughout the period of this case study (2007–2017), the public controversy surrounding shale development had taken prominence on the global stage. Public opposition to fracking was rising in many countries, and the documentary *Gasland* by Josh Fox, capturing the image of tap water lighting on fire, had caught attention. Controversies surrounding scientific research on the impacts of hydraulic fracturing remained evident (Dunlop et al., 2021)—experts disagreed on the causes of methane in groundwater or seismic activity linked to fracking. These are considerations that featured in the public debate in NB about permitting shale development to take place on a large scale in the province. This research is interested in how theory helps us understand the implications of risk for decision-making in the process of engaging the public. Specifically, this study is interested in the role of trust and lay knowledge, and their impact on learning among policy actors.

Multidisciplinary literature asserts the importance of public engagement in different ways. Different strands of the literature aimed at understanding the association between public perceptions and expert risk assessment, question the privileged role of expert views over lay knowledge. The "deficit" model under the Public Understanding of Science (PUS) frame, which asserts that expert knowledge needs to be used to educate lay people to make rational decisions, has been challenged very broadly. Science and Technology Studies (STS), where the role of expert knowledge is privileged, has also seen a turn in the literature toward participatory approaches emphasizing the role of communicative action and lay knowledge.

There has also been recognition of the role of citizens in the processes of risk communication and management (Fiorino, 1990; Slovic, 1993). Advocates argue that democratic legitimacy and trust can be improved by including the public in government decision-making regarding controversial technological innovations (Renn and Levine, 1991). Stern and Fineberg (1996) argue that from a governance perspective, "the instrumental rationale for broad public participation is that it may decrease conflict and increase acceptance of or trust in decisions by government agencies" (p. 24). This is especially the case in the context of governing risks, where distrust can create conflict (Stern and Fineberg, 1996). While the approach to the model of the individual may differ among the various approaches, the idea that involving people in a deliberative process can create new understandings of problems, and potentially more effective ways to address them, is a common driving thread. This is also evident in the literature on environmental planning and management, where the role of citizens in deliberating on the regulatory and policy decisions that impact them is emphasized (Fischer, 2000; Coburn, 2003).

The concept of public engagement has evolved from one of informing people about science and technical expertise to make rational decisions (a deficit model), to one about recognizing the knowledge that people hold and their views regarding risks that can be incorporated into the process (Petts and Brooks, 2006). Studies point to the importance of learning among participants in decision-making forums and processes to bring public values and perspectives to the table with technical expertise. Learning writ large is posited as a process of updating preferences based on new information or experience that can potentially lead to more effective outcomes. This is not exclusive to just citizens learning: the idea that experts can engage and learn as well is an important consideration of this approach.

The policy literature has also been challenged with "argumentative" and "deliberative" turns underlining the importance of engaging the public in government processes. Theoretical frameworks that seek to explain policy outcomes have also tended to privilege the role of experts and technical information. Therefore, in trying to assess democratization and its role in risk management in the case of shale development in NB, this study is focused on how policy actors (including experts and engaged citizens) learn and whether it influences policy change.

The Advocacy Coalition Framework (ACF) is a major theoretical framework to consider when it comes to understanding policy change (Sabatier and Jenkins-Smith, 1993), and it has been widely applied in the literature to study energy and environment subsystems (Litfin, 2000; Elliott and Schlaepfer, 2001; Bomberg, 2013; Heikkila et al., 2014). For the ACF, the basic unit of study is a policy subsystem that consists of advocacy coalitions and, according to the framework, is the most useful unit of analysis to understand how policy changes over a period of 10 years or more (Jenkins-Smith and Sabatier, 1994).

Coalitions are mainly composed of policy elites[1] that have high levels of interest and knowledge in the respective subsystem. These coalitions compete to dominate the subsystem, whereby they succeed in translating their beliefs into policy (Sabatier and Jenkins-Smith, 1999). For this to occur, they need to gather allies, formulate strategies, and share resources. The central argument that forms the basis of the framework is that policy participants will partner with those who share similar beliefs, and that if they partake in a "nontrivial degree of coordination," they form an advocacy coalition (Sabatier and Weible, 2007). As such, beliefs play an important role in their ability to bring participants together and form the interests that they then aim to transform into policy.

The ACF contends that there are four major paths to policy change: policy-oriented learning, external shocks, internal shocks, and a hurting stalemate (Sabatier and Jenkins-Smith, 1999). In this chapter, the focus is on policy-oriented learning, a path to policy change that occurs as a result of belief change within a coalition. Importantly, policy-oriented learning

[1] Policy brokers are another set of actors worth mentioning within the ACF; they are those whose main concern is to maintain the degree of political conflict within reasonable limits and find acceptable answers to the issues (Sabatier, 1988).

relative to policy participants is defined as "relatively enduring alterations of thought or behavioural intentions that result from experience and/or new information and that are concerned with the attainment or revision of policy objectives" (Sabatier and Jenkins-Smith, 1999, p. 123). Thus, knowledge plays an explicit role in the change of beliefs that ultimately creates a change in policy through the process of policy-oriented learning.

However, for this to happen, the ACF specifies a set of conditions and a requirement for knowledge to be evidence-based "scientific and technical information" (Sabatier and Weible, 2007). As such, the ACF puts forward a very "rationalistic, technocratic understanding of learning" (Fischer, 2003, p. 109). The ACF argues that expert debate is what stimulates learning between the participants, and argues that certain technical capacities and specific contexts are a requisite for this to take place. The role of experts is privileged in the ACF, and they play a significant role in the dissemination and modification of ideas, rendering the model a technocratic one that emphasizes policy elites. Furthermore, the approach neglects the social aspects associated with learning and the socially constructed foundations of expertise (Fischer, 2003).

Many scholars have argued against such a technocratic understanding of policy-oriented learning and make calls for its revision (see, for example, Fischer, 2003; James and Jorgensen, 2009; Hajer and Wagenaar, 2003). They emphasize the role of discourse and lay knowledge. Shulock (1999) argues that it is insufficient to only consider technical and scientific information, and underscores the importance of an approach that includes knowledge from those affected by policy. Fischer (2000) argues that the conception of the type of knowledge that is used in policymaking needs to change. This research points to the importance of integrating the role of interactive knowledge in learning in the ACF. "Expert-knowledge generation is institutionalised and exclusive and shared through peer-review processes, whereas lay knowledge is embedded in the world around and directly impacting on individuals" (Petts and Brooks, 2006, p. 1046). Knowledge in a societal context is shared and developed informally among people; the emphasis is on contextualized knowledge in local settings. This is an interactive form of knowledge that is built on experience and interaction (Bryant, 2001).

The case of NB presents an opportunity to assess learning based on the integration of lay knowledge, where an extensive set of mechanisms for public participation[2] regarding shale development was implemented between 2010 and 2015. In the context of a high degree of uncertainty and perceived risk associated with hydraulic fracturing, the role of lay knowledge in learning and its effect on change can therefore be examined.

According to the ACF, beliefs are the core explanatory factor in understanding coalitions. It is on the basis of pre-existing beliefs that actors perceive the world in the ACF and therefore coordinate to influence policy. Beliefs are layered and start fundamentally with (a) the deep core connected to worldviews (e.g., ranking of values such as freedom and security based on cultural orientations and political ideologies), (b) the policy core of key policy positions based on worldview perspectives (e.g., government intervention versus market drivers based on problem severity and cause), and (c) secondary beliefs associated with instruments and tools to achieve policy objectives (instrumental means for achieving policy core ends or goals). For the ACF, cognitive components identify the basic causes of problems and favoured solutions (Jenkins-Smith et al., 2014). Ultimately, cognitive and normative elements are the ways in which actors find the "most appropriate" means to attain specified values and objectives. If changes happen to the policy core of a coalition, this can lead to policy change (Sabatier, 1998). Furthermore, there is evidence to suggest that if enough changes occur at the secondary level, this can also trigger a change of policy core beliefs (Weible and Nohrstedt, 2012).

The ACF posits that the parameters of learning are differentiated by cognitive and normative perceptions that can drive change, where cognitive precepts have a higher degree of influence on altering behavior or thought (Weible and Nohrstedt, 2012). Interested in the role of lay knowledge, this study aims to test a proposed hypothesis that interactive knowledge can serve to inform normative beliefs, which can in turn lead to belief change and therefore policy change. To do so, the research looks at the case of New Brunswick where a policy change in 2014 introduced a moratorium on hydraulic fracturing, after years of engagement on the opportunities of shale gas development in the province.

[2] This research does not focus on the efficacy of the engagement mechanisms themselves and is more interested in understanding how learning takes place in the context of increased citizen participation where the mechanisms allow for lay knowledge to interact with institutionalized knowledge.

New Brunswick Case Study and Engagement Overview

In 2009, the government of NB leased out approximately one-fifth (1.5 million hectares) of the province's landmass for exploration to several energy corporations. In 2011, members of the New Brunswick Energy Commission (Bill Thompson and Jeannot Volpe) were appointed by the government to lead the consultations on the future of energy policy in the province. Their report, published in 2011, identified natural gas as an energy transition fuel, as a part of the province's energy mix that could support the transition to a low carbon economy. While shale development was not exclusively discussed in detail, the potential through unconventional gas became an obvious opportunity for production. The government aimed to engage the public in NB around the development of a regulatory framework for shale gas in the event that it would be technically and economically feasible to advance in the province. There were ten major public engagement initiatives related to fracking in New Brunswick starting in 2012.[3] They include[4]:

- Two-person panel in 2010 to highlight key areas of concern for a provincial energy strategy through public engagement and recommend the establishment of an energy commission for the province;
- NB Energy Commission to engage New Brunswickers on the province's long-term energy plan over the course of 2010/11;
- Stakeholder Forum in 2011 hosted by the Natural Gas Steering Committee (NGSC);
- A public discussion period based on the Natural Gas Group's (NGG) 2011 engagement to develop a policy and regulatory framework for oil and gas development;
- An engagement tour led by Dr. Louis LaPierre, a professor from the University of Moncton, to seek further input on the shale gas regulatory recommendations across nine different locations in the province in mid-2012;

[3] Based on information through the Right to Information and Protection of Privacy Act request as well as key public documents.

[4] There is very limited information on engagement mechanisms 3, 4, and 6 listed here. These are not well documented publicly, and the researcher had limited success through the Right to Information and Privacy request and interviews.

- Meetings with the Assembly of First Nations Chiefs of New Brunswick held by both industry and the NGG in 2012;
- Virtual town halls held by the Department of Energy and Mines where questions were posed to the Minister and experts in 2012;
- New Brunswick Energy Institute Roundtable evidence-based series in 2013;
- Over 70 presentations by the NGG to various groups such as municipal associations, chambers of commerce, and other government departments, with more than 2,200 people that attended between 2011 and 2012.
- The New Brunswick Commission on Hydraulic Fracturing (NBCHF) in 2014, led by John McLaughlin, Cheryl Robertson, and Marc Léger to study the issue of fracking and determine whether five conditions set out by the government could be met.

Methodology

This research is focused on a single in-depth case study of policy change in the province of NB, Canada. This approach helps develop a rich account of the case, and it is designed to account for the real-world context and deep understanding of decision-making for the policy actors involved (Yin, 2009). The study uses theory testing formulating a mechanism within existing theory to test causality. It adopts a mechanistic understanding of causality and employs process tracing, where outcomes are produced through "interactions of a series of parts that transmit causal forces from X to Y" (Beach and Pedersen, 2013, p. 176). The major mechanism under examination here is policy-oriented learning, a mechanism through which a change in beliefs leads to policy change. The key policy change for this study took place in 2014, when a moratorium on hydraulic fracturing was applied and then indefinitely extended in 2017.

To assess beliefs and strategies, 17 semi-structured interviews were conducted in order to understand policy actors' perceptions. To triangulate, a media review was also conducted as well as an extensive review of relevant documentation and available public materials. Also, through the Right to Information and Protection of Privacy Act (RTIPPA), the researcher submitted requests to the Department of Energy and Mines and that of Environment and Local Government in NB. For the media review, the researcher searched the Canadian Major Dailies database for

Telegraph-Journal[5] articles, with the terms "shale gas," "hydraulic fracturing," and "fracking." The search resulted in 439 articles, and by employing the methodology outlined by Lodge and Matus (2014), the articles were coded for key claims by actors and associated dates of the claims. The media analysis was used to assess key coalitional actors and their policy core and secondary arguments. Public documents, national media, and interviews were used to bolster this analysis and to assess deep core beliefs.

Analysis

This section presents the empirical findings from this analysis and details the normative and causal claims related to hydraulic fracturing in NB to determine the nature of policy learning among key coalitional actors. This research is focused on the normative and causal dimensions of policy core beliefs as outlined by the ACF, and hypothesizes that normative claims as defined earlier can lead to policy change.[6] The research hypothesizes that preferences are determined and altered through social interaction and that they have an effect on normative components of the policy core. This is particularly the case where the process of decision-making has opened up the policy process to a broader segment of engaged citizens. A secondary hypothesis is a supporting notion that contrary to what is proposed in the ACF, it is not the causal components[7] that are more likely to change, but rather, it is the normative components that have a significant effect through information from the social world and experience.

The basis of the ACF is coalitions, and in the case of NB, two coalitions involving a wide range of actors, including officials from government (both elected officials and officials in the public service), interest group leaders, environmental and business organizations, experts, the media and citizens, competed to advance their objectives regarding shale development. The media analysis identified a pro-development and an anti-shale

[5] The Telegraph-Journal is the largest newspaper in circulation at the provincial level.

[6] As noted earlier, the ACF hypothesizes that casual claims have higher potential to lead to change, a notion that this chapter critiques.

[7] According to the ACF—"Empirical policy core beliefs are more susceptible to verification and refutation from scientific and technical information—and therefore are more likely to exhibit change compared to normative policy core beliefs" (Weible and Sabatier, 2009, p. 197).

gas coalition, with distinct policy statements regarding the environmental risks versus the economic benefits of shale gas. The pro-development coalition, composed of elected officials, industry representatives, experts, journalists, and citizens, argued broadly at the policy core level for the economic benefits of shale development to the province. Their more nuanced policy positions included claims for economic development in the context of promoting good paying jobs, ensuring a natural gas supply for the region, as well as the potential for investment attraction. A small minority of claims argued that fracking is fundamentally safe. The claims are represented in Fig. 6.1 by frequency.

The anti-shale gas coalition was composed of elected officials (including members of the opposition and municipal representatives), government officials, experts, environmental organizations, Indigenous representatives, citizens, labour organizations, and journalists. They argued at the policy core level against shale gas development predominantly in relation to the environmental risks associated with the practice. More nuanced policy positions included claims about the environmental risk of shale development in the context of climate change more broadly and Indigenous representatives asserted their land rights against the risks

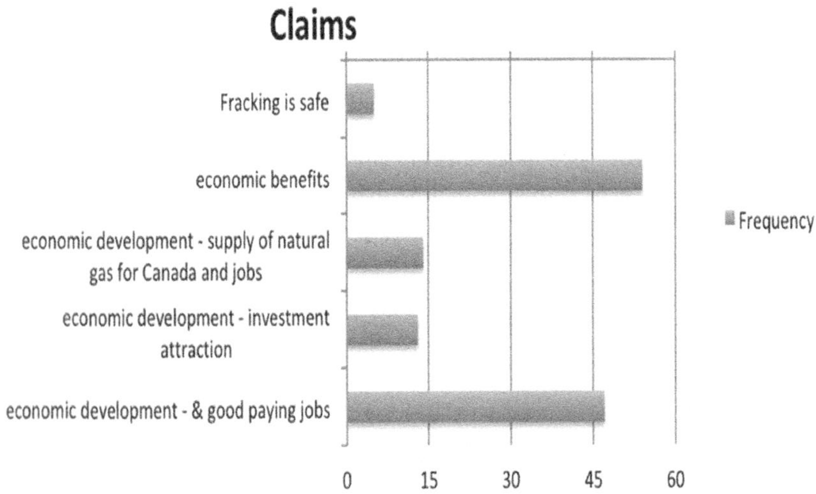

Fig. 6.1 Frequency of claims: economic benefits (2007–2017) (*Source* Author's own source)

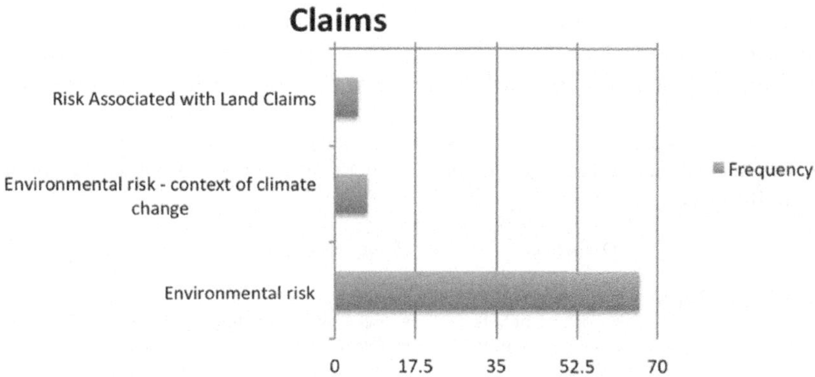

Fig. 6.2 Frequency of claims: environmental risks (2007–2017) (*Source* Author's own source)

of fracking. The claims of the anti-shale coalition are demonstrated in Fig. 6.2 by frequency of claims.

The analysis differentiated between problem severity and causal arguments versus normative claims. The pro-development coalition relied on arguing that shale gas could be produced safely with the right regulatory measures, thereby managing the risks associated with the practice to profit from the economic benefits. The anti-shale gas coalition argued that more information about the impacts of fracking was needed for decisions to be made. They raised concerns about the uncertainty associated with the practice and made calls for the implementation of a moratorium on fracking, until the risks could be better understood. In doing so, the anti-shale gas coalition questioned the assertion of the pro-coalition that regulations would truly protect the people and the land in NB if fracking were to proceed on a large scale.

The media analysis assisted in sketching out some of the key differentiations at that level of policy core beliefs. However, it is limited in providing a deeper understanding of the context. Interviews and document analysis provide a much clearer understanding of the impact of engagement and risk on the nature of policy learning. Thus far, this section has established the coalitional positions and some of the basic claims made by each. The next section will highlight the findings according to the categories of assertions outlined in the methodology. This includes claims about information, risk, and uncertainty, as well as public engagement. A summary of the key beliefs that emerged from thematic examination of the data from the interviews in addition to the data from the media analysis is provided in Table 6.1. The findings will be discussed in the following discussion.

Table 6.1 Belief systems of pro-development and anti-shale gas coalitions in New Brunswick

	Definition	*Pro-development*	*Anti-shale gas*
Policy core beliefs	Normative and empirical beliefs concerning policy subsystems	General goals: • Economic benefits • Economic development: jobs, natural gas supply, investment attraction Positions on general policy solutions and instruments: • Regulation and resource revenue Problem severity and cause: • Risks can be managed with stringent regulations Normative precepts and role of government: • Create domestic supply of natural gas • Promote eastern region's growth and self-reliance • Create and keep jobs for New Brunswickers to stay home	General goals: • Protect the environment • Promote sustainable energy alternatives Positions on general policy solutions and instruments: • More information needed to decide Problem severity and cause: • Uncertainty regarding the impacts of fracking (particularly in the case of water) Normative precepts and role of government: • Public engagement needs to be meaningful; people should feel heard and included in policy process • Don't trust that regulations will be enforced • Should be advancing sustainable energy development alternatives

(continued)

Table 6.1 (continued)

	Definition	*Pro-development*	*Anti-shale gas*
Secondary beliefs	Instrumental beliefs or beliefs about a subset of a policy subsystem	• Regulatory regime to ensure safe operations • Royalty regime to ensure economic benefits earned for resources	• Implement a moratorium • Extend the moratorium/apply an indefinite ban

Source Author's own source

Claims About Information

Interview participants from both coalitions claimed that in the early stages of learning about fracking (2007–2011), there was very little information available. On the one hand, the members of the pro-development coalition relayed their experiences in trying to ensure that information was aggregated and disseminated in a reliable manner (interviews 2, 6, 14, 15). As one industry representative put it: "the interest in hydraulic fracturing was just getting underway so it's something new to people, new to the government and the industry, and there wasn't a lot of information available in particular to the public" (interview 5). On the other hand, anti-shale gas coalition members noted the difficulties in finding information to understand the impacts of fracking.[8]

During this initial four-year period, a strategy that was adopted by both coalitions was to engage with people who experienced fracking. The pro-coalition did this by arranging a fact-finding mission to jurisdictions in the United States (US) and Canada with a recent history of fracking activities. In some cases, joined by members of the anti-coalition, they spoke with industry representatives and farmers working in and experiencing shale gas development. An elected official noted: "I wanted to see it [fracking process] with my own eyes and hear it with my own ears. And that's why we went to Texas, Houston, we went to Pennsylvania, we went to the northern part of BC and went to Calgary and then northern Alberta to where operations have been in process" (interview 2).

[8] Confidential interviews 3, 7, 11, 12.

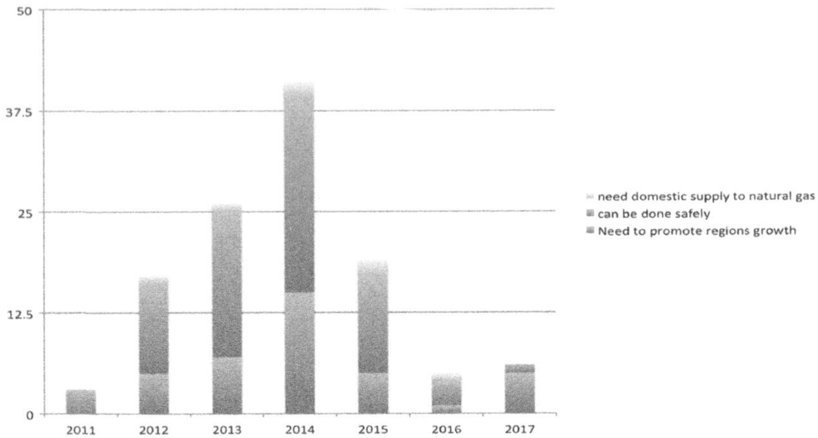

Fig. 6.3 Distribution of economic benefit claims by date (2007–2017) (*Source* Author's own source)

Members from the anti-shale gas coalition likewise were interested in hearing first-hand about the experience. As one environmental advocate noted, "the reality is that people want to hear from people. I want to hear stories, and I want to hear what's your experience with this" (interview 1). To counter the approach, they arranged for bringing knowledge in from the US as well. As one environmental advocate noted: "initially, in the beginning it was about bringing experience of the United States from a civil society perspective and expertise to those who would be involved in policy development and regulation [in NB]" (interview 3). During this period, the public debate about shale gas development was only beginning to emerge, and the claims recorded for or against the resource's extraction were not as frequent (see Figs. 6.3 and 6.4).[9]

The coalitions also turned to the experiences within the province to understand and learn about fracking. Experiences were shared from residents in the Penobsquis area in Kings County who lost their well water starting in 2004, where Potash Corporation operated a large mining

[9] Notably, in 2012, approximately 20,000 New Brunswickers signed a petition asking for a ban on shale gas exploration and development in the province (Dudziak and D'Arcy, 2017).

project. The residents cited loss of water in direct relation to the seismic[10] and drilling activity that Potash Corporation and Corridor Resources (a proponent in the case of fracking) undertook in the area. About 60 wells went dry, and for several years, the provincial government supplied the community with water that was trucked in and rationed per head. Several interview participants cited these concerns with water loss and a lack of appropriate remediation[11] as potentially negative experiences to be had with the resource industry. An environmental advocate noted: "the courts were allowing them to do that so [...] the whole legal system was geared to protect the industry and not the people, and so what happens if we allow that to happen here?" (interview 8). While the causes of water loss are disputed, this experience is one that associated risk in a province where many households depend on well water for their daily consumption. The province estimates that there are upwards of 100,000 private water wells in NB (GNB, 2020). A First Nations representative noted: "NB has a lot of groundwater, we have a lot of people who are living in rural areas, and so [there was] the concern that that this would somehow impact the groundwater" (interview 4).[12]

Risk to water resources from hydraulic fracturing activity emerged as a key concern in the analysis. Public reports summarizing various phases of the public engagement process identify water protection as a key concern (OCMOH, 2012; LaPierre, 2012; NBCHF, 2016). This is a core issue identified in the interviews as well, where the anti-coalition pointed to the lack of baseline water testing in the province and the repeal of wetland protections in 2012. They argued that without these basic measures it would be difficult for the government to monitor and enforce measures to protect water in the province. They also pointed to experience with the province's capacity to ensure adequate oversight. As one government official noted, people "are not trusting the government and those that apply the regulations. Because in other industries there is not a sufficient amount of people to ensure that it is followed [regulation] by the letter,

[10] Seismic activity is deployed in exploration activities for hydraulic fracturing to assess the geology of the earth.

[11] The community actively sought financial compensation from Potash Corp. and took their case before the Mining Commissioner rather than advancing a legal case through the courts—the process took several years.

[12] A concern repeated in several other interviewees from the anti-shale coalition (3, 8, 9, 10, 11).

and so when they see that in other types of industries or sectors, they ask why would this sector have more resources?" (interview 9).

The anti-shale gas coalition was also concerned about the required capacity and infrastructure to support wastewater management from fracking operations. NB is largely composed of unincorporated regions governed by Local Service Districts (LSDs) with unique governance at the local level but no policy or financial levers.[13] Incorporated regions also have limited authority in terms of municipal policy and regulatory functions, and in a province where municipal water treatment systems are small scale, anti-coalition members from local government questioned the province's capacity to support effective water treatment. The Chief Medical Health Officer (CMHO) reaffirmed this when a key recommendation in a CMHO report raised the need to establish capacity to manage the infrastructure around fracking activities. Members of the anti-coalition relied on their experiences with local government to question the effectiveness of regulating a large-scale industry like shale gas in their province.

Claims About Risk and Uncertainty

As more scientific research about hydraulic fracturing emerged in the period between 2011 and 2014, the frequency of claims for or against hydraulic fracturing increased significantly (see claims in Figs. 6.3 and 6.4). This was also accompanied by the rollout of several public engagement mechanisms in NB and a rise in the controversies surrounding development in the province. As noted earlier, the government released public discussion documents in 2012. One key document in May 2012, titled *Responsible Environmental Management of Oil and Gas Activities in New Brunswick—Recommendations for Public Discussion*, proposed 116 changes to the regulatory framework governing oil and gas production (GNB, 2012). To engage people on the proposed rules, the government appointed Dr. LaPierre with a mandate to solicit feedback from citizens, record public sessions, review submissions, and ultimately provide a report that summarized the issues related to the regulations for shale gas development in New Brunswick. Within the same time frame, NB's CMHO

[13] Elected advisory committees oversee LSDs and operate under delegated authorities to provide services in communities (GNB, 2017).

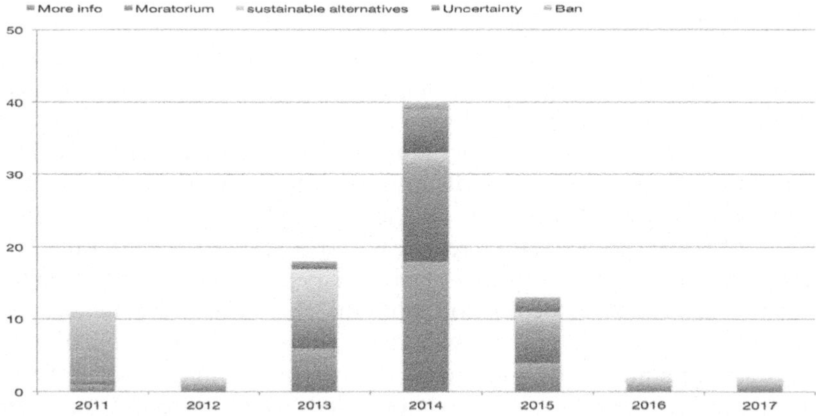

Fig. 6.4 Distribution of environmental risks claims by date (2007–2017) (*Source* Author's own source)

Dr. Eilish Cleary was mandated with developing recommendations for protecting public health in light of potential shale development activities.

Public opposition to fracking in the province was on the rise in 2012, and this was evidenced by the organization of public rallies, blockades, and petitions. In May 2012, before the engagement tour was slated to take place, there was a large parade organized in May in Fredericton outside of the legislature, and many community groups spoke out against shale development in the province (NBEN, 2012a). In early fall of 2012, controversy emerged around the CMHO's report as it became clear that the report had been written but not released to the public. It is not clear how this information emerged, but the opposition Liberals and many members of the anti-shale gas coalition demanded the report's release, and the media controversy surrounding the government's lack of commitment to releasing the report stimulated public outcry. Anti-shale members encouraged people to take civil action, including writing to their MLAs to request the report's release (NBEN, 2012b).

During this period, despite sustained engagement and the government's effort to democratize the decision-making process, opposition intensified against shale development in NB. This emerged as the pro-development coalition's narrative that hydraulic fracturing could be done safely solidified. The engagement tour concluded and in October 2012 released a report with recommendations for shale gas development, titled

The Path Forward[14] (LaPierre, 2012). Advancing a path for shale gas development to proceed based on what was heard, the report identified a number of key areas for work aside from just addressing the proposed changes to regulation of oil and gas.

The report also claimed that industry and government had failed at providing information to New Brunswickers to understand the implications of shale gas development (LaPierre, 2012, p. 26). On the same day in October, the CMHO's report was also released and it contained several recommendations from a health perspective. A key recommendation that gained a lot of traction with the anti-shale coalition was the establishment of health impact assessments alongside environmental ones, and more monitoring to understand the cumulative impacts of fracking on health. Furthermore, with a section titled "What we know and what we don't know," the report also made clear that there was more information needed for decision-making about shale development. In hindsight, an industry representative asserted that: "one of the key things would be the fear of the unknown. Looking back over the course of four years the number one thing that kept coming from the public at large was we don't have enough information" (interview 6).

Several interview participants noted their disappointment in the *Path Forward* report. It appeared to carve out a way forward when they felt they had been vocal during engagement processes about their views for not moving ahead with shale gas development just yet. This was evident in a public release by the anti-shale gas coalition that stated that the report "did not reflect the will of the people as expressed at public meetings held in 2012, and Dr. LaPierre did not gather evidence over the course of the public meetings to support his opinion finding that a moratorium on shale gas development was or was not warranted" (NBEN, 2012a). Participants also accorded a significant degree of credibility to the CMHO's report, citing the need to have more information to protect public health. More broadly, the coalition asserted that the province could not proceed without more information, particularly in terms of regulating unknowns.

[14] Recommendations based on what the report identified as top concerns including: government integrity, water contamination, well integrity, jobs, chemicals, health risks, security of fresh water supply, baseline testing, regulation, and royalties among many other concerns.

Claims About Public Participation

Claims about environmental risks rose, as demonstrated in Fig 6.4. A robust narrative emerged from the anti-shale gas coalition claiming that there was not enough information to make a decision to proceed with development. The calls for a moratorium from the anti-shale gas coalition also grew between 2013 and 2014. The pro-development coalition argued that the regulations that they developed were among the most stringent in North America. They released an *Oil and Gas Blueprint* that highlighted the key planks to move forward with shale gas development. Additionally, Dr. LaPierre was appointed to head up the New Brunswick Energy Institute (NBEI) to provide reliable scientific data to inform decision-making around shale. Figure 6.3 shows the increase in claims that fracking is a practice that can proceed safely if well managed for all the risks associated with the practice.

The anti-coalition undermined this narrative, questioning how this would be done. As one environmental advocate noted, there were "a lot of unknowns, so how can you regulate something that is still evolving? And then there's the bigger questions around climate change. Whether this is safe or not is one question but then there's a whole other context of should we do it even if it is safe?" (interview 1). They also advanced their perspective that they did not trust the government to regulate the resource, and noted limited capacity in NB at the provincial and municipal levels to provide an adequate degree of oversight to ensure that the regulations were enforced.

In terms of public participation, many members noted that they felt unheard and were frustrated to see the province move ahead with regulations for fracking. The issue became even more contentious when it was revealed that Dr. LaPierre had falsified his credentials and stepped down from the NBEI. The trust of people in the government and in the engagement process emerged as a strong theme in the analysis. As one environmental advocate mentioned "so I mean that kind of a thing where government isn't even respecting the process [...] that is when the trust is really broken because it became apparent they have no interest in trying to get some sort of measure of approval or listen to what was going on" (interview 8). Public protest and civic action rose in NB in opposition to shale gas development and culminated in the clashes at Rexton, NB, where images of police cars on fire made national and international news. An industry representative noted that "people started to mistrust the process itself – public consultation ceased to be about the mechanism anymore – people don't trust it and why would they?" (interview 6).

In 2014, the Liberals ran on a campaign to put a moratorium on hydraulic fracturing in the province. They won in October of that year, defeating the Progressive Conservative government that had been in power since 2010. The new Premier, Brian Gallant, implemented a moratorium on fracking along with five conditions that needed to be met before fracking could proceed in NB: (a) that a "social licence" be established through consultations to lift the moratorium; (b) clear and credible information on the impacts on air, health, and water so a regulatory regime could be developed; (c) a plan to mitigate impacts on public infrastructure and address issues such as waste water disposal; (d) that a process be in place to fulfill the province's obligation to consult with First Nations; and (e) that a proper royalty structure be established to ensure benefits are maximized for New Brunswickers (CBC, 2014). To address these conditions, the government appointed three thought leaders in the province, John McLaughlin, Marc Léger, and Cheryl Robertson, to lead the New Brunswick Commission on Hydraulic Fracturing (NBCHF). They were mandated with engaging New Brunswickers and providing their recommendations on whether the conditions could be met.

The NBCHF toured the province, but rather than engaging on shale development itself, they broadened the discussion to one regarding the future of energy in NB (interview 15). The engagement process itself was well received by the anti-shale gas coalition, and in its findings, it reported that the government needed to work on strengthening public trust and instituting an independent regulator to oversee resource development. It does not appear that the government acted on their recommendations, and it is not clear to what extent the rest of the conditions were implemented/advanced. The government extended the moratorium on hydraulic fracturing indefinitely in 2016.

Discussion and Conclusion

The key consideration of this analysis is assessing belief change. As Wellstead (2017) states: "changing a coalition's core policy belief would eventually alter the basic perception of policy problems as well as the general policy prescription of an issue" (p. 551). Evaluating whether policy-oriented learning led to change requires identifying who learns, what they learn, and to what effect (see Table 6.2, which builds on Bennett and Howlett, 1992).

The research indicates that the increase in engagement mechanisms allowed the anti-shale gas coalition to find many opportunities to engage in decision-making. The anti-shale coalition was able to contest the goals and means of the pro-development coalition's strategies. By drawing on the normative claims linked to their policy core beliefs, the anti-fracking coalition was able to successfully challenge the dominant coalition's policy and push for change. They drove the assertion that there was not enough information to proceed with hydraulic fracturing development in New Brunswick. As the main strategy, the anti-coalition appealed to this argument to undermine the very underpinnings of the pro-development coalition's claims that it was safe and could be developed safely to expand resource production in the province based on evidence and regulatory best practice.

From the analysis presented here, it is clear that the learning was isolated to the instruments and settings related to shale development in NB. In the case of NB, it is not clear that policy change was due to the alteration of beliefs on behalf of either coalition. Rather, the problem definition was altered to become a question of whether

Table 6.2 Learning by category

Type of learning	*Who learns*	*Learns what*	*To what effect*
Policy-oriented learning	Advocacy coalitions; actors include: bureaucrats, government officials, regulators, journalists, industry and environmental representatives, members of the public	Policy core policy preferences—involving beliefs about the role of government, severity of problem cause and goals for policy (including based on scientific expertise and experience) around shale development	Goals, instruments, and settings related to shale development (regulation predominantly)—policy change

Source Adapted from Bennett and Howlett (1992)

fracking was socially acceptable, rather than how to develop it safely and maximize economic benefit. By evaluating risk differently, the coalition challenged the technocratic management of risk regulation of an emerging technology.

Furthermore, the case demonstrates the importance of trust in risk management. The lack of trust in public authorities in NB meant that the public did not trust that they would be protected from the risks posed by hydraulic fracturing. The process of engagement appears to have amplified that mistrust, but what the case also demonstrates is the role of lay knowledge in decision-making. Members of the anti-shale gas coalition had a deep understanding of their context that they brought to the engagement exercises. Their local knowledge focused more on how policy and regulation would impact their lives and less on the technical aspects of risk governance.

References

Beach, D., & Pedersen, R. B. (2013). *Process-tracing methods: Foundations and guidelines*. University of Michigan Press.

Bennett, C. J., & Howlett, M. (1992). The lessons of learning: Reconciling theories of policy learning and policy change. *Policy Sciences, 25*(3), 275–294.

Bomberg, E. (2013). The comparative politics of fracking: Networks and framing in the US and Europe. In *APSA 2013 Annual Meeting Paper*.

Bryant, T. (2001). *The social welfare policy change process: Civil society actors and the role of knowledge*. Doctoral Thesis, Faculty of Social Work, University of Toronto.

Canadian Broadcasting Corporation (CBC). (2014, December 18). Shale gas moratorium details unveiled by Brian Gallant. *CBC News*. Retrieved from: https://www.cbc.ca/news/canada/new-brunswick/shale-gas-moratorium-details-unveiled-by-brian-gallant-1.2877440

Coburn, J. (2003). Bringing local knowledge into environmental decision making: Improving urban planning for communities at risk. *Journal of Planning Education and Research, 22*(4), 420–433.

Council of Canadian Academies. (2014). *Environmental impacts of shale gas extraction in Canada: The expert panel on harnessing science and technology to understand the environmental impacts of shale gas extraction* (pp. 1–262, Rep.). Ottawa, ON: Council of Canadian Academies.

Dudziak, S., & D'Arcy, M. (2017). Reinventing democracy through public advocacy: The case of the anti-shale gas movement in New Brunswick. In *The shifting terrain: Non-profit policy advocacy in Canada*, eds. N. Mulé & G. De Santis. Montréal and Kingston: McGill-Queen's University Press.

Dunlop, L., Atkinson, L., & Turkenburg-van Diepen, M. (2021). The environment and politics in science education: The case of teaching fracking. *Cultural Studies of Science Education, 16*, 557–579. https://doi.org/10.1007/s11422-021-10017-z

Elliott, C., & Schlaepfer, R. (2001). The advocacy coalition framework: Application to the policy process for the development of forest certification in Sweden. *Journal of European Public Policy, 8*(4), 642–661.

Fiorino, D. J. (1990). Citizen participation and environmental risk: A survey of institutional mechanisms. *Science, Technology, & Human Values, 15*(2), 226–243.

Fischer, F. (2000). *Citizens, experts, and the environment: The politics of local knowledge*. Duke University Press.

Fischer, F. (2003). *Reframing public policy: Discursive politics and deliberative practices*. Oxford University Press.

Government of New Brunswick (GNB). (2010). Citizen Engagement Unit to be established [News release]. Retrieved from: https://www2.gnb.ca/content/gnb/en/news/news_release.2010.10.1664.html

Government of New Brunswick (GNB). (2012). Responsible environmental management of oil and gas activities in New Brunswick: Recommendations for Public Discussion. New Brunswick Natural Gas Group. Retrieved from: https://www2.gnb.ca/content/dam/gnb/Corporate/pdf/ShaleGas/en/RecommendationsDiscussion.pdf

Government of New Brunswick (GNB). (2017). A Guide to the Local Governance Act. Retrieved from: https://www2.gnb.ca/content/dam/gnb/Departments/lg-gl/pdf/ReferenceGuideToTheLocalGovernanceAct.pdf

Government of New Brunswick (GNB). (2020). Water Well Testing. Department of Environment and Local Government. Retrieved from: https://www2.gnb.ca/content/gnb/en/services/services_renderer.3877.Water_Well_Testing.html

Hajer, M. A., & Wagenaar, H. (Eds.). (2003). *Deliberative policy analysis: Understanding governance in the network society*. Cambridge University Press.

Heikkila, T., Pierce, J. J., Gallaher, S., Kagan, J., Crow, D. A., & Weible, C. M. (2014). Understanding a period of policy change: The case of hydraulic fracturing disclosure policy in Colorado. *Review of Policy Research, 31*(2), 65–87.

James, T. E., & Jorgensen, P. D. (2009). Policy knowledge, policy formulation, and change: Revisiting a foundational question. *Policy Studies Journal, 37*(1), 141–162.

Jenkins-Smith, H. C., & Sabatier, P. A. (1994). Evaluating the advocacy coalition framework. *Journal of Public Policy*, 175–203.

Jenkins-Smith, H. C., Nohrstedt, D., Weible, C. M., & Sabatier, P. A. (2014). The advocacy coalition framework: Foundations, evolution, and ongoing research. In *Theories of the policy process*, 3rd ed. Boulder: Westview Press, pp. 183–224.

LaPierre, L. (2012). The Path Forward. Government of New Brunswick (GNB). Retrieved from: http://www2.gnb.ca/content/dam/gnb/Corporate/pdf/ShaleGas/en/ThePathForward.pdf

Litfin, K. T. (2000). Advocacy coalitions along the domestic foreign frontier: Globalization and Canadian climate change policy. *Policy Studies Journal, 28*(1), 236–252.

Lodge, M., & Matus, K. (2014). Science, badgers, politics: Advocacy coalitions and policy change in bovine tuberculosis policy in Britain. *Policy Studies Journal, 42*(3), 367–390.

New Brunswick Commission on Hydraulic Fracturing (NBCHF). (2016). Commission's Final Reports. Retrieved from: https://www.nbchf-cnbfh.ca/documents/

New Brunswick Environmental Network (NBEN). (2012a). Walk for a ban on fracking—Stop fracking with our water and air. Retrieved from: https://nben.ca/index.php/en/news-from-groups/344-walk-for-a-ban-on-fracking-stop-fracking-with-our-water-and-air

New Brunswick Environmental Network (NBEN). (2012b). An Open Letter to MLAs for the Immediate Release of the Complete Health Study on Shale Gas Sector. Retrieved from: https://nben.ca/en/component/tags/tag/dr-eilish-cleary

NRCan. (2020, November 5). *Government of Canada*. Natural Resources Canada. Retrieved January 22, 2022, from: https://www.nrcan.gc.ca/our-natural-resources/energy-sources-distribution/clean-fossil-fuels/natural-gas/shale-tight-resources-canada/17669

Office of the Chief Medical Officer of Health (OCMOH). (2012). Chief Medical Officer of Health's Recommendations Concerning Shale Gas Development in New Brunswick. New Brunswick Department of Health. Retrieved from: https://www2.gnb.ca/content/dam/gnb/Departments/h-s/pdf/en/HealthyEnvironments/Recommendations_ShaleGasDevelopment.pdf

Petts, J., & Brooks, C. (2006). Expert conceptualisations of the role of lay knowledge in environmental decisionmaking: Challenges for deliberative democracy. *Environment and Planning A, 38*(6), 1045–1059.

Renn, O., & Levine, D. (1991). Credibility and trust in risk communication. In *Communicating risks to the public*. Dordrecht: Springer, pp. 175–217.

Sabatier, P. A. (1988). An advocacy coalition framework of policy change and the role of policy-oriented learning therein. *Policy Sciences, 21*(2–3), 129–168.

Sabatier, P. A. (1998). The advocacy coalition framework: Revisions and relevance for Europe. *Journal of European Public Policy, 5*(1), 98–130.

Sabatier, P. A., & Jenkins-Smith, H. C. (1993). *Policy change and learning: An advocacy coalition framework*. Boulder: Westview.

Sabatier, P. A., & Jenkins-Smith, H. C. (1999). The advocacy coalition framework: An assessment. In *Theories of the policy process*, ed. P. A. Sabatier. Boulder (Colorado): Westview Press, pp. 117–166.

Sabatier, P. A., & Weible, C. M. (2007). The advocacy coalition framework. Innovations and clarifications. In *Theories of the Policy Process*, ed. P. A. Sabatier, 2nd ed. Boulder (Colorado): Westview Press, pp. 189–220.

Shulock, N. (1999). The paradox of policy analysis: If it is not used, why do we produce so much of it? *Journal of Policy Analysis and Management*, *18*(2), 226–244.

Slovic, P. (1993). Perceived risk, trust, and democracy. *Risk Analysis*, *13*(6), 675–682.

Stern, P., & Fineberg, H. V. (1996). *Understanding risk: Informing decisions in a democratic society.* Committee on Risk Characterization and Commission on Behavioral and Social Sciences and Education, National Academy Press, Washington, DC.

Weible, C. M., & Nohrstedt, D. (2012). Coalitions, learning and policy change. *Routledge Handbook of Public Policy*, 125–137.

Weible, C. M., & Sabatier, P. A. (2009). Coalitions, science, and belief change: Comparing adversarial and collaborative policy subsystems. *Policy Studies Journal*, *37*(2), 195–212. https://doi.org/10.1111/j.1541-0072.2009.00310.x

Wellstead, A. (2017). Plus ça change, plus C'est La Même chose? A review of Paul Sabatier's "an advocacy coalition framework of policy change and the role of policy-oriented learning therein". *Policy Sciences*, *50*(4), 549–561.

Yin, R. (2009). *Case study research: Design and methods* (4th ed.). Sage Publications.

CHAPTER 7

Carbon Capture, Utilization, and Storage: Public Confidence in Risk Decision-Making

Patricia Larkin, Monica Gattinger, and Stephen Bird

Introduction

Canada has developed extensive expertise and experience in point-source carbon capture, utilization, and storage (CCUS). The country's four large-scale integrated projects include carbon dioxide (CO_2) capture at a coal-fired electricity generating facility, upstream oil production facilities, and a fertilizer plant. Depending on the project, CO_2 injection and storage occurs as sequestration in a deep saline aquifer geologic formation or for enhanced oil recovery (EOR) operations. Carbon capture, transformation, and conversion, also under the umbrella of CCUS, is an important option for emissions-intensive and trade-exposed industries (EITE) such as cement, steel, and chemical manufacturing. In the

P. Larkin · M. Gattinger (✉)
Institute for Science, Society and Policy, University of Ottawa, Ottawa, ON, Canada
e-mail: mgatting@uottawa.ca

S. Bird
Clarkson University, Potsdam, NY, USA

M. Gattinger (ed.), *Democratizing Risk Governance*,
https://doi.org/10.1007/978-3-031-24271-7_7

last fifteen years, the CCUS industry has emerged as an important CO_2 mitigation option, but it has not reached its potential.

Despite Canadian expertise, and the fact that CCUS is a key component of many global emissions reductions scenarios (IPCC 2005; IEA 2019a), the technology faces challenges when it comes to public confidence in decision-making across a range of socioeconomic and political risk issues. This includes concerns for the adequacy of regulatory oversight and controversies over carbon issues more broadly. Technologies such as CCUS that extend or continue fossil fuel extraction or use can be controversial because of concerns over the degree of actual carbon reductions.

This study aims to identify and propose recommendations to mitigate the key risk issues driving public confidence in CCUS and government decision-making processes that govern and support it. The authors undertook a comprehensive review of academic, industry, and government publications, as well as in-depth interviews with decision-makers representing a variety of sectors involved in CCUS policy and implementation. Risk issues related to public confidence were categorized into thirteen categories identified in the risk management literature. Recommendations to mitigate the identified risk issues were developed using a slightly modified version of the REACT framework for risk management and population health (Krewski et al. 2007). The study suggests that a wide variety of risk management actions is needed in order for CCUS to make the contribution to climate mitigation that continues to be envisioned for large industrial sites.

The chapter proceeds in four sections. The first provides a primer on CCUS. The next delineates the research objectives, analytical approach, and methodology. The third section delves into the research findings, while the final section offers a discussion and recommendations for risk management.

A Primer on Carbon Capture, Utilization, and Storage

Carbon capture, utilization, and storage technologies may be applied to CO_2 emissions at point-source fossil energy electricity generation (coal, natural gas) and heavy industry sites (including oil and gas facilities). Demonstrated applications include carbon capture with saline aquifer sequestration (CCS) and carbon capture for enhanced oil recovery

(CCUS/EOR). Carbon capture and conversion in emissions-intensive and trade-exposed (EITE) industries focuses especially on cement, steel, and chemical manufacturing. Additionally, the term "CarbonTech" has been used to encompass all carbon capture technologies and technological processes that reduce CO_2 emissions (CMC Research Institutes and Canadian Business for Social Responsibility 2019). The acronym CCUS will be used for the remainder of this chapter, unless CCS is highlighted specifically.

CCS and CCUS large-scale integrated projects (LSIPs) include four activities: capture, transport, deep well-head injection, and storage. Globally, LSIP CO_2 capture may be undertaken using pre-combustion, post-combustion, and oxy-fuel technologies (Gale et al. 2015). The capture activity also includes compression of the CO_2 emissions into a supercritical state, with the CO_2 concentration approaching 99% pure. This substance is usually transported by pipeline to the injection site.

Beginning in the early 1970s, supercritical CO_2 was injected into depleted oil reservoirs to improve miscible flood operations for enhanced tertiary oil recovery purposes. At the time, this was not conceived as a climate mitigation strategy because CO_2 procured for EOR was seen as a cost to be reduced while at the same time enhancing oil production.

The IPCC *Special Report on carbon dioxide capture and storage* (2005) put a spotlight on CCS as a climate change mitigation option, with CO_2 sequestration in saline aquifer formations 800–1200m deep underground. For its part, the International Energy Agency (IEA) has consistently included CCS as a lowest cost GHG emission reduction solution for point-source emissions sites through 2050. However, the projected CCS contribution to mitigation has been in decline under a variety of emissions reduction scenarios proposed by the IEA since 2009. This is principally due to slower than anticipated near-term deployment of the technology and also because of improvements in renewable technologies, particularly wind and solar.

There has instead been a propensity towards more CCUS/EOR projects (Larkin et al. 2019) with CCS and EOR reframed as *carbon capture utilization and storage* (CCUS) beginning in 2012 (Markusson et al. 2017). Given the high costs of CO_2 capture, however, Dixon et al. (2015) argue that CO_2 sales for use in EOR projects have been critical to demonstrating the concept and verifying storage longevity. For example, the Weyburn-Midale EOR project in Saskatchewan involving CO_2 sales was established in 2000 and was subject to a decade of biosphere and geosphere monitoring programmes (Bowden et al. 2013a, b).

With respect to the EITE sector, the current emphasis is on carbon conversion via chemical or biological processes rather than underground sequestration/storage. This emerging era is focused on the use of CO_2 emissions within an industry, such as for cement manufacturing, or offered as a valued carbon feedstock in the downstream industry marketplace, such as chemicals, plastics, or fuels (Jones et al. 2017).

Public authorities, CCUS companies, and CCUS advocacy organizations across Canada are among the global leaders in support and development of this mitigation technology. They have substantial expertise in policy, regulatory, and technological innovation. Despite this expertise, the technology still faces significant socioeconomic and political challenges and risks that we outline in the next section.

In terms of Canadian emissions, upstream oil and gas development is the country's largest source of GHG emissions, accounting for approximately 27% of emissions in 2017 and projected to grow to 32% by 2030 (Government of Canada 2019). Shell's Quest CCS project in Alberta, operating since 2015, is a showcase LSIP using geological sequestration for emissions sourced at an oil sands upgrader. A portion of the capacity of a second LSIP, the Alberta Carbon Trunk Line for EOR purposes, uses emissions from the North West Redwater Sturgeon refinery.

Electricity generation is Canada's fourth-largest source of GHG emissions (about 10%) and emissions are projected to decline to 4% of the total by 2030 (Government of Canada 2019). This is primarily due to the federal Reduction of Carbon Dioxide Emissions from Coal-fired Generation of Electricity Regulations (Environment Canada 2018) and Regulations Limiting Carbon Dioxide Emissions from Natural Gas-fired Generation of Electricity (Government of Canada 2018). Currently, CCS is the only functioning technology that can reduce emissions from fossil fuel-fired power plants (Canadian Electricity Association 2020). SaskPower's Boundary Dam coal-fired electricity plant is Canada's only LSIP operating in this domain.

With respect to the application of CCUS for heavy industry outside oil and gas production, this is another area with strong potential. Approximately 11% of GHG emissions originated from heavy industry in 2017 and emissions are projected to grow to approximately 13% by 2030 (Government of Canada 2019). The IEA (2019b) suggests that emissions reductions in iron and steel, cement, aluminium, and chemical industries remain particularly difficult. Alberta's Carbon Trunk Line, noted above, uses CO_2 sourced in part from the Nutrien fertilizer plant. Carbon conversion may be applied to a greater extent for the EITE sector, as demonstrated by the cement industry (Carbon Cure 2021). Canada's existing large-scale projects (Shell Quest, Boundary Dam and Alberta Carbon Trunk Line/North West Refiner/Nutrien) could eliminate up to 4 megatons of CO_2 equivalent per year or 6% of Canada's emissions reductions through 2030 (Larkin et al. 2021).

Research Objectives, Analytical Approach, and Methodology

This study focuses primarily on decision-making processes for CCUS technologies. As noted previously, for over 15 years this climate change mitigation technology has been identified as an important option for CO_2 emissions reductions at large point-sources such as fossil-based electricity generation and heavy industry sites (IEA 2019a; IPCC 2005). And yet, the technology has not been deployed as rapidly as envisioned. This study aims to help understand why, with a specific focus on identifying the key risk issues driving public confidence (or lack thereof) in CCUS and government decision-making processes that govern and support it.

We use the term public confidence broadly to indicate the overall support and comfort that the public and private sector actors like investors have for a given energy system and its associated regulatory scheme. A key aspect of public confidence is social acceptance. When it comes to energy transition technologies like CCUS, social acceptance has become one of the most policy-relevant concerns (Upham et al. 2015; Gaede and Rowlands 2018). Two issues are particularly important for new technologies. First, social acceptance can be thought of as a continuum, a range of positive and negative responses for both the outcome of a decision-making process and the process itself. As suggested by Batel et al. (2013), societal responses may take the form of a simple lack of opposition, or they may reflect stronger, positive reactions such as support, interest, or even admiration. On the negative side, rejection can include degrees of uncertainty, resistance, or apathy. A second important issue is that societal responses are not static: stakeholders' views and innovation contexts evolve throughout the public policy cycle (Busse and Siebert 2018). For example, at the level of individual projects, local context matters a great deal (i.e., for communities adjacent to facilities). Alternately, social acceptance can also manifest at the national level, with a focus on a particular technology in the context of national policymaking and goals. Moreover, individual reactions may increase or decrease the risk perceptions of others, in what Kasperson et al. refer to as the social amplification of risk (1988).

The challenges of public confidence and social acceptance exist within a broader context of other factors described by Cleland and Gattinger (2019) that have fundamentally transformed the context for energy decision-making. These include lower levels of public trust in institutions of various sorts, greater expectations on the part of citizens and communities to be involved in decisions that affect them, greater political fragmentation and tendencies towards polarization, the need for adaptation and resilience in the energy system itself, and growing levels of economic, political, and technological uncertainty.

We use a risk-based framework for the analysis (Rothstein et al. 2013). Risk assessment and risk management (RA/RM) have been applied for decades to health and environmental protection and public safety, as well as to issues in banking, insurance, and organizational management. If done well, RA/RM provides a systematic, open, and transparent process for stakeholders and decision-makers to follow. Similarly, it can be used as a framework for analysis, as we do here.

The risk issues selected for analysis build on the findings of L'Orange Seigo et al. (2014), where the technology acceptance framework of Huijts et al. (2012) was applied to public perception of CCS. The list is supplemented by risk issues identified by Leiss and Krewski (2019) as being "most likely to attract wide public attention and thus … likely to have, in the long run, significant influence on the public acceptance of CCS" (p. 239).

Specifically, we categorize the risk issues that affect public confidence in CCUS decision-making into three groups (see Table 7.1). The first category (seven risks) comprises ***cross-cutting*** factors for all stakeholders (government, industry, and the public): worldviews, problem perception, trust, energy context, knowledge/information provision, tolerable costs, and distributive justice. The second category relates to ***governance*** factors (three risks): policy and regulatory stability, inter-jurisdictional challenges, and procedural justice. The third category focuses on ***industry*** factors (three risks): willingness and/or capacity to act, pace/demonstration of technological feasibility, and market competitiveness/international trade. Each of these issues is described in the following section of the chapter.

The study's analysis draws on academic literature and government documents, as well as fourteen interviews with decision-makers that have

Table 7.1 Risk issues related to public confidence in decision-making for CCUS

Risk issue	*Definition*
1. Cross-cutting factors for government, industry, public	
*Worldviews (10)	The sets of assumptions, beliefs, and experiences that inform attitudes (stakeholders, public, etc.) towards CCUS
*Problem perception (11)	Awareness of problems related to energy systems
Trust (5)	Trust in technical/scientific information, industry, regulatory competence, implementation
*Energy context (10)	Trends in implementation of energy alternatives in decision-making jurisdiction
*Knowledge/Information provision (12)	Awareness, common understanding, distribution of information
*Tolerable costs (12)	Financial outlay to implement and maintain a project
*Distributive justice (11)	Distribution of costs, risks, benefits
2. Governance factors	
Policy and regulatory stability (7)	GHG emissions reductions goals and measures that could support (directly or indirectly) CCUS implementation
*Inter-jurisdictional challenges (9)	Decision-making process and outcomes that involve two or more jurisdictions
Procedural justice (5)	Transparent, engaged, accountable decision processes, including competent regulatory oversight
3. Industry factors	
*Willingness and/or capacity to act (10)	Planning, preparedness, agreement to implement CCUS
*Pace/Demonstration (12)	Technological feasibility and implementation
Market competitiveness/International trade (7)	Economic opportunity/export of technologies

Notes in table (#) represents how many of the fourteen interview participants mentioned the risk issue. *Issues marked with an asterisk emerged as higher risk issues (more than half of interview participants mentioned them). List of risk issues adapted from Leiss and Krewski (2019) and L'Orange Seigo et al. (2014)

varying degrees of shared interests in advancing carbon capture mitigation options and outcomes. Within this CCUS ecosystem, the participants were chosen to provide a range of perspectives spanning policy development and implementation from the federal government, research and funding institutions, the private sector (fossil and trade-exposed industries, including technology developers and users), and environmental and industry advocacy NGOs.

Scholars characterize this kind of network as an "epistemic community" or knowledge-based network of recognized experts. Members of these networks typically have common "principled and causal beliefs but also have shared notions of validity and a shared policy enterprise" (Haas 1992 cited in Stephens et al. 2011, p. 379). Such a community is usually focused on "risks to" the advancement of the technology as opposed to "risks of" the technology, the latter often being the public's concern (Stephens et al. 2011).

Semi-structured interviews were undertaken by telephone in May and June 2019. The findings are organized in part by distinguishing between group types. This may be based on the organizational type (government, industry, non-government); industry type (oil, gas, and coal, or trade-exposed); or those working directly with the technology ("implementers").

Research Findings: Public Confidence Risks Related to CCUS Decision-Making

Risk issues related to CCUS occur at two levels: (1) CCUS as a climate mitigation technology at the international/national/provincial levels (hereafter, the *policy level*) and (2) specific CCUS projects and related government decision-making processes at the provincial/regional/local levels (hereafter, the *project level*). Risks at both of these levels influence the extent of policy support for CCUS, as well as final project-level investment decisions. These risks affect all sectors attempting to manage GHG reductions: large industrial emitters and electricity generation (our study's focus), as well as transmission, transportation, the built environment, agriculture, forestry, waste, and government operations (Specific Mitigation Opportunities Working Group 2016).

As shown in Table 7.1 and detailed below, nine of the thirteen risk issues noted above were mentioned by more than half of participants. We categorize these as "higher risk" issues. Of note, with the exception of trust, all *cross-cutting* risk issues were mentioned more frequently.

The three risk areas mentioned by the largest number of interviewees (12 of 14) include:

(i) inadequate knowledge and information provision,
(ii) the need to reduce costs, and
(iii) inadequate pace of effective project demonstration.

As explained below, participants noted that CCUS is vulnerable in a public and decision-maker context marked by inadequate awareness and understanding of the industry and the broader energy system. Costs (and by extension financial support) were noted as a key concern, especially in terms of being able to make CCUS technologies cost-effective enough for wide-spread implementation. Finally, interviewees noted that the timeline to effective project demonstration is critical in the context of rapid clean energy technology development and climate mitigation solutions. Feasibility and successful demonstration of CCUS technologies need to occur at a pace fast enough to provide solutions.

Cross-Cutting Factors for Government, Industry, the Public

Worldviews (higher risk: mentioned in 10 of 14 interviews). Worldviews refer to the sets of assumptions, beliefs, and experiences that inform attitudes towards CCUS. Worldviews fundamentally affect attitudes towards climate change and energy technologies, as well as risk perceptions and preferences for actions that address climate change (Kahan et al. 2011). In so doing, they can shape government policy and regulatory responses, which can have implications for the nature and speed of emissions reductions.

Studies about the development and future of CCS technology have found an uneasy coalition of supportive actors with a variety of viewpoints from industry, government, NGOs, and civil society (Markusson et al. 2012). Opinions at the policy level vary across a range of issues, including how effective the technology is for long-term storage or sequestration, and whether it perpetuates fossil fuel production and use. At the project level, worldviews (assumptions, beliefs, and experiences that inform attitudes) and their relationship to beliefs about local benefits and safety seem to have the largest impact on social acceptance of CCS (Krause et al. 2014; Warren et al. 2014).

The interviews supported much of this existing literature. Respondents noted that for some people, using CCUS/EOR means the technology should not be defined as "clean tech." This relates to a worldview that affects CCUS acceptance because it is seen to perpetuate fossil fuel production and use and is perceived to represent unacceptable risks to the environment. As one interviewee noted, "[There is] kind of a moral hazard problem of proceeding with CCS ... ultimately that by buying into CCS, you are accepting a lesser solution for decarbonization in the energy sector."

Participants agreed that a variety of actions, such as information provision and a focus on CCUS as part of the solution to climate change, particularly for hard to reach sectors, would be most likely to help lessen this challenge. In the words of one participant, "Canada can [...] serve as a leader to other countries in the development of cleaner technologies for oil and gas. In other words, in addition to providing product, Canada can provide solutions to the world for the development of oil and gas resources with lower environmental impact." Other suggestions included developing a common GHG reduction vision in Canada and demonstrating the technology's relevance beyond conventional fossil fuel applications such as EOR.

Problem perception (higher risk: mentioned in 11 of 14 interviews). This issue is an extension of the risk of different worldviews. It refers to problem perceptions varying across different groups or belief systems when it comes to climate change and the place of carbon capture as a mitigation option. Study participants emphasized this issue as very important.

Transition expectations can fundamentally affect perceptions of CCUS as a solution or a problem. For example, previous research has identified two expectations for energy transition among the general public and among energy and environmental leaders in Canada: one focused on a gradual process of change, and the other focused on aggressive emissions reductions (Beck 2020; Bird et al. 2019). Survey research reveals that 73% of Canadians expect at least a moderate pace of reducing GHG emissions with expectations for substantive change in 25 years or less. Within that group, 43% prefers a more aggressive pace of 10 years or less (Bird et al. 2019). A small minority (27%) prefers a much slower pace of 50 years or longer. These differences in views can dramatically shape perceptions of CCUS.

As noted earlier, CCS scholarship has found CCUS/EOR to be controversial because it does not address the production of downstream GHG emissions or look at alternative (non-fossil fuel) energy sources (Einsiedel et al. 2013). Indeed, interviewees' remarks regarding problem perception suggested that discussions of CCUS technologies should not focus on capture, but about what is done with the CO_2, including what else can usefully be done to reduce emissions in sectors with limited technological options to abate emissions. In the EITE sector, differences in problem perception may be muted because the question of carbon storage includes the potential for conversion into a resource (rather than storage as a waste or increasing fossil fuel production).

Key suggestions to manage differences in problem perception included the development of carbon capture with permanent storage and/or conversion destinations, the need for tax incentives to mobilize the EITE sector and accelerate the pace of CCUS technology beyond EOR, and better outreach and communications for the actions being taken. Respondents believed that progress in these areas would help to address the risk posed by differences in problem perception.

Energy context alternatives (higher risk: mentioned in 10 of 14 interviews). This issue refers to the challenge of trade-offs and opportunity costs of developing one technology over another, especially at the provincial and local level. Existing literature suggests that public confidence in decision-making for energy alternatives can be strengthened where new technologies are discussed within the broader energy context. For example, Lock et al. (2014) assessed participant trade-offs between CCS and renewable energy sources in situations where one technology is developed at the expense of the other. They found that making these decisions in the context of broader conversations about energy use improved trust and perceptions of legitimacy in government decisions about technology. Stated another way, public confidence in these decisions is affected by peoples' perceptions of fairness in decision-making processes and their assessments of collective and individual costs and benefits. This applies for all forms of energy and energy projects, from oil and gas through to renewable energy (Nourallah 2016; Cleland and Gattinger 2017).

Participants in this study noted that CCUS has the potential to achieve multi-billion dollar markets internationally, but it is challenged by competition from increasingly affordable natural gas, wind, and solar energy technologies. To address potential trade-offs in the energy context, interviewees recommended including more coherent and comprehensive approaches to decision-making at all jurisdictional levels. Such approaches would presumably make clear some of the underlying benefits of CCUS in comparison with other technologies. Suggestions to highlight the value of CCUS included ongoing community education and outreach for CCUS science and safety, research to make the technology more affordable, and efforts to better understand the potential role of CCUS in contributing to net-zero emissions. As one interviewee noted, "[CCUS] work that's been done in utilities and [the] oil and gas sector will be tremendously beneficial [...] across a broader range of sectors that we know are going to be here to stay."

Lack of trust (lower risk: mentioned in 5 of 14 interviews). Lack of public trust in project developers, public authorities, and decision-making processes can be a significant impediment to public confidence in energy project decisions. Research demonstrates that trust is a critical factor in social acceptance of energy project decisions (Cleland et al. 2016, 2018; Nourallah 2016), including for CCS (Einsiedel et al. 2013). This is due in part to levels of trust in new technologies; communities can be sceptical of non-established science and infrastructure.

Interestingly, interviewees emphasized the importance of trust to a lesser degree than other risk factors, but they did raise it as an issue. Respondents noted the critical importance of trust in science. They also highlighted the importance of trust in industry, particularly if CCUS pursuits are seen as self-serving and not a response to community or broader needs. In addition, they suggested that policy longevity and stability are essential to promote trust in government, particularly industry trust in government. Industry participants also noted that individual actions by their own sector could undermine trust in the entire CCUS endeavour. In the words of one participant, "[There can be] suspicion, skepticism [of] industry ... where [a technology] is pushed by industry – [people think] there's got to be a catch. If industry tends to be self-serving rather than serving a social good, 'How can this be a good thing?'".

Tolerable costs (higher risk: mentioned in 12 of 14 interviews). Concerns over cost emerged as one of the most important risk issues for all participants, but they took a variety of forms. The cost issue begins with initial investments in the technology without knowing the outcome. By extension, this means high levels of financial risk. Here, participants noted that arguments can be made for public money to be spent instead on renewables, nuclear, or direct air capture of CO_2. As for private spending, it tends to focus on lowest cost solutions, which also represents a risk that investment dollars won't flow in sufficient volume to CCUS.

There are also regional dimensions to the cost issue. Previous research has shown that energy and environmental leaders are concerned that the costs and opportunities of transition are unlikely to be distributed equally across Canada (Beck 2020). In this study, participants also noted that variations in government funding between jurisdictions can have different regional cost/benefit impacts. Further, participants noted that if capture innovation is subsidized by government it could lead to negative public perceptions because of concern over government favouritism of fossil fuels.

Interviewee suggestions for managing these risks included avoiding punitive regulations or generous grants/subsidies, and instead focusing on more moderate programmes of public support through tax incentives, supportive policy, and research support via effective demonstration and pilot projects.

Inadequate knowledge/information provision (higher risk: mentioned in 12 of 14 interviews). Participants noted that inadequate knowledge sharing and information provision slow down or block CCUS acceptance at the policy and project levels. While scientific and engineering expertise was underlined as a positive attribute in the Canadian context, participants noted that public knowledge of the underlying technology and functions of CCUS infrastructure remains low. They expressed similar concerns over limited knowledge levels among politicians (as compared to the working level bureaucracy), regulatory leaders, and environmental stakeholders. Participant concerns mainly focused on the degree of knowledge about market risks, challenges, and specific attributes of the technology. One participant noted, "[the technology is] not that well understood actually. There's a risk that policymakers and governments – and I've seen this – are kind of interested but they don't know what to do with it. ... Especially at the higher policy levels of the government they don't really understand it, even though government scientists may understand it fairly well."

Suggestions to mitigate risks related to knowledge and information included developing a supportive narrative, improving industry outreach, and better information and resource sharing among technology developers. In addition, participants emphasized that there may be stronger support for CCUS as an emergent technology for the EITE sector.

Distributive justice (higher risk: mentioned in 11 of 14 interviews). This issue encompasses policy and project decision-making that involves trade-offs and allocation of costs and benefits among different groups. In general, policy discussions in Canada and elsewhere have emphasized that options for climate change mitigation should not unfairly impact vulnerable or minority populations. At the project level, the concern is whether impacts are distributed equitably ***across the whole of a community*** and that the community is not unfairly impacted ***compared to other communities.*** Concerns for environmental impacts related to post-combustion technologies or to pipeline and CO_2 leakage to the surface have the potential to impact specific areas or jurisdictions and may be distributed inequitably within or across communities.

Study participants suggested that risk mitigation measures could include socializing costs across local and provincial jurisdictions, ensuring strong and effective regulatory standards, and improving information using lifecycle analysis. Participants also noted the importance of better communicating health and safety standards, and more effectively identifying and supporting stakeholders who stand to lose if industries shut down.

Governance Factors

Lack of policy and regulatory stability (lower risk: mentioned in 7 of 14 interviews). Policy and regulatory stability for GHG emissions reductions can affect support (directly or indirectly) for CCUS implementation. While consistent policies for CCUS mitigation technologies are important, participants emphasized them less than other factors. Lack of policy stability is problematic because it creates mixed signals for industry and other stakeholders, and because it increases uncertainty in a policy regime in which there are already high levels of political, economic, and social risk. This is a particular concern when policies are implemented by a government and then reversed when a new government comes into power. When this happens, participants noted that it increases mistrust and risk, and weakens the investment climate. Interviewees noted that the largest concern for CCUS is variability in provincial and federal carbon policies. Other factors include differences in policy instruments, for instance, using taxes, levies, or performance standards. In the words of one interviewee, "Stable climate change policy: people are hungry for it."

Almost all participants noted that a stable price on carbon is essential to mitigate risk. Respondents emphasized the need for cross-partisan agreements both within and between jurisdictions to provide a clear and consistent direction for CCUS technology. They also noted the importance of clear funding models to support innovation, research and development, and investment.

Inter-jurisdictional challenges (higher risk: mentioned in 9 of 14 interviews). This risk issue concerns decision-making that involves two or more jurisdictions. Study participants voiced strong concern over inter-jurisdictional issues and tensions between provincial governments and between national and provincial jurisdictions. Similar to policy instability, the challenge arises when multiple jurisdictions are inconsistent and unaligned in their approaches to CCUS. For example, Saskatchewan remains committed to coal-fired electricity, but the federal government committed to phasing out unabated coal-fired power by 2030. There are misalignments between provinces as well. Not all provinces have a regulatory framework for CCUS, and others may include additional reviews of CCUS projects by municipal or Indigenous authorities, creating a hodgepodge of regulatory approaches across jurisdictions. Participants noted that inter-jurisdictional challenges tend to play out in political and partisan contexts, rather than at the project, bureaucratic, or regulatory levels. As noted by one interviewee, "One of the reasons why I don't think we've seen as much uptake on carbon capture is that we collectively never moved forward in an effective way on pricing carbon. We'd always pushed for that consistent price on carbon on a North American-wide basis. We're not there – instead now we're in a federal-provincial quagmire on this issue."[1] Industry participants also worried that government consultation

[1] The Supreme Court of Canada decision affirming the constitutionality of federal carbon pricing legislation followed the interview research.

with industry to develop more consistent policies may be slow to materialize, and that negative public views about a lack of movement to address challenges will fall on industry.

All risk management options suggested by participants involved improving and accelerating cooperation and coordination between governments.[2]

Procedural justice (lower risk: mentioned in 5 of 14 interviews). Risk issues related to procedural justice are focused on decision-making processes, including policy and regulatory decision-making that is transparent, engaged, and accountable. There is an extensive literature underscoring that policy processes perceived as open, transparent, and unbiased are much more likely to result in public support for both policies and projects (Cleland et al. 2016; Simard 2018; Frank and Lindsay 2020). Interestingly, this topic did not garner a lot of attention from study participants, but those who mentioned it pointed to risks at both the policy and project levels. According to one interviewee, "[It is important to] always start with the regulations and policy. Society feels comfortable and protected through regulations and policy. Listen to their concerns and factor that into how we develop and deploy the technology as well so […] you're bringing […] society into the technology, their involvement and the raising of concerns. [There needs to be a] desire and willingness to listen to stakeholders about their concerns."

Interviewees also noted that in some parts of Canada there are no specific regulations for risk management review. This has the potential to impact public confidence in individual project decisions and implementation. Of note, interviewees did not highlight the need for transparency in the determination of costs and benefits or in lifecycle assessments. It may be that this issue has less "play" with participants because there are so few CCUS projects in Canada or because many of the large-scale projects exist in Alberta, where regulatory provisions are the most developed.

Suggestions to mitigate risk for this issue included improving transparency and information-sharing, incorporating broad lifecycle perspectives into industry and project analyses, and third-party reviews of applications to government funding programmes.

[2] For a review of energy-environment federalism in Canada, see Bratt (2021).

Industry Factors

Willingness/capacity to act (higher risk: mentioned in 10 of 14 interviews). This emerged as a relatively important risk issue and refers to the tension between industry actors that are able and willing to move forward on technology implementation and those that are not. Tension can be heightened by public perceptions that industry is lagging when it comes to vigorously moving forward with a clear commitment to finding emissions abatement solutions. A number of participants stated that some companies are in favour of the status quo and that the speed of the slowest is advantageous. Others noted that the challenge is exacerbated by different approaches taken for different sectors. For example, new building requirements related to carbon inputs could affect the cement industry more significantly than the steel industry. Participants held diverse opinions on this issue.

Beyond the need for government to provide a clearer path on GHG emissions reductions, most recommendations for risk management focused on industry actions, including CEO leadership and coordination, higher investment and cost reductions, and greater commitment to innovation in the project demonstration phase.

Pace and demonstration of technological feasibility (higher risk: mentioned in 12 of 14 interviews). Study participants emphasized this issue strongly and noted the inability to meet technological feasibility expectations in any area of CCUS. Some participants argued that expectations were simply unrealistic and lacked appropriate timelines. The issue of pace is directly related to many of the other concerns discussed above. Jurisdictional issues, differences in worldviews, alternative technological options, and lack of consistent carbon pricing and policy all play a role in driving pace to a slow grind. Several participants raised the importance of scaling up the technology to a level that has a meaningful emissions impact.

Participants noted that addressing this challenge will require action by industry and government in concert. Recommendations included increasing policy and funding stability, improving cost reductions, strengthening existing partnerships and research networks, and creating new international partnerships.

Market competitiveness and international trade (lower risk: mentioned in 7 of 14 interviews). This area was one of the few bright spots for participants, who characterized it as a strength. Government documents and interviewees emphasized that Canada could be well-positioned to benefit from international markets and to emerge as a leader in this technology space. Some participants noted that Canada is already considered to be a global leader in the development of CCUS. Suggested risk mitigation options included demonstrating and showcasing investment, having coherent government policies, building export market opportunities, and developing Canada's role as a global leader in CCUS.

The following section categorizes participant recommendations to mitigate the above-noted risks using the REACT framework of risk assessment and risk management.

Discussion, Conclusion, and Risk Management Options Emerging from the Research

The nations that lead in policy and project support for CCUS include Canada, the United States, Norway, the United Kingdom, and Australia. Other nations score lower on a 2021 "readiness index" (GCCSI 2022). Overall, however, global implementation of CCUS is not on track to meet mitigation projections (IEA 2019c). Specifically with respect to CCS, a variety of reasons explain limited progress. These are generally identified as technical, economic, political-institutional, social, and international (Viebahn and Chappin 2018; Markusson et al. 2017; Gaede and Meadowcroft 2016). Many of the challenges facing CCS and CCUS are not unique to Canada.

But who should do what and how to address this? We apply a lightly modified version of the REACT framework for risk management and population health (Krewski et al. 2007, 2014). We use REACT to categorize the various approaches and tools participants recommended and highlighted for each risk issue discussed above (see Table 7.2). These fall under the purview of policymakers, regulators, and industry, often working in concert.

Policy/regulatory options involve government policy, legislation, regulations, guidelines, permits, or approvals for action. Interestingly, participants did not suggest any risk management options under the sole purview of industry players. They viewed implementation as a joint government/industry climate change mitigation endeavour, which suggests the need for a national vision for CCUS in the context of Canadian climate policy. Further, participants underscored the need for

Table 7.2 Risk management options for policymakers, regulators, and industry

Purview of action		
Policy	*Regulation*	*Large point-source industry*
	Policy/regulatory options	
Policy clarity and certainty for climate change/GHG	Develop a national vision for CCUS	
	Regulatory clarity and certainty for climate change/GHG	
	Clear and coherent climate change and GHG reduction plans	
Federal/provincial policy and regulatory collaboration		
	Economic/financial options	
Government/industry cost sharing	Carbon pricing to create value proposition	Government/industry cost sharing; ***plus*** industry cost reductions
	Advisory/communications options	
Information/education regarding CCUS, energy systems, mitigation alternatives	Government, industry and public analyse CCUS alongside alternative mitigation options	Information/education regarding CCUS, energy systems, mitigation alternatives
Increased development of international networks		Increased development of international networks
Cooperation and engagement in knowledge sharing, including international networks		Cooperation and engagement in knowledge sharing, including international networks
	Community-based options	
	Transparency and engagement in information/ technological options	
	Collaborative learning/engagement with the public	
	Technological options	
	Broaden CCUS uses beyond fossil applications	

(continued)

Table 7.2 (continued)

Purview of action		
Policy	*Regulation*	*Large point-source industry*
	Expand storage strategies (CO_2 destination point, monitoring)	
Government and industry demonstration for export market		Government and industry demonstration for export market

Source Authors' own source

stable, detailed, and coherent climate policy and GHG reduction plans to signal opportunities for investors, reduce policy risk and variability, and clarify the need for the technology. This includes carbon pricing, an economic/financial measure discussed below. Interviewees noted that industry and individual company climate plans also need to be detailed and coherent. In addition, participants noted that clear and stable climate policy and carbon pricing hinge on federal/provincial cooperation to foster policy stability and reduce risk. Much of the industry still requires "green industrial development," which requires a shared vision among governments and industry.

Economic/financial tools refer to insurance, levies, and other cost structures designed as incentives to take action. Interviewees emphasized that carbon pricing is a critical component for CCUS to help achieve tolerable costs as well as create opportunities for venture capital and investment. Carbon prices need to be reasonable, predictable, and robust to provide adequate economic incentives for CCUS development. Participants also recommended using cost sharing between government and industry to further encourage industry to be creative, entrepreneurial, and successful. In the period since the interviews were undertaken, the US has extended its 45Q carbon sequestration tax incentive programme (US Code 2022) and the Government of Canada has initiated creation of a CCUS investment tax credit (Government of Canada 2022b).

Advisory/communications tools encompass communications, education, and awareness activities. Participants recommended deepening and broadening the analysis of CCUS to demonstrate potential value. This could include, notably, comparing various CCUS technologies to other mitigation options using lifecycle analysis. Similarly, interviewees noted the importance of improving understanding of CCUS technologies, approaches, and uses across energy systems and industrial contexts. They suggested providing policymakers and the general public with information and education on CCUS more often and more effectively. This would include, critically, the potential of CCUS to reduce carbon intensity in operational contexts beyond fossil fuel use and production. They likewise proposed better communication of cost improvements to clearly demonstrate and communicate progress on the economics of CCUS to policymakers, stakeholders, and the public. Finally, they suggested increasing knowledge sharing and demonstrations in international export markets to increase opportunities for Canadian leadership.

Community-based tools range from those targeting the CCUS industrial ecosystem to engagement with communities where CCUS projects are located. Here, interviewees noted the importance of building transparent learning and engagement with all stakeholders and the general public to foster public confidence in the technology and decisions surrounding it.

Technological tools refer to advances in technology. The key recommendation here was to broaden the potential uses of CCUS, notably for EITE industries. Interviewees noted that technology assessments should be broadened to explore more potential uses across all energy systems and industry contexts.

CCUS is and will be an essential component of climate mitigation efforts in Canada and globally. As noted above, the Canadian and US federal governments have announced additional investment tax credit proposals for sequestration projects (Government of Canada 2022a; US Code 2022), and Canada has announced grants and contributions for technology development (Government of Canada 2022b). However, much additional analysis is required to identify how to best support CCUS technology development and deployment, including how to strengthen public confidence in decision-making. Looking forward, it will also be important to better understand the motivations and concerns of potential opponents to CCUS, and to better assess whether there is common ground between proponents and detractors to build public confidence in decision-making. This could include better understanding the views of those who oppose CCUS because of concerns for fossil fuel lock-in, or, alternately, better understanding the regional and local concerns of communities near CCUS infrastructure. This research suggests that advancing understanding in these areas, along with implementing the recommendations emerging from this study, will help to build public confidence in CCUS decision-making and position CCUS technology to make the contribution to climate mitigation envisioned for it over the past fifteen years.

Acknowledgments This chapter draws on Larkin, Bird, and Gattinger 2021. Larkin was supported by postdoctoral research funding from the Natural Sciences and Engineering Research Council (NSERC CREATE Training Program in Carbon Capture). The authors would like to thank Michael Cleland and Brendan Frank from the University of Ottawa's Positive Energy Research Team for their insightful comments on the research. They would also like to thank Dr. Phil De Luna (then Director, Materials for Clean Fuels Challenge Program, National Research Council Canada) and Professor Margot Hurlbert (Canada Research Chair in Climate Change, Energy and Sustainability Policy, University of Regina) for their helpful comments. Finally, thanks are owed to those who gave generously of their time and expertise participating in interviews for this study. As is customary, any errors of fact or interpretation are the responsibility of the authors.

References

Batel, S., Devine-Wright, P., & Tangeland, T. (2013). Social acceptance of low carbon energy and associated infrastructures: A critical discussion. *Energy Policy*, 58, 1–5. https://doi.org/10.1016/j.enpol.2013.03.018.

Beck, M. (2020). *What is "transition"? The two realities of energy and environmental leaders in Canada*. Ottawa: Positive Energy, University of Ottawa. Retrieved 8 April 2020, from https://www.uottawa.ca/positive-energy/content/what-transition-two-realities-energy-and-environmental-leaders-canada.

Bird, S., Lachapelle, E., & Gattinger, M. (2019). *Polarization over Energy and Climate in Canada: Survey results*. Ottawa: Positive Energy, University of Ottawa. Retrieved 14 May 2020, from https://www.uottawa.ca/positive-energy/sites/www.uottawa.ca.positive-energy/files/polarization_survey_final.pdf.

Bowden, A. R., Pershke, D. F., & Chalaturnyk, R. (2013a). Biosphere risk assessment for CO_2 storage projects. *International Journal of Greenhouse Gas Control*, 16, S291–S308. https://doi.org/10.1016/j.ijggc.2013.02.015.

Bowden, A. R., Pershke, D. F., & Chalaturnyk, R. (2013b). Geosphere risk assessment conducted for the IEAGHG Weyburn-Midale CO_2 Monitoring and Storage Project. *International Journal of Greenhouse Gas Control*, 16, S276–S290. https://doi.org/10.1016/j.ijggc.2013.02.014.

Bratt, D. (2021). *Energy-Environment Federalism in Canada: Finding a path for the future*. Ottawa: Positive Energy, University of Ottawa. Retrieved 17 May 2021, from https://www.uottawa.ca/positive-energy/content/energy-environment-federalism-canada-finding-path-future.

Busse, M., & Siebert, R. (2018). Acceptance studies in the field of land use—A critical and systematic review to advance the conceptualization of acceptance and acceptability. *Land Use Policy*, 76, 235–245. https://doi.org/10.1016/j.landusepol.2018.05.016.

Canadian Electricity Association. (2020). *Carbon Capture Storage*. Retrieved 15 June 2020, from https://electricity.ca/learn/future-of-electricity/carbon-capture-storage/.

Carbon Cure. (2021). *Clean Tech Company, CarbonCure Wins NRG COSIA Carbon XPRIZE*, April 19. Retrieved 17 May 2021, from https://www.carboncure.com/news/clean-tech-company-carboncure-wins-nrg-cosia-carbon-xprize/.

Cleland, M., Bird, S., Fast, S., Sajid, S., & Simard, L. (2016). *A Matter of Trust: The role of communities in energy decision-making*. Ottawa: Positive Energy, University of Ottawa, and Canada West Foundation. Retrieved 31 October 2019, from https://www.uottawa.ca/positive-energy/sites/www.uottawa.ca.positive-energy/files/mattertrust_report_24nov2016-1_web.pdf.

Cleland, M., & Gattinger, M. (2017). *System under Stress: Energy decision-making in Canada and the need for informed reform*. Ottawa, ON: University of Ottawa. Retrieved 2 July 2018, from https://www.uottawa.ca/positive-energy/sites/www.uottawa.ca.positive-energy/files/2_positive_energy-system_under_stress-cleland_and_gattinger.pdf.

Cleland, M., & Gattinger, M. (2019). *Canada's energy future in an age of climate change*. Ottawa: Positive Energy, University of Ottawa. Retrieved 4 November 2019, from https://www.uottawa.ca/positive-energy/sites/www.uottawa.ca.positive-energy/files/canadas_energy_future_design_rd_web.pdf.

Cleland, M., Gattinger, M., Aguirre, R., & Beck, M. (2018). *Durable balance: Informed reform of energy decision-making in Canada*. Ottawa: Positive Energy, University of Ottawa. Retrieved 2 July 2017, from https://www.uottawa.ca/positive-energy/sites/www.uottawa.ca.positive-energy/files/180418-db-report-final.pdf.

CMC Research Institutes and Canadian Business for Social Responsibility. (2019). *CarbonTech: A primer on carbon capture, conversion, utilization and storage technologies*. Retrieved 7 September 2022, from https://cmcghg.com/wp-content/uploads/2019/05/CBSR-CarbonTech_Report_final.pdf.

Dixon, T., McCoy, S. T., & Havercroft, I. (2015). Legal and regulatory developments on CCS. *International Journal of Greenhouse Gas Control*, 40, 431–448. https://doi.org/10.1016/j.ijggc.2015.05.024.

Einsiedel, E. F., Boyd, A. D., Medlock, J., & Ashworth, P. (2013). Assessing socio-technical mindsets: Public deliberations on carbon capture and storage in the context of energy sources and climate change. *Energy Policy*, 53, 149–158. https://doi.org/10.1016/j.enpol.2012.10.042.

Environment Canada. (2018). *Reduction of Carbon Dioxide Emissions from Coal-Fired Generation of Electricity Regulations, SOR/2012–167*. Ottawa, CA: Queen's Printer. Retrieved 20 September 2018, from http://ec.gc.ca/lcpe-cepa/eng/regulations/detailReg.cfm?intReg=209; http://laws-lois.justice.gc.ca/eng/regulations/SOR-2012-167/index.html.

Frank, B., & Lindsay, S. G. (2020). *Addressing polarization: What works? Case study: The just transition task force*. Ottawa: Positive Energy, University of Ottawa. Retrieved 18 February 2021, from https://www.uottawa.ca/positive-energy/sites/www.uottawa.ca.positive-energy/files/just_transition_final.pdf.

Gaede, J., & Meadowcroft, J. (2016). Carbon capture and storage demonstration and low-carbon energy transitions: Explaining limited progress. *The Palgrave Handbook of the International Political Economy of Energy*, 319–340. https://doi.org/10.1057/978-1-137-55631-8_13.

Gaede, J., & Rowlands, I. H. (2018). Visualizing social acceptance research. *Energy Research & Social Science*, 40, 142–158. https://doi.org/10.1016/j.erss.2017.12.006.

Gale, J., Abanades, J. C., Bachu, S., & Jenkins, C. (2015). Special issue commemorating the 10th year anniversary of the publication of the Intergovernmental Panel on Climate Change Special Report on CO_2 Capture and Storage. *International Journal of Greenhouse Gas Control*, 40, 1–5. https://doi.org/10.1016/j.ijggc.2015.06.019.

GCCSI. (2022). *CO_2RE policy, readiness and requirement indicators*. Retrieved 7 September 2022, from https://co2re.co/Policies.

Government of Canada. (2018). *Regulations limiting carbon dioxide emissions from natural gas-fired generation of electricity*. Retrieved 17 September 2018, from http://gazette.gc.ca/rp-pr/p1/2018/2018-02-17/html/reg4-eng.html.

Government of Canada. (2019). *Canada's Fourth Biennial Report to the United Nations Framework Convention on Climate Change*. Environment and Climate Change Canada. Retrieved 10 January 2020, from https://www.canada.ca/en/environment-climate-change/services/climate-change/greenhouse-gas-emissions/fourth-biennial-report-climate-change.html.

Government of Canada. (2022a). *Additional design features of the investment tax credit for carbon capture, utilization, and storage: Recovery mechanism, climate risk disclosure, and knowledge sharing*. Retrieved 7 September 2022, from https://www.canada.ca/en/department-finance/news/2022a/08/additional-design-features-of-the-investment-tax-credit-for-carbon-capture-utilization-and-storage-recovery-mechanism-climate-risk-disclosure-and-k.html.

Government of Canada. (2022b). *Energy Innovation Program—Carbon capture, utilization and storage RD&D Call*. Retrieved 7 September 2022, from https://www.nrcan.gc.ca/science-and-data/funding-partnerships/funding-opportunities/funding-grants-incentives/energy-innovation-program/energy-innovation-program-carbon-capture-utilization-and-storage-stream/23815.

Huijts, N., Molin, E., & Steg, L. (2012). Psychological factors influencing sustainable energy technology acceptance: a review-based comprehensive framework. *Renewable Sustainable Energy Review*, 16, 525–531.

IEA. (2019a). *Carbon capture, utilisation and storage: A critical tool in the climate energy toolbox*. Retrieved 21 August 2019, from https://www.iea.org/topics/carbon-capture-and-storage/.

IEA. (2019b). *Material efficiency in clean energy transitions*. https://webstore.iea.org/download/direct/2454?fileName=Material_efficiency_in_clean_energy_transitions.pdf.

IEA. (2019c). *Tracking clean energy progress*. Retrieved 17 October 2019, from https://www.iea.org/tcep/.

IPCC. (2005). *IPCC Special Report on Carbon Dioxide Capture and Storage*. Cambridge, UK: Cambridge University Press. Retrieved 16 August 2016, from http://www.ipcc.ch/pdf/special-reports/srccs/srccs_wholereport.pdf.

Jones, C. R., Olfe-Kräutlein, B., Naims, H., & Armstrong, K. (2017). The social acceptance of carbon dioxide utilisation: A review and research agenda. *Frontiers in Energy Research*, 5. https://doi.org/10.3389/fenrg.2017.00011.

Kahan, D. M., Jenkins-Smith, H., & Braman, D. (2011). Cultural cognition of scientific consensus. *Journal of Risk Research*, 14(2), 147–174. https://doi.org/10.1080/13669877.2010.511246.

Kasperson, R. E., Renn, O., Slovic, P., Brown, H. S., Jacque, E., Goble, R., Kasperson, J. X., & Ratick, S. (1988). The social amplification of risk: A conceptual framework. *Risk Analysis*, 8, 177–187.

Krause, R. M., Carley, S. R., Warren, D. C., Rupp, J. A., & Graham, J. D. (2014). "Not in (or under) my backyard": Geographic proximity and public acceptance of carbon capture and storage facilities. *Risk Analysis*, 34(3), 529–540. https://doi.org/10.1111/risa.12119.

Krewski, D., Hogan, V., Turner, M. C., Zeman, P. L., McDowell, I., Edwards, N., et al. (2007). An integrated framework for risk management and population health. *Human and Ecological Risk Assessment: An International Journal*, 13(6), 1288–1312. https://doi.org/10.1080/10807030701655798.

Krewski, D., Westphal, M., Andersen, M. E., Paoli, G. M., Chiu, W. A., Al-Zoughool, M., et al. (2014). A framework for the next generation of risk science. *Environ Health Perspect*, 122(8), 796–805. https://doi.org/10.1289/ehp.1307260.

Larkin, P., Bird, S., & Gattinger, M. (2021). *Carbon capture, Utilization and Storage: Polarization, public confidence and decision-making*. Ottawa: Positive Energy, University of Ottawa. Retrieved 28 March

2021, from https://www.uottawa.ca/positive-energy/sites/www.uottawa.ca.positive-energy/files/ccus_final_web_1.pdf.

Larkin, P., Leiss, W., Arvai, J., Dusseault, M., Fall, M., Gracie, R., Heyes, A., & Krewski, D. (2019). An integrated risk assessment and management framework for carbon captureand storage: A Canadian perspective. *International Journal of Risk Assessment and Management*, 22(3/4), 464–508.

Leiss, W., & Krewski, D. (2019). Environmental scan and issue awareness: Risk management challenges for CCS. *International Journal of Risk Assessment and Management*, 22(3/4), 234–253.

Lock, S. J., Smallman, M., Lee, M., & Rydin, Y. (2014). "Nuclear energy sounded wonderful 40 years ago": UK citizen views on CCS. *Energy Policy*, 66, 428–435. https://doi.org/10.1016/j.enpol.2013.11.024.

L'Orange Seigo, S., Dohle, S., & Siegrist, M. (2014). Public perception of carbon capture and storage (CCS): A review. *Renewable and Sustainable Energy Reviews*, 38, 848–863. https://doi.org/10.1016/j.rser.2014.07.017.

Markusson, N., Dahl Gjefsen, M., Stephens, J. C., & Tyfield, D. (2017). The political economy of technical fixes: The (mis)alignment of clean fossil and political regimes. *Energy Research & Social Science*, 23, 1–10. https://doi.org/10.1016/j.erss.2016.11.004.

Markusson, N., Shackley, S., & Evar, B. (Eds.). (2012). *The social dynamics of carbon capture and storage: Understanding CCS representations, governance and innovation*. Oxon, UK: Routledge.

Nourallah, L. (2016). *Communities in perspective: Literature review of the dimensions of social acceptance for energy development and the role of trust*. Ottawa: Positive Energy, University of Ottawa. Retrieved 2 July 2017, from https://www.uottawa.ca/positive-energy/sites/www.uottawa.ca.positive-energy/files/positive_energy-community_social_acceptance_literature_review_0.pdf.

Rothstein, H., Borraz, O., & Huber, M. (2013). Risk and the limits of governance: Exploring varied patterns of risk-based governance across Europe. *Regulation & Governance*, 7(2), 215–235. https://doi.org/10.1111/j.1748-5991.2012.01153.x.

Simard, L. (2018). *How to decide—Engagement: Information and capacity*. Retrieved 2 July 2017, from https://www.uottawa.ca/positive-energy/sites/www.uottawa.ca.positive-energy/files/pe_louis_simard_final.pdf.

Specific Mitigation Opportunities Working Group. (2016). *Final Report*. Ottawa, ON: Government of Canada. Retrieved 20 September 2018, from https://www.canada.ca/content/dam/eccc/migration/cc/content/6/4/7/64778dd5-e2d9-4930-be59-d6db7db5cbc0/wg_report_specific_mitigation_opportunities_en_v04.pdf.

Stephens, J. C., Hansson, A., Liu, Y., de Coninck, H., & Vajjhala, S. (2011). Characterizing the international carbon capture and storage community.

Global Environmental Change, 21(2), 379–390. https://doi.org/10.1016/j.gloenvcha.2011.01.008.

Upham, P., Oltra, C., & Boso, À. (2015). Towards a cross-paradigmatic framework of the social acceptance of energy systems. *Energy Research & Social Science*, 8, 100–112. https://doi.org/10.1016/j.erss.2015.05.003.

US Code. (2022). *26 USC 45Q: Credit for carbon oxide sequestration*. Retrieved 7 September 2022, from https://uscode.house.gov/view.xhtml?req=(title:26%20section:45Q%20edition:prelim).

Viebahn, P., & Chappin, E. (2018). Scrutinising the gap between the expected and actual deployment of carbon capture and storage—A bibliometric analysis. *Energies*, 11(9), 2319. https://doi.org/10.3390/en11092319.

Warren, D. C., Carley, S. R., Krause, R. M., Rupp, J. A., & Graham, J. D. (2014). Predictors of attitudes toward carbon capture and storage using data on world views and CCS-specific attitudes. *Science and Public Policy*, 41(6), 821–834. https://doi.org/10.1093/scipol/scu016.

PART III

Genomics

CHAPTER 8

Public Inclusion and Responsiveness in Governance of Genetically Engineered Animals

Jennifer Kuzma and Teshanee T. Williams

Introduction

In 2016, Canada approved the first genetically engineered (GE) animal for sale on the food market, the AquAdvantage Salmon (AAS). AAS is genetically engineered to grow to adulthood in half the time in comparison to the wild-type Atlantic salmon. Developers inserted genes for the growth hormone gene from Chinook salmon and a promoter gene from the ocean pout to achieve this more rapid growth. The Canadian AAS case is the first time that a GE animal has been approved for human consumption and sold in the marketplace.

GE animals in the food supply are even more controversial than GE plants, and the AAS continues to face significant opposition from

J. Kuzma (✉)
North Carolina State University, Raleigh, NC, USA
e-mail: jkuzma@ncsu.edu

T. T. Williams
University of North Carolina at Chapel Hill, Chapel Hill, NC, USA

M. Gattinger (ed.), *Democratizing Risk Governance*,
https://doi.org/10.1007/978-3-031-24271-7_8

consumers, nongovernmental organizations (NGOs), and retailers (Halliday 2020). Surveys suggest that the public is more significantly concerned about GE animals for food than GE crops (Frewer et al. 2014; Cuite et al. 2005; Hoban 1998; Halliday 2020; Martin-Collado et al. 2022). GE animals are viewed as the most negative of various food technologies—even more so than pesticides and hormones (Henson et al. 2008). Deeply held attitudes, values, and beliefs often underlie this negativity. For example, measures of "disgust sensitivity" are strongly correlated with resistance to GE animals (Scott et al. 2016), and genetically engineering animals provoke diverse ethical concerns outside of scientific safety (Thompson 1997). Nep and O'Doherty (2013) also found in focus groups that there is significant distrust among Canadian consumers with regard to governance of GE salmon by companies and government.

GE foods are not labeled in Canada and there is a lack of public awareness that AAS has entered the market. The lack of transparency and labeling of GE foods derived from animals may fuel consumer distrust (Nep & O'Doherty 2013). In the words of one biotechnology expert, "There's a lack of transparency across the board in the industry...ordinary consumers don't really understand what genetic engineering is all about...animals make for a whole other layer of complexity...This biotechnology in food has arrived without any sort of social consent provided by consumers" (S. Charlebois, quoted in Halliday 2020). An NGO representative summarizes the shortcomings in public participation in decision-making about AAS as "There's no consultation with the public...no consultation with fisherfolk or farmers. They don't look at the questions, 'Do we need or want this technology?' The regulatory system looks only at the question of safety and excludes those questions" (L. Sharratt, as quoted in Halliday 2020).

In this chapter, we examine the regulatory decision-making for AAS up until its market introduction in 2017. We seek to understand whether, where, when, and how there are opportunities for public participation and values-discussions within the Canadian governance system for GE animals. Where there are opportunities to articulate values in the Canadian governance system, we examine how decision-makers respond to and incorporate broader concerns about AAS. The AAS case serves as a current example for governance of GE food animals and an instructive case for future governance of GE and gene-edited animals and their food products.

We frame our evaluation on two principles of responsible innovation (RI)—*inclusion and responsiveness*—the public-facing endeavors of RRI. First, we look at the regulatory approval process for AAS to examine when there were opportunities for public and stakeholder participation in decision-making (*inclusion*). Second, we report on findings from our study which utilized textual analysis of one public participation window—a series of Parliamentary hearings associated with GE animal oversight in Canada in 2016. Here, we examine whether decision-makers incorporated the diverse stakeholder perspectives and concerns voiced at the hearings into their final reports (*responsiveness*). Finally, we identify barriers to putting inclusion and responsiveness into practice in risk governance of GE organisms and discuss ways to overcome these barriers to facilitate responsible innovation practices in oversight systems for emerging technologies.

Responsible Research and Innovation

The framework of responsible research and innovation (RRI) may provide a way forward for biotechnology developers to act on their desires for greater public trust and legitimacy (Kuzma 2018) and to address the public concerns about lack of consultation in GE approval processes and the AAS case more specifically (Halliday 2020). RRI has been proposed in the last decade to better align science and technological development with democratic engagement processes, public values, and societal goals (e.g., Gardezi et al. 2022; Owen et al. 2012, 2013; Stilgoe et al. 2013). RRI arose out of a longer history of work on the ethical, legal, and social implications/aspects of scientific research and technology development (Felt 2018).

Although RRI as a framework is continually evolving, its most-cited conception centers around four principles: anticipation, inclusion, reflexivity, and responsiveness (Stilgoe et al. 2013). *Reflexivity* moves governance of science and technology away from solely a risk-based approach to one that encompasses reflection on the underlying goals, motivations, limits of knowledge, assumptions, and alternative framings of problems. *Anticipation* incorporates a forward-looking dimension where potential consequences are explored and analyzed before technologies are fully developed in order to anticipate downstream potential risks and impacts. *Inclusion* involves citizens and publics, in addition to stakeholders, in governance of research and innovation, opening up processes of reflexivity and anticipation to voices beyond those of subject-matter experts. Finally, ***responsiveness*** requires a capacity to change shape or direction of innovation in response to stakeholder and public values (discovered by anticipation, inclusion, and reflexivity), as well as new or changing information or circumstances. The RRI framework based on these 4 principles is "deemed to be characteristic of a more responsible vision of innovation" than other frameworks centering on research ethics, diversity, and inclusion in STEM fields, and interdisciplinarity (Wittrock et al. 2021, p. xi) and has been "operationalized by national funding bodies" and "integrated in research practice" in the EU (Wittrock et al. 2021, p. xi).

We evaluate the oversight process for AAS according to two of these four principles—inclusion and responsivity. We choose these two as they are more public-facing endeavors of RRI, putting public engagement and the incorporation of societal values into the process of biotechnology innovation. Below we consider whether the government approval processes for AAS in Canada provided opportunities for meaningful, bi-directional public engagement and input (e.g., as those suggested in NASEM 2016 for gene drives).

Inclusion in Risk Governance for GE Animals

In late 2013, AAS was approved for commercial production in Canada, and in 2016, AAS was also approved for human consumption in Canada (Fig. 8.1). In 2017, AquaBounty technologies announced that it had already sold 4.5 tons of AquAdvantage Salmon (AAS) (Waltz 2017). Current labeling laws in Canada are based on voluntary labeling standards, so much of the Canadian public was and remains unaware that salmon on the market could be genetically engineered (Halliday 2020). For the initial production of AAS, eggs were fertilized in a facility on Prince Edward Island (PEI) and then shipped to the Panamanian highlands for "grow-out" where the GE salmon were grown to full size in a land-locked location using recirculating aquaculture tanks. Once AAS were grown to full size in Panama, they were transported to food distributors in Canada for sale in food markets. The initial parameters for Canadian approval of AAS were specific to being grown out in Panama, but since, AAS has been approved for grow-out in both the U.S. and Canada (AquaBounty 2019; Callegari & Mikhailova 2021). Below we

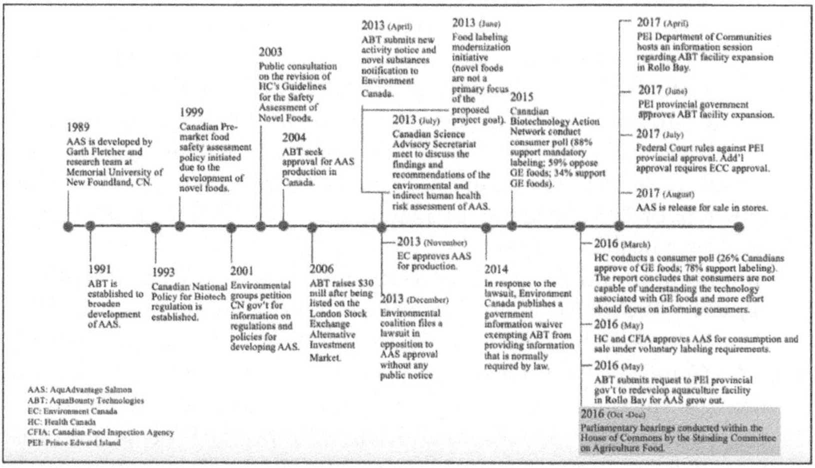

Fig. 8.1 Timeline of GE Salmon in Canada

describe the regulatory approval process from 2013 to 2017 for AAS in Canada to investigate places where the public was informed or engaged.

AAS Regulatory Approval: Living Organisms

The Canadian Environmental Protection Act 1999 (CEPA 1999), administered by Environment Canada (EC) and Health Canada (HC), is the key authority for novel organisms like the AAS. The regulatory process for novel organisms focuses extensively on the assessment of risks to human health and the environment, as the Government of Canada ensures that all new substances, including organisms, are assessed for their potential harm to the environment and human health. The New Substances Notification Regulations (Organisms) [NSNR (Organisms)] under CEPA 1999 prescribe the information that must be provided to Environment Canada (EC) prior to the import to or manufacture in Canada of new organisms that are living products of biotechnology, including fish like the AAS (Government of Canada 2005, 2010).

The Department of Fisheries and Oceans Canada (DFO), EC and HC signed a Memorandum of Understanding to implement the NSNR (Organisms) for fish (Department of Fisheries and Ocean 2013). DFO assists by conducting an environmental and indirect human health risk assessment for GE fish like AAS and recommending any necessary measures to manage risks. The risk assessments evaluate whether the notified fish product of biotechnology is "CEPA toxic": a substance is toxic if it may enter the environment and (a) have or may have an immediate or long-term harmful effect on the environment or its biological diversity; (b) constitute or may constitute a danger to the environment on which life depends; or (c) constitute or may constitute a danger in Canada to human life or health.

A notification under the NSNR (Organisms) was submitted to EC by AquaBounty Technologies in April 2013. DFO conducted an environmental and indirect human health risk assessment to make recommendations to EC and the Minister of the Environment about any necessary risk management measures for the AAS. A review of the DFO risk assessment was conducted under the National Science Response Process, Canadian Science Advisory Secretariat (CSAS) in July 2013. The purpose of this meeting was to peer-review the conclusions presented in DFO's preliminary comprehensive Environmental and Indirect Human Health Risk Assessment for AAS. However, only 3 of 23 participants were from outside of the government sector. Sixteen out of 23 participants were from DFO itself, 3 from other federal government. Agencies (HC and EC), 1 from the PEI Provincial government, and only 3 from outside government (1 consultant, 1 academic, and 1 from the Atlantic Salmon Foundation) (Department of Fisheries and Oceans 2013).

Both the risk assessment process under DFO and the meeting to review the risk assessment were generally closed to the public. No public comment period was conducted, and decision-making was conducted between the federal agencies and the developers of AAS. The public was not directly solicited for input on the decision-making process for the approval of AAS production under the NSNR (Organisms) process. Up until this point, there would be little if any information available to the public on the approval of AAS in Canada.

The federal Ministers of the Environment and Health ultimately approved the commercial production of AAS eggs in a notice published in the Canada Gazette on November 23, 2013 (Goldenberg 2013). The decision allowed AquaBounty to proceed with the production of the GE salmon eggs in PEI, Canada for shipping to Panama for grow-out and processing. Once the approval under CEPA and the NSNR process was made, however, Ecology Action Centre (EAC) and Living Oceans (LOS), took the federal government to court in 2014 over substantive portions of the review and legal requirements. This opposition to AAS approval included the argument that the review did not include an assessment of "whether the genetically engineered salmon could become invasive, potentially putting ecosystems and species such as wild salmon at risk" (Wristen 2014). These NGOs also objected to the permits EC granted for "unassessed uses" of AAS at the time such as its grow-out in Canada (Wristen 2014). One NGO leader involved in the suit also bemoaned the lack of transparency and public consultation in the decision-making process stating that "this is the world's first genetically modified food animal to go into production...this was done without any public debate at all and under circumstances that look like a deliberate attempt to prevent public comment. Canadians have a right to know about decisions like this in advance of them being made" (Wristen 2014). In the end, the court ruled in the favor of the federal government's approval, but at the time, it also restricted AquaBounty egg production to a single facility in PEI and did not grant permission to grow out the GE salmon at other locations (note: grow-out was ultimately approved in Rollo Bay, PEI in 2019—see Evans 2019). The court also ruled that the federal government's current practices of issuing waivers for grow-out without public notification could not be continued.

In spring 2018, we interviewed several decision-makers in Canada involved in the assessment and approval of AAS who confirmed the lack of public consultation on the approval.

In contrast to the Canadian approval, the U.S. has requirements under the Administrative Procedures Act (APA) for notice and public comment in rulemaking and these were invoked for the AAS approval under the U.S. Food and Drug Administration. The draft environmental assessment was available for public comment in the U.S. prior to the approval decision being made. In addition, the U.S. FDA convened its Veterinary Medicine Advisory Committee, an external advisory committee (no government employees) to review the assessment of AAS and deliberate in an open public meeting. Although the U.S. process for AAS was far from the ideal forms of public engagement discussed in the scholarly literature and suggested by the RRI principle of inclusion, there were multiple windows of opportunity to obtain and incorporate public feedback for the AAS decision (Stilgoe et al. 2013; Callegari & Mikhailova 2021). In contrast, in Canada, the public was not solicited for input on the decision-making process for the approval of AAS production under the NSNR (Organisms) process. The process lacked public transparency and no public information was available until the 2013 final AAS approval was published in the Canada Gazette and NGOs announced they were taking the federal government to court for the approval.

AAS Regulatory Approval: Food and Feed

For food and feed approval of AAS, two additional separate assessments were conducted. In addition to the DFO assessment for EC's NSNR regulation, Health Canada assessed the safety and nutrition of AquAdvantage Salmon for use as food (Health Canada 2016a) and the Canadian Food Inspection Agency (CFIA) assessed the safety and nutrition of AquAdvantage Salmon for use as a livestock feed (CFIA 2016). In Canada, genetically engineered foods like AAS fall under the category of "novel foods." According to the Food and Drugs Act (Health Canada 1985), novel foods are described as food that have been produced through new processes, that do not have a history of safe use as a food, or that have been modified by genetic manipulation (Health Canada 2006). The guidelines for the safety assessment of novel foods stipulate that approval requests for the production and sale of novel foods be made to the Health Products and Food Branch (Health Canada 2006). These foods are subjected to a food safety and nutrition assessment that is based on substantial equivalence. Substantial equivalence is the argument that the novel food product is equivalent to conventional foods in terms of safety and nutritional standards (Health Canada 2006). Health Canada's assessment of AAS was conducted based on the Guideline for the Conduct of Food Safety Assessment of Foods Derived from Recombinant-DNA Animals (UN FAO 2008). CFIA assessed the safety and nutrition of AquAdvantage Salmon for use as a livestock feed to establish the safety of feed ingredients derived from AAS.

Both HC and the CFIA published the summaries of their AAS assessments on May 19, 2016, in which it was determined AAS were as safe and nutritious as other farmed Atlantic salmon and that there were no feed safety concerns (CFIA 2016; Health Canada 2016a) (Fig. 8.1). Once again, however, there were not opportunities for public comment or open meetings through which to solicit public feedback on AAS approval for the Canadian food market. It is notable that the complete risk assessments for the CEPA NSNR process, the HC novel foods and CFIA novel feeds were not published online for public viewing, but rather just summaries are posted on the HC and EC websites. In contrast, in the U.S., one can obtain the risk assessment documents for biotech products online in most cases, and thus see how the data is interpreted (e.g., Meghani & Kuzma 2018). In the U.S., there are also mandates for public comment periods on regulatory decisions, sometimes involving the approval of individual GE food animals, in addition to broader policies on how to regulate. This is a key weakness in transparency for Canada—external academics and stakeholders with subject-matter knowledge, and other interested publics, cannot judge the data, its interpretation, and risk conclusions from just a summary. One could argue that this lack of *external* peer review and scrutiny is harmful to both public legitimacy and the scientific enterprise. Without external eyes and drawing upon a wide expertise of various publics and stakeholders, scholars have argued that the analyses will suffer from a deficit of important standpoints (Meghani & Kuzma 2011)—quality may suffer, and public legitimacy of decisions may decrease. Ultimately, the lack of inclusive processes, as suggested by the RRI framework, may jeopardize the future of GE animal-based foods.

Barriers to Inclusion for AAS Approval

In informal interviews with biotech stakeholders,[1] we found that at the time of the AAS approval, regulators struggled with broader goals to engage the public in decision-making and increase transparency for biotechnology product decisions. Agency staff recognized that GE animal products are controversial and that there will soon be an explosion of them as gene-editing and CRISPR make genetic modification on animals easier to perform. They agreed that greater public inclusion should be a goal. However, agency managers also noted significant barriers to increasing public transparency and participation in GE animal approvals. One is the need to protect confidential business information (CBI) in biotech product submissions. At the time of the AAS approval, NSNR (Organisms) notifications and assessments for GE animals did not have to be posted prior to approval (unlike for new chemical substances under the same law). The rationale in designing the closed process for NSNR (Organisms) was that there would need to be a higher degree of intellectual property (IP) protection given the novelty of genetically engineered

[1] One author spent 4 months in Ottawa Canada in Spring 2018 meeting and speaking with regulators, innovators, trade organization representatives, government leaders, NGOs, and other stakeholders associated with GE animal policies.

animals (compared to chemicals). Protecting CBI was deemed as important, although it came at the expense of public transparency during regulatory review. Balancing the protection of IP and CBI with openness was considered a significant challenge for GE animals and AAS oversight.

The second challenge Canadian regulators noted is that there is little capacity and experience among the regulatory staff with the public comment and rulemaking process. Regulators expressed concern about comments falling outside of their jurisdiction and worried about the possible high volume and low relevance of public comments. The agency staff felt constrained in not being able to consider "non-scientific" concerns and expected most comments to contain information not relevant to the technical scope of their decision-making which centers around plausible health and environmental risks. Broader public concerns about social, cultural, or economic harms or social values about "naturalness" and sustainability are likely to be expressed in public comment or input processes, yet lie outside the authorities granted by the regulations public administrators are tasked with administrating.

Shortly after the AAS approval, the agencies were being asked by Parliament to implement public participation processes and operate under greater transparency. For example, the Standing Committee on Environment and Sustainable Development (ENVI) in the Parliament's House of Commons published a report in June 2017 with two specific recommendations for NSNR (Organisms) (House of Commons 2017): (1) Recommendation 26: "The Committee recommends that CEPA be amended to establish a more open, inclusive and transparent risk assessment process that better enables public participation in the evaluation of new living modified organisms" and (2) Recommendation 28: "The Committee recommends that CEPA be amended to ensure that provisions that set out a requirement for consultation with the provinces and territories also require consultation with Indigenous peoples." While waiting for CEPA to be amended by Parliament (which it ultimately was in April 2021), regulators expressed plans to design a voluntary process for developers to notify the public of NSNR (Organisms) submissions for GE animals. In this process, developers would have the option to publish a non-confidential summary of their product on the EC website when they submit a GE animal or microbe to EC NSNR division. EC NSNR would then solicit public comments on what types of environmental and health risks are of concern to people for that GE animal or microbial product.

After the AAS approval, HC placed increasing emphasis on openness and transparency for novel foods like those derived from AAS. These are reflected in Health Canada's Regulatory Transparency and Openness Framework (Health Canada 2019). They have adopted digital methods (i.e., online consultation, online comment submissions) to foster public participation. This has provided novel approaches to allow Canadians to provide public input and interact with the Government of Canada. Health Canada and the Public Health Agency of Canada also published Guidelines on Public Engagement to strengthen their public engagement efforts and capacity by providing staff with guidance on sound principles, practical tools and templates, best practices, and the use of innovative technologies for engagement (Health Canada 2016b).

AAS Post-approval Inclusion

Despite a lack of public participation during the federal regulatory process for AAS, there was a significant point *after* the regulatory approval in which public and stakeholder input was sought via Parliamentary hearings focused on Canadian policy toward GE animals (Fig. 8.1). In the next section, we use this window of participation to explore public concerns and values associated with risk governance and oversight for AAS and GE animals. As an indicator of RRI's principle of responsiveness, we also examine whether decision-makers incorporated the diverse perspectives expressed in the hearings in their final reports.

Responsiveness in Risk Governance for GE Animals

Soon after CFIA and HC made their decisions on feed and novel-food approval for AAS, Parliament's House of Commons Committee on Agriculture and Agri-Food was asked by the Minister of Agriculture and Agri-Food to examine the legal and regulatory framework around GE animals more generally and their increasing availability for human consumption. On 1 June 2016, the Committee agreed to investigate GM animals for food and the issues around regulatory approval in Canada beyond health and safety, as well as steps to provide the public with information about the market entry of GM animal-based products (House of Commons 2016). The Committee was to release its report to the House by December 8, 2016 (ibid.).

The Committee held four public hearings in fall 2016, soliciting input from representatives of the agriculture and agri-food sector, regulatory authorities, and civil society about the issues raised by the arrival of GE animals for human consumption. Stakeholders were invited to present their viewpoints at these committee hearings. In December 2016, Parliament's summary report was released and became available online along with full transcripts of the meetings (House of Commons 2016). In April 2017, the government agencies involved in AAS approvals and GE animal regulation more broadly responded to the report.

These reports and transcripts of the meetings provided a window of participation into regulatory policy-setting for GE animals in Canada. Although it occurred after the formal regulatory decision to approve AAS, given the lack of other venues for public comment, we used it to examine narrative frames and stakeholder attitudes toward governance policy for GE animals and how those concerns were considered in the final reports written by either Parliament or by the federal agencies in response to the hearings. Here, we use the reflection of public and stakeholder narratives and concerns in the Parliament and agency reports as evidence for "policy uptake" or "responsiveness" to public and stakeholder concerns. Below, we briefly summarize our findings, and more details on the methodologies and results can be found in Williams (2019).

Stakeholder Comments and Cultural Worldviews

Our textual analysis of the Parliamentary hearings involved categorizing each comment by stakeholder group and cultural worldview (Williams 2019). Cultural worldviews were assessed using cultural theory (Douglas & Wildavsky 1982) which has a long history of explaining how people perceive risks from emerging technologies (Finucane & Holup 2005; Jones & Song 2014; Kahan et al. 2011). The four cultural types identified by cultural theory are egalitarian, hierarchical, individualistic, and fatalistic (Douglas & Wildavsky 1982; Kahan et al. 2011; Jones & Song 2014; Thompson et al. 2018). Previous research has used these four cultural types to examine the way that issues concerning risk are framed as narratives (i.e., stories) depending on the cultural worldview (Jones & Song 2014). Figure 8.2 describes how we translated this prior work and applied it to identify cultural narratives in the Parliamentary hearings for the GE salmon and GE animals (for more details, see Williams 2019). Table 8.1 shows the results of the comments displayed during

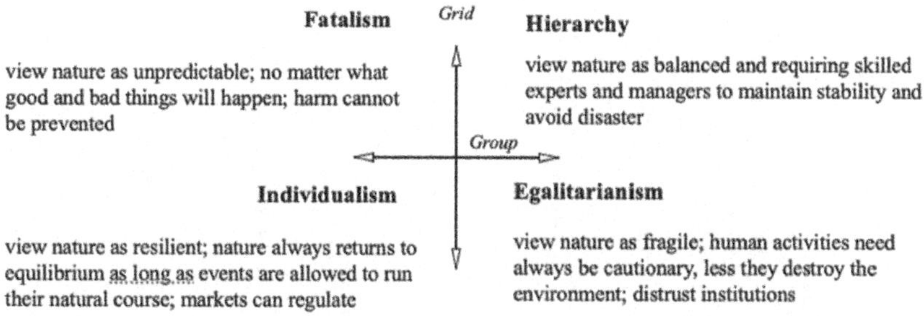

Fig. 8.2 Four cultural archetypes and narratives about technology, risk, and governance

the Parliamentary hearings by stakeholder affiliation and these cultural worldviews. Notably, those with greater hierarchical worldviews tended to be in the federal agency positions, positions of authority and decision-making for AAS, while those with egalitarian views were in the advocacy sector. Members of Parliament, industry, and trade groups displayed a more even breakdown of hierarchical (H), individualistic (I), and egalitarian (E) worldviews. Fatalistic worldviews were not identified in the hearings.

Comments by "Science-Based" or "Science-Plus"

In addition to tracking the cultural worldviews and stakeholder groups, we probed whether the points and issues raised in the hearings are related to strictly "science-based" concerns (such as direct toxicity or allergenicity of the product) or "science plus" which includes broader concerns about potential impacts or issues (including social, economic, cultural, political, or indirect ecosystem impacts). We found that egalitarian and individualistic commentators were much more likely to expand the scope of the issues beyond "science-only" to "science plus"; however, hierarchs were more likely to contract the scope of GE animals to "science only" (Table 8.2) (see also Williams 2019; Williams & Kuzma 2022).[2]

We also found differences in the cultural groups as to *how* "science-only" or "science-plus" arguments were used (Williams 2019; Williams &

[2] These relationships were statistically significant at $p < 0.05$ as reported in Williams (2019) and Williams and Kuzma (2022).

Table 8.1 Parliamentary hearing comments according to stakeholder and cultural groups

	Stakeholder Group	*Cultural Group* Hierarchical n = *121(45%)*		Individualist n = *65 (24%)*		Egalitarian n = *83 (31%)*		*Total* n = *269 (100%)*
Liberal, Conservative, NDP	Member of Parliament	53	(45%)	25	(22%)	39	(33%)	117 (100%)
Canadian Cattlemen's association, BioteCanada	Alliance/Coalition	29	(37%)	30	(39%)	19	(24%)	78 (100%)
AquaBounty Technologies, Canadian Aquaculture Industry Alliance	Fishing industry	5	(31%)	3	(19%)	8	(50%)	16 (100%)
Department of Agriculture and Agri-Food, Canadian Food Inspection Agency, Department of Health	Agency official	30	(83%)	6	(17%)	0	(0%)	36 (100%)
Canadian Biotechnology Action Network, CropLife Canada, Ecology Action Centre, Vigilance OGM	Advocacy group	2	(11%)	0	(0%)	17	(89%)	19 (100%)

	Stakeholder Group	*Cultural Group* Hierarchical n = *121(45%)*		Individualist n = *65 (24%)*		Egalitarian n = *83 (31%)*		*Total* n = *269 (100%)*
University professor/Board member of AquaBounty	Researcher/Scientist	2	(67%)	1	(33%)	0	(0%)	3 (100%)

Reading across the rows, for that stakeholder group, the number and % of comments falling into each cultural group is shown. Across all stakeholder groups (*n* = 269), cultural group representation in the comments is Hierarchical 45% (*n* = 121), Individualist 24% (*n*= 65), and Egalitarian 31% (*n* = 83)

Kuzma 2022). Egalitarian comments tended to be against GE animal approval ($n = 18$ comments for approval and $n = 59$ comments against approval) and used "science plus" arguments more frequently than "science-only" arguments both to support their positions and to refute opposing positions (Table 8.2). In contrast to egalitarians, individualists were generally in favor of GE animal approval ($n = 60$ comments for and $n = 3$ comments against approval). However, like egalitarians, individualists also used "science plus" more frequently than "science-only" both to support their arguments and to refute opposing arguments (Table 8.2). In contrast, hierarchical comments, which were mostly in favor of GE animal approval ($n = 109$ comments for and $n = 9$ comments against), shifted the use of "science plus" arguments or "science-only" arguments depending on whether they were arguing for their own position in favor of GE animals or refuting other's arguments against GE animals. As shown in the highlighted cells in Table 8.2, hierarchs with a pro-GE position tended to use "science-only" to refute arguments against the release of GE salmon ($n = 16$ "science plus" vs $n = 48$ "science-only") but expanded to "science plus" when arguing in favor of GE animal release ($n = 32$ "science plus" vs $n = 13$ "science-only").

The last finding was interesting to us in the context of prior literature that describes the marginalization of perspectives that are not "science-based" in the biotechnology debates (e.g., see Thompson 2007; Meghani & Kuzma 2011). The predisposition of those in positions of regulatory authority (largely hierarchs—see Table 8.1) is to go beyond the science when arguing *for* the approval of GE animals by making appeals to the economy, markets, or sustainability, but to refute arguments of those who oppose GE animals by forcing them to stick to the scientific risks (i.e., objections to GE animals must be "science-based"). In other words, hierarchs, and those in favor of GE animals seem to reject arguments against GE animal approval that appeal to the economy, cultural, or sustainability, yet use these same appeals to support their position for GE animals.

Policy Uptake of Stakeholder Concerns by Science/Science-Plus and Cultural Worldview

To get at responsiveness to diverse public concerns, we next analyzed whether the Parliamentary hearing final report and the federal agency response reports reflected the arguments we saw in the hearing transcripts.

Table 8.2 Use of science-plus and science-only arguments by cultural type in Parliamentary hearings

		To support their own position		*To refute opponent's position*		
Cultural type	*Scope*	Comment with: Pro-GE stance	Anti-GE stance	Comment with: Pro-GE stance	Anti-GE stance	Total
Egalitarian $n = 77$	*Science PLUS*	13	45	3	14	75 (97%)
	Science ONLY	1	0	1	0	2 (3%)
Individualist $n = 63$	*Science PLUS*	31	1	14	0	46 (73%)
	Science ONLY	4	1	11	1	17 (27%)
Hierarchical $n = 118$	*Science PLUS*	32	5	16	4	57 (48%)
	Science ONLY	13	0	48	0	61 (52%)

Source Authors. Note: the hierarchical row shows the shift that hierarchs, who are largely in favor of GE animals, make from "science-plus" in arguing for their position in comparison with "science-only" to arguing against an opponent's position, as discussed in the text

We examined the two reports for the cultural worldviews expressed in the reports and the appeals to "science-only" or "science-plus" concerns to compare them to these appeals in the hearing transcripts. Figure 8.3 shows that although a significant number of comments in the hearings went beyond science and were "science-plus" (69%), both the Parliament report (49%) and to a greater extent the federal agency report (13%) reduced these appeals, focusing more on "science-only" issues. We interpret this result as an indicator of low responsivity to concerns outside of direct scientific risk in the public policy process for GE animals.

Likewise, we tracked and compared the cultural types of arguments among the hearings, Parliament report, and federal agency response report. Figure 8.4 shows that the uptake of cultural worldviews in Parliament's report after the hearings largely reflected the comments in the hearings. However, the federal agency report tended to overemphasize hierarchical worldviews at the expense of individualistic worldviews. As expected, the results in Table 8.1 indicate that hierarchical worldviews

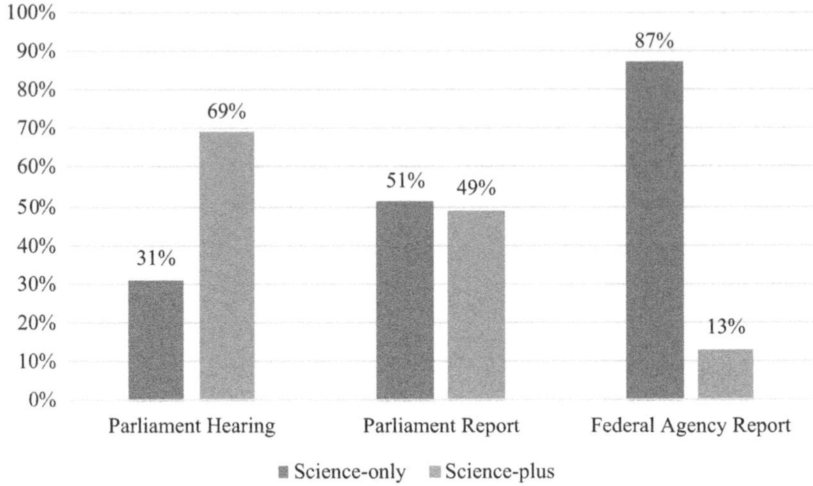

Fig. 8.3 Responsiveness to science-only versus science-plus concerns—uptake in final reports

were more dominant in federal agency representative comments. Egalitarian worldviews shared a similar percentage across the hearings and both reports. We interpret this result as a better indication of responsivity to diverse cultural perspectives in comparison to the minimal uptake of science-plus concerns in the final reports.

Although there are limitations to textual analysis, we present it here as a potential novel way to look at responsivity or policy-uptake from participatory events. Future research could build upon and validate such approaches.

Barriers and Opportunities for Inclusion and Responsiveness

As previously mentioned, in our conversations with decision-makers for GE animals in Canada, they highlighted two key barriers to public inclusion in the regulation of GE animals: protecting intellectual property and lack of capacity for dealing with public comments. Previous work in the U.S. on stakeholder attitudes to RRI in biotech innovation systems

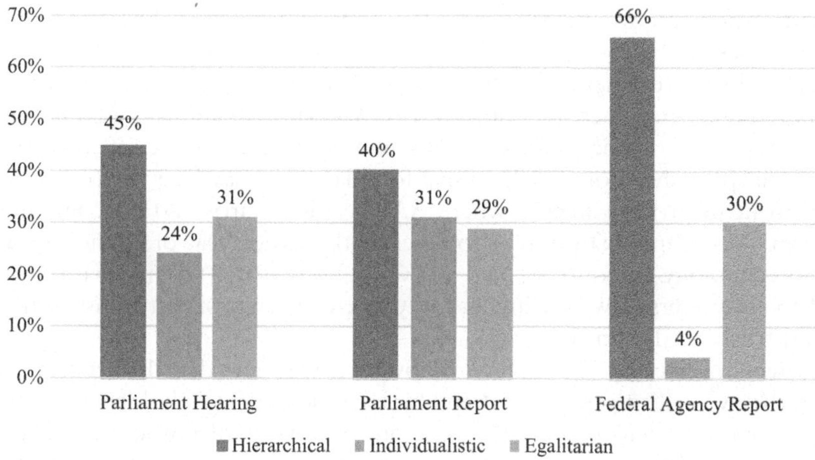

Fig. 8.4 Responsiveness to cultural worldviews—uptake in final reports

also uncovered barriers to RRI, especially for the principles of inclusion and responsiveness. Roberts et al. (2020) analyzed the attitudes of different biotechnology stakeholders toward principles and practices of RRI with a mixed-method approach. Homogenous focus groups (by stakeholder affiliation) and pre- and post-focus group surveys were used to measure sector attitudes toward RRI. Significant differences were found in stakeholder reactions to practices to implement RRI. In comparison to government and consumer groups, industry, trade organizations, and academics had more negative reactions to RRI practices that relinquish control to people outside of technology development, namely practices of inclusion and responsiveness (Roberts et al. 2020). Qualitative analysis of focus-group conversations revealed barriers to RRI associated with institutional goals and cultures. Regarding inclusion and responsiveness, innovators were cynical about including external voices in innovation pathways due to the inflexibility of funding programs which leads to constraints on their work, and they were also concerned about these RRI practices causing potential delays to innovation given the highly competitive national and international environments for financing and patents (Roberts et al. 2020).

Our conversations with Canadian regulators and innovators from industry also expressed these fears and barriers to greater public inclusion and responsiveness and fell into a few general categories. In addition to the issues of IP protection and capacity previously discussed, government and industry representatives mentioned that greater public inclusion and transparency may increase public fear of AAS and GE animal-based foods. In the words of one industry representative, greater public engagement or transparency through GE animal food labeling may lead to "picketing in front of grocery stores" where AAS is sold. Marris (2015) coins the term "biotechphobia-phobia" to describe this expert fear of public fear of biotechnology. In contrast, Nep and O'Doherty (2013) found that Canadian consumers view labeling as a way to enhance consumer transparency and trust in GE salmon.

Second, like the US biotech innovators interviewed in Roberts et al. (2020), Canadian government and biotech industry stakeholders for AAS were motivated to protect the pace and standards of innovation. Their views were optimistic about the contribution of genetic engineering to society (techno-optimistic), and they expressed concern about the delays to innovation that would likely result from greater public inclusion and responsiveness. Finally, they pointed out that there would be threats to science-based decision-making if broader socioeconomic and cultural perspectives were incorporated (as public responsiveness a la Stilgoe et al. 2013 would likely require).

In prior work, Callegari and Mikhailova (2021) also explore RRI as a framework in investigating AAS governance, but in the U.S. They found that companies adopt practices "entirely opposite to those being advocated within the RRI framework" and focus on "exclusive communication with the scientific community and legal authorities" (p. 1). They conclude that these practices are "undesirable from the perspective of both the organizations involved and society at large" (p. 1). Both they and Roberts et al. (2020) recognize that fulfilling RRI ideals from the scholarly literature (e.g., Stilgoe et al. 2013) may present too many practical barriers to innovators and regulators, and they suggest compromise in taking smaller steps toward public and stakeholder inclusion and responsiveness. Roberts et al. (2020) suggest the co-design of RRI pathways that include biotech innovators and other stakeholders to consider the very practical limitations that innovators face (e.g., with respect to CBI, IP, and competitiveness of innovation and funding systems). They also propose that institutional incentives for incorporating RRI practices be developed to encourage government regulators and innovators to adopt greater inclusion and responsiveness. Callegari and Mikhailova (2021) suggest that as a step toward RRI, stakeholder engagement should be "strategic and selective" (p. 14), and at first, limited to stakeholders that are willing to compromise to accommodate the goals of biotech innovators. Both Roberts et al. (2020) and Callegari and Mikhailova (2021) recognize that these accommodations may not be considered true to the scholarly visions of RRI, however, if RRI is to advance at all beyond an academic set of ideal principles, incremental steps should be the near-term focus.

In the context of regulatory decision-making and GE animals, these incremental steps could include: (1) the formation of a supra-agency federal body that convenes stakeholders and publics in dialogue about GE animal foods, allowing for the consideration of concerns and benefits that go beyond individual and narrow federal regulatory jurisdictions to include socioeconomic issues, indirect risks, and cultural impacts (see for example NASEM 2017, p. 9); (2) federal incentives provided to innovators for incorporating RRI practices in upstream innovation of GE animals to encourage openness and transparency with the public and stakeholders; and (3) funding and capacity-building for government regulators in Canada to open up public comment periods on every federal regulatory decision for GE animals, engage in public hearings on a regular basis, and convene public and stakeholder workshops and focus groups. These steps will not achieve the pure vision of RRI articulated by scholars, but will go a long way toward building greater public legitimacy and trust, even if opposition to GE animal foods is likely to remain.

References

AquaBounty (2019). *Aquabounty Technologies, Inc. environment and climate change Canada approves Rollo Bay facility for commercial production of Aquadvantage Salmon*. Retrieved May 19, 2021, from https://www.globenewswire.com/news-release/2019/04/02/1794998/0/en/AquaBounty-Technologies-Inc-Environment-and-Climate-Change-Canada-Approves-Rollo-Bay-Facility-for-Commercial-Production-of-AquAdvantage-Salmon.html.

Callegari, B., & Mikhailova, O. (2021). RRI and corporate stakeholder engagement: The Aquavantage salmon case. *Sustainability*, 13(4), 1820.

Canadian Environmental Protection Act (1999). Retrieved May 19, 2021, from http://ec.gc.ca/lcpe-cepa/default.asp?lang=En&n=CC0DE5E2-1&toc=hide.

Canadian Food Inspection Agency (2016). *Decision document DD2016–117 determination of the safety of Aquabounty Technologies Inc's salmon (Salmo salar l.) event*. Retrieved May 19, 2021, from http://www.inspection.gc.ca/plants/plants-with-novel-traits/approved-under-review/decision-documents/dd2016-117/eng/1463076782568/1463076783145#a4.

Canadian Science Advisory Secretariat (2017). *Proceedings of the national science response process for the environmental and indirect human health risk assessment of the AquAdvantage Salmon*. Retrieved May 19, 2021, from http://publications.gc.ca/collections/collection_2017/mpo-dfo/Fs70-4-2017-007-eng.pdf.

Cuite, C. L., Aquino, H. L., & Hallman, W. K. (2005). An empirical investigation of the role of knowledge in public opinion about GM food. *International Journal of Biotechnology*, 7, 178–194.

Department of Fisheries and Oceans Canada (2013). *Summary of the environmental and indirect human health risk assessment of Aquadvantage Salmon. Canadian Science Advisory Secretariat science response 2013/023*. Retrieved May 19, 2021, from http://waves-vagues.dfo-mpo.gc.ca/Library/361091.pdf.

Douglas, M., & Wildavsky, A. (1982). How can we know the risks we face? Why risk selection is a social process. *Risk Analysis*, 2(2), 49–58.

Evans, O. (2019, April 2). Canada approves production of genetically modified salmon. *Salmon Business*. https://salmonbusiness.com/canada-approves-production-of-genetically-modified-salmon/#:~:text=Now%20Canada's%20federal%20authorities%20at,grow%2Dout%20of%20AquAdvantage%20Salmon.

Felt, U. (2018). Responsible research and innovation. In S. Gibbon, B. Prainsack, S. Hilgartner, & J. Lamoreaux (Eds.), *Handbook of genomics, health and society* (2nd ed., pp. 108–116). Routledge.

Finucane, M. L., & Holup, J. L. (2005). Psychosocial and cultural factors affecting the perceived risk of genetically modified food: An overview of the literature. *Social Science & Medicine*, 60(7), 1603–1612.

Frewer, L. J., Coles, L., Houdebine, D., & Kleter, A. G. (2014). Attitudes towards genetically modified animals in food production. *British Food Journal*, 116, 1291–1313.

Gardezi, M., Adereti, D. T., Stock, R., & Ogunyiola, A. (2022). In pursuit of responsible innovation for precision agriculture technologies. *Journal of Responsible Innovation*, 1–24.

Goldenberg, S. (2013, November 25). Canada approves production of GM salmon eggs on commercial scale. *The Guardian*. https://www.theguardian.com/environment/2013/nov/25/canada-genetically-modified-salmon-commercial.

Government of Canada (2005). *New substances notification regulations (organisms), SOR 2005–248 §*. Retrieved May 19, 2021, from https://laws-lois.justice.gc.ca/PDF/SOR-2005-248.pdf.

Government of Canada (2010). *Guidelines for the notification and testing of new substances: Organisms—Pursuant to the new substances notification regulations (organisms) of the Canadian Environmental Protection Act, 1999*. Retrieved May 19, 2021, from https://www.canada.ca/en/environment-climate-change/services/managing-pollution/evaluating-new-substances/biotechnology-living-organisms/guidelines.html.

Halliday, M. (2020, July 2). The world's first bioengineered salmon hits the market. *Pivot Magazine*. https://www.cpacanada.ca/en/news/pivot-magazine/2020-07-02-aquabounty-bioengineeredsalmon#:~:text=With%20a%20new%20farm%20in,history%20of%20commercial%20food%20production.

Health Canada (1985). *Food and Drugs Act, R.S.C., 1985, c. F-27 § (1985)*. Retrieved May 19, 2021, from http://laws-lois.justice.gc.ca/eng/acts/F-27/index.html.

Health Canada (2006). *Guidelines for the safety assessment of novel foods*. Retrieved May 19, 2021, from https://www.canada.ca/en/health-canada/services/food-nutrition/legislation-guidelines/guidance-documents/guidelines-safety-assessment-novel-foods-derived-plants-microorganisms/guidelines-safety-assessment-novel-foods-2006.html.

Health Canada (2016a). *Novel food information—AquAdvantage Salmon*. Retrieved May 19, 2021, from https://www.canada.ca/en/health-canada/services/food-nutrition/genetically-modified-foods-other-novel-foods/approved-products/novel-food-information-aquadvantage-salmon.html.

Health Canada (2016b). *Health Canada and the Public Health Agency of Canada—Guidelines on public engagement*. Retrieved May 19, 2021, from https://www.canada.ca/en/health-canada/services/publications/health-system-services/health-canada-public-health-agency-canada-guidelines-public-engagement.html?_ga=2.162702462.318742135.1513101816-1240815627.1513101816.

Health Canada (2019). *Regulatory transparency and openness*. Retrieved May 19, 2021, from https://www.canada.ca/en/health-canada/corporate/transparency/regulatory-transparency-and-openness.html.

Henson, S., Annou, M., Cranfield, J., & Ryks, J. (2008). Understanding consumer attitudes toward food technologies in Canada. *Risk Analysis*, 28, 1601–1617.

Hoban, T. J. (1998). Trends in consumer attitudes about agricultural biotechnology. *AgBioForum* 1, 1–7.

House of Commons (2016). *Genetically modified animals for human consumption*. Standing Committee on Agriculture and Agri-food. Number 022. 1st session. 42nd parliament. Retrieved May 19, 2021, from https://www.ourcommons.ca/Committees/en/AGRI/StudyActivity?studyActivityId=9004589.

House of Commons (2017). *Healthy environment, healthy Canadians, healthy economy: Strengthening the Canadian environmental protection act, 1999*.

Retrieved May 19, 2021, from http://publications.gc.ca/collections/collection_2017/parl/xc50-1/XC50-1-1-421-8-eng.pdf.

Jones, M. D., & Song, G. (2014). Making sense of climate change: How story frames shape cognition. *Political Psychology*, 35(4), 447–476.

Kahan, D. M., Jenkins-Smith, H., & Braman, D. (2011). Cultural cognition of scientific consensus. *Journal of Risk Research*, 14(2), 147–174.

Kuzma, J. (2018). Regulating gene edited crops. *Issues in Science and Technology*, 35(1), 80–85.

Marris, C. (2015). The construction of imaginaries of the public as a threat to synthetic biology. *Science as Culture*, 24, 83–98.

Martin-Collado, D., Byrne, T. J., Crowley, J. J., Kirk, T., Ripoll, G., & Whitelaw, C. B. A. (2022). Gene-edited meat: Disentangling consumers' attitudes and potential purchase behavior. *Frontiers in Nutrition*, 9, 856491.

Meghani, Z., & Kuzma, J. (2011). The "revolving door" between regulatory agencies and industry: A problem that requires reconceptualizing objectivity. *Journal of Agricultural and Environmental Ethics*, 24(6), 575–599.

Meghani, Z., & Kuzma, J. (2018). Regulating animals with gene drive systems: Lessons from the regulatory assessment of a genetically engineered mosquito. *Journal of Responsible Innovation*, 5(S1), 203–222.

NASEM (2016). National Academies of Science, Engineering and Medicine. *Gene drives on the horizon*. National Academy Press.

NASEM (2017). National Academies of Science, Engineering, and Medicine. *Preparing for future products of biotechnology*. National Academy Press.

Nep, S., & O'Doherty, K. (2013). Understanding public calls for labeling of genetically modified foods: Analysis of a public deliberation on genetically modified salmon. *Society & Natural Resources*, 26(5), 506–521.

Owen, R., Macnaghten, P., & Stilgoe, J. (2012). Responsible research and innovation: From science in society to science for society, with society. *Science and Public Policy*, 39(6), 751–760.

Owen, R., Stilgoe, J., Macnaghten, P., Gorman, M., Fisher, E., & Guston, D. (2013). A framework for responsible innovation. In R. Owen, J. Bessant, & M. Heintz (Eds.), *Responsible innovation: Managing the responsible emergence of science and innovation in society* (pp. 27–44). Wiley.

Roberts, P., Herkert, J., & Kuzma, J. (2020). Responsible innovation in biotechnology: Stakeholder attitudes and implications for research policy. *Elementa: Science of the Anthropocene*, 8(1), 47. https://doi.org/10.1525/elementa.446.

Scott, S. E., Inbar, Y., & Rozin, P. (2016). Evidence for absolute moral opposition to genetically modified food in the United States. *Perspectives on Psychological Science*, 11(3), 315–324.

Stilgoe, J., Owen, R., & Macnaghten, P. (2013). Developing a framework for responsible innovation. *Research Policy*, 42, 1568–1580.

Thompson, P. B. (1997). Science policy and moral purity: The case of animal biotechnology. *Agriculture and Human Values*, 14(1), 11–27.

Thompson, P. B. (2007). *Food biotechnology in ethical perspective*. Springer.

Thompson, M., Ellis, R., & Wildavsky, A. (2018). *Cultural theory*. Routledge.

UN FAO (2008). *United Nations Food and Agricultural Organization: Guideline for the conduct of food safety assessment of foods derived from recombinant-DNA animals (No. CAC/GL 68–2008)*. Retrieved May 19, 2021, from http://www.fao.org/fao-who-codexalimentarius/sh-proxy/en/?lnk=1&url=%3A%2F%2Fworkspace.fao.org%2Fsites%2Fcodex%2FStandards%2FCAC%20GL%2068-2008%2FCXG_068e.pdf.

Waltz, E. (2017). First genetically engineered salmon sold in Canada. *Nature News*, 548(7666), 148.

Williams, T. (2019). *Narrative policy framework: Examining policy narratives as tools of influence in the regulatory review process of bioengineered salmon in the United States and Canada*. Doctoral Thesis. NC State University (Under the direction of Dr. Jennifer Kuzma and Dr. RaJade Berry James). Retrieved May 19, 2021, from https://repository.lib.ncsu.edu/bitstream/handle/1840.20/36795/etd.pdf?sequence=1.

Williams, T., & Kuzma, J. (2022). Narrative policy framework at the macro level—Cultural theory-based beliefs, science-based narrative strategies, and their uptake in the Canadian policy process for genetically modified salmon. *Public Policy and Administration*. 09520767211065609.

Wittrock, C., Forsberg, E. M., Pols, A., Macnaghten, P., & Ludwig, D. (2021). Implementing responsible research and innovation: Organisational and national conditions. *Springer briefs in ethics* (eBook). https://doi.org/10.1007/978-3-030-54286-3.

Wristen, K. (2014). *Environmental groups take federal government to court for permitting manufacture of genetically modified salmon in Canada*. Retrieved May 19, 2021, from http://www.livingoceans.org/media/releases/environmental-groups-take-federal-government-court-permitting-manufacture-genetically.

CHAPTER 9

Decision-Making About Newborn Screening Panels in Canada: Risk Management and Public Participation

Marisa Beck, Brendan Frank, Sara Minaeian, and Stuart G. Nicholls

Newborn Bloodspot Screening: An Under-Studied Risk Issue

The World Health Organization defines screening as "the presumptive identification of unrecognized disease in an apparently healthy, asymptomatic population by means of tests, examinations or other procedures that can be applied rapidly and easily to the target population" (Wilson & Junger, 1968). Population-based screening programs exist for different stages of life, from prenatal screening of the developing fetus, through newborn screening, to screening of adults.

M. Beck (✉) · B. Frank
Ottawa, ON, Canada
e-mail: mpenslar@gmail.com

S. Minaeian
University of Oxford, Oxford, United Kingdom

M. Gattinger (ed.), *Democratizing Risk Governance*,
https://doi.org/10.1007/978-3-031-24271-7_9

Initiated in the 1960s, and with programs now existing in all continents of the world (Therrell et al., 2015), newborn bloodspot screening (NBS) is the largest and longest running example of a population screening program internationally (Nicholls et al., 2014). However, programs vary in size and scope across jurisdictions.

NBS detects rare diseases in asymptomatic neonates. The process begins with taking a small blood sample 24–72 hours after birth, usually through a heel prick or heel lance. The sample is then screened for a range of biomarker targets that indicate elevated risk for a number of conditions. Newborns who screen positive for a condition undergo further testing to either confirm or rule out a diagnosis. If a diagnosis is confirmed, patients receive treatment from specialized healthcare providers. In Canada, screening is offered to all children as standard of care (Nicholls et al., 2014), predicated on early identification and early intervention, to ameliorate or prevent disease symptoms (Canadian Agency for Drugs and Technologies in Health, 2011). While provided as standard of care, parents can opt out of the screening program, in which case their child would not be tested for any of the conditions.

Over the past two decades, new and disruptive technologies have made it possible to include an increasing number of targets in screening panels at relatively low cost. In particular, the advent of tandem mass spectrometry technologies marked a step change for NBS, allowing for simultaneous detection of biomarkers for multiple disorders at minimal incremental costs (Levy, 1998). Progress in whole genome screening technologies could similarly trigger a sudden and substantive expansion of screening panels (Botkin & Rothwell, 2016; Bailey et al., 2021; Watson et al., 2022). But absent such technological breakthroughs, technical capacity for screening is only one factor of many in decisions about whether to add or remove conditions from screening panels.

S. G. Nicholls
Clinical Epidemiology Program, Ottawa Hospital Research Institute, Ottawa, ON, Canada
e-mail: snicholls@ohri.ca

The Expansion of Newborn Screening: An Exercise in Risk Governance

The universality of screening and the growing number of targets have sparked a discussion regarding accepted principles that underpin decision-making. For example, while decisions regarding the addition of targets have focused on the benefits to the individual child, there is a debate in the literature regarding what constitutes a benefit (Cornel et al., 2020). In some instances, there may be better health outcomes for the child as a result of early treatment of diagnosed conditions, but families may benefit in a number of ways following the diagnosis of a rare condition as well, such as the psychological comfort of avoiding the "diagnostic odyssey" and better knowledge to inform future reproductive decision-making (Bailey et al., 2006; Buchbinder & Timmermans, 2011; Potter et al., 2009; Bombard & Miller, 2012). Nevertheless, the US body responsible for NBS stopped considering benefit to the family in the nomination and review process for adding conditions to the panel (Watson et al., 2022). Early diagnosis and treatment of rare diseases can also significantly reduce long-term costs to the healthcare system (Sims et al., 2007; Shih et al., 2021). As such, NBS may provide benefits to both the individual and the healthcare system and, ultimately, society as a whole.

From a broader health system perspective, decisions about adding conditions to NBS panels must weigh several additional factors that may present challenges. Decisions to expand the list of conditions for newborn screening are decisions about benefits, but also risks. We use the term risk broadly to refer to the consequences—whether intended or unintended—of an event or activity for something that people value, including health, property, nature, beliefs, social institutions, and cultural practices (Renn, 2008; Beck, 1992; Stern & Fineberg, 1996). Risk is determined by two essential parameters: (1) the likelihood or probability that a consequence occurs and (2) the severity of the consequence for human health, well-being, or the natural environment.

In the context of NBS, the associated risks may accrue at different levels to different stakeholders. For example, screening tests commonly require decisions around thresholds. Setting this threshold requires a fine balance; too low and there may be many 'false positive' results—children that test positive but do not have the condition—but too high and there may be too many 'false negative' results—children who have the condition but screen negative. Including a condition where testing is insufficiently accurate can create personal and social costs or risks. There are also ethical risks related to privacy and sample storage and short-term economic risks related to budget constraints. Indeed, any decision made within the context of a finite envelope of funds also involves the opportunity costs of alternate services that do not receive those funds (UK National Screening Committee, 2000; Ulph et al., 2017; Rogowski et al., 2014). Decisions about NBS panel expansions affect at least four distinct groups, two of which are less obvious: healthcare professionals who deliver the screening, babies and their families likely to benefit from any expansion, people who receive screening but who are unlikely to benefit from the expansion, and people who might lose access to healthcare resources that are now directed to newborn screening but could have been allocated elsewhere.

We thus contend that the decision-making process regarding the addition of a target to newborn screening panels is one of risk governance (Renn, 2008), where the goal is to reduce or prevent risks but do so "while taking into account social, cultural, ethical, political, and legal considerations" (Presidential/Congressional Commission on Risk Assessment Risk Management, 1997, 8).

Implications of Different Value Judgments: Why Public Participation Is Key to Effective Risk Governance

While conventional risk analysis quantifies all possible outcomes and multiplies them by their respective probabilities to arrive at a single indicator of risk, there are several challenges posed with this approach. First, people may value consequences differently. As such, risk assessment varies depending on whose perspectives are included (Zikmund-Fisher et al., 2007). Second, while the terminology of 'risk' assumes that we have sufficiently certain knowledge of potential outcomes and/or their associated probabilities, this knowledge is incomplete in many decision-making situations (Stirling, 2007). Indeed, in the context of rare diseases—the focus of newborn bloodspot screening—the scientific evidence may be limited (Watson et al., 2022). When available analysis or scientific knowledge is unable to reliably identify outcomes and/or probabilities, subjective judgments play an important role in risk assessment. In this context, highlighting the need to examine whose views and judgments are included in risk governance becomes even more crucial (Stirling, 2007).

Over the last two decades, scholars in risk governance have drawn attention to the crucial ways in which the opinions of scientific experts may differ from other stakeholders, such as those affected by the decisions made, and have emphasized the need for public involvement in public policy decision-making (Renn, 2008; Jardine et al., 2009; Webler & Tuler, 2018). Public involvement generally refers to the engagement of multiple, diverse social groups in the formation of public policymaking or regulatory decision-making to address societal issues. While some authors such as Fiorino (1990), have argued that public participation is imperative for moral reasons (because it is the right thing to do), the engagement of groups affected by the decision may also have instrumental effects such as driving more publicly acceptable outcomes relative to decisions based on expert knowledge alone.

The inclusion of public(s) in policy decision-making challenges traditional notions about science and politics that underlie models of evidence-based decision-making. First, it problematizes the notion that science and politics—or facts and values—are separate and need to stay separate. Second, it undercuts the position that effective decision-making about risk should rely on scientific and expert knowledge alone. In reality, the two are intricately linked: not only does scientific evidence inform and shape political discourse but science itself is infused with politics and values—and legitimately so (see Douglas in this edited volume): involving citizens in risk governance may expose implicit value-judgments embedded in expert assessments (Nicholls et al., 2016; Kuzma, 2016). For example, empirical research shows that risk perceptions of experts frequently differ from those of the general population (Krewski et al., 2012) and that people's risk perception is strongly driven by their value commitment and cultural identity (Kahan, 2012). As a consequence, other forms of knowledge, including people's life experiences, 'local' and cultural knowledge are legitimate and valuable in risk governance and decision-making (Corburn, 2005).

The Need to Better Understand Decision-Making for Newborn Bloodspot Screening

Despite the acknowledged benefits and risks to the expansion of newborn screening panels and documented variation between programs internationally (Jansen et al., 2016), very little work has explored how decisions are made regarding the inclusion or exclusion of targets within newborn screening programs. To date, descriptions regarding the structures that support the decision-making process in NBS are essentially non-existent. Furthermore, despite established principles for population screening, there is little if any examination of *how* criteria are applied (Jansen, 2017). A recent exception to this has been work by Jansen et al. (2016) who provide an overview of the decision-making process in the Netherlands (Jansen et al., 2021) as well as a brief description by Shone (2019) regarding the process in North Carolina.

This lack of data is problematic for several reasons. First, it precludes examination of the process and whether the decisions are fair or equitable; justifying the choice of diseases in an NBS program requires balancing the costs and benefits for society. This requires consideration of the broader population who, as taxpayers and recipients of healthcare services, are affected by decisions concerning funding and distribution of these services (van der Burg & Oerlemens, 2018). Second, it offers fewer opportunities to learn and understand the constraints placed on these decision processes as well as ways to improve them. Finally, it obfuscates the reasons for differences between provinces and territories, which may depend as much on value judgments and resource availability as they do on evidence (Nicholls et al., 2016).

This chapter reports on research findings about the process to expand NBS panels in Canada. Specifically, it focuses on how decision-making processes for NBS panel additions address risks, including how the public is involved in the process. To answer these questions, we draw on document analysis and interviews with key informants.

This chapter proceeds as follows. The next section, "Analytic Frameworks and Methods" introduces the analytical frameworks and methodology used in this study. The following section, "Who Decides How About NBS Panel Additions in Canada?" describes empirical results, and the Section, "Economic and Advisory Risk Management Tools Dominate" discusses these findings in the context risk management. The final section, "Where from Here? Avenues Forward for Decision-Makers and Scholars" offers concluding thoughts and identifies fruitful avenues for future research.

Analytic Frameworks and Methods

In this chapter, we examine the decision processes for NBS panel additions through a risk governance lens, with specific focus on the work of scientific advisory bodies. We consider these decision processes to address—if largely implicitly—the risks associated with NBS, including economic risks, health risks, and ethical risks identified above. In our discussion, we draw on the typology of risk management tools identified in the REACT framework (Krewski et al., 2007, 2014) presented previously in Chapters 5 and 7. This framework provides an organizing structure for risk management tools that public authorities (governments and regulators) may choose to apply, and includes regulatory, economic, advisory, technological, and community interventions.

Given our focus on the democratization of decision-making, and public involvement specifically, we also apply the public participation spectrum developed by the International Association for Public Participation (IAP2) as a framework when determining the quality of public participation in decision-making. The IAP2 framework sets out levels of engagement that gradually transfer increasing amounts of agency to the public: (1) inform the public about the problems, alternatives, and solutions; (2) consult the public and ask feedback on assessments and alternative solutions; (3) involve the public to effectively incorporate perspectives and concerns; (4) collaborate with the public on every aspect of decision-making, (5) empower the public to have final decision-making authority (Fig. 9.1).

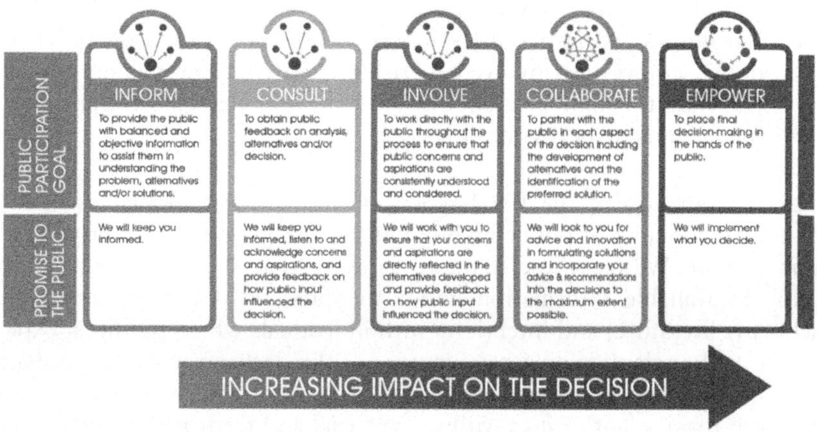

Fig. 9.1 The International Association for Public Participation (IAP2) Spectrum of Public Participation (© International Association for Public Participation www.iap2.org, retrieved from https://www.iap2canada.ca/foundations/ 22 September 2022. Reproduced with permission)

Choosing and designing appropriate means for public participation is highly dependent on local contexts and resource availability (Webler & Tuler, 2018). Involving the public constructively in risk decision-making can be expensive; it requires effort, skill, and learning.

Regarding the democratic quality of public involvement in risk governance, the introduction of this edited volume identified four principles:

> ***Transparency*** concerns the ease with which stakeholders can access information about risk-related decision processes and outcomes.
>
> ***Inclusiveness and representativeness*** focus on whether those who are impacted or concerned by risk issues have formal opportunities to make their voices heard in decision-making about these risks (inclusiveness). This principle also refers to whether the range of stakeholders involved, including marginalized social groups, is representative of potentially affected or concerned populations (representativeness).
>
> ***Deliberative quality*** refers to the 'how' of public engagement: is there a genuine opportunity for members of the public to engage in dialogue and exchange? Are their voices heard and seriously considered in the deliberations?
>
> ***Accountability of decision-makers*** focuses on the accountability of public authorities involved in risk-related decision-making toward citizens (through elected officials) or non-elected officials (e.g., bureaucrats, expert committees).

The combination of the IAP2 spectrum and the four principles of the democratic quality of public involvement in risk governance is original to this chapter. We use these frameworks as analytic structures to analyze: publicly available information about the process of test addition, the scholarly literature, and interviews with individuals who have direct experience at the scientific advisory juncture of the decision-making processes within Canadian newborn screening programs. Interviews with final decision-making authorities within provincial and territorial governments were out of scope for this research.

We conducted documentary analysis (of websites, publicly available materials) as well as semi-structured interviews with eight participants involved in five different NBS programs across Canada. We made great efforts to speak with participants from all Canadian screening programs, but only representatives from Saskatchewan, Alberta, British Columbia, the Maritimes (Nova Scotia, Prince Edward Island, and New Brunswick), and Quebec accepted our invitation to participate in this study.[1] Some of our interviewees have experience with both the medical and administrative components of NBS but are not involved in the final decision. Interviews were conducted over the phone and focused on the governance structures and decision-making processes concerning additions to NBS panels.

Who Decides How About NBS Panel Additions in Canada?

Canadian NBS Programs: Great Variety, Little Transparency

In Canada, jurisdiction over NBS programs and screening panel composition lies with provinces and territories and there is no central organizing body. While all babies born in Canada today have access to screening, the number of conditions included in the screening panels differs between jurisdictions (Potter et al., 2008) and not all provinces/territories have their own screening facilities (Table 9.1). Prince Edward Island and New Brunswick share a regional facility with Nova Scotia; Yukon sends its samples to British Columbia; the Northwest Territories and Nunavut share facilities with Alberta (Kitimeot) and Ontario (Baffin). Provinces and territories that use the same facility share the same screening panels. The number of conditions screened for in Canada ranges from 11 in Québec to 40 in Manitoba.

The lack of standardization across Canada not only means that populations in different regions have unequal access to testing but also that decisions regarding the composition of screening panels likely differ across the country. At the same time, and consistent with the broader newborn screening literature, publicly available information about these processes is scarce and uneven across programs. In most provinces, it is close to impossible for members of the public to learn about the evidence that decisions are based on and the mechanisms that operate when decisions are made. For example, while all NBS programs have websites, the available resources about NBS are targeted toward expectant parents and

[1] Since British Columbia and Alberta provide testing services to Nunavut, the Northwest Territories, and the Yukon, our participants are effectively involved in NBS programs in 10 provinces/territories across Canada.

healthcare professionals and provide little information about the decision-making process itself. Ontario's NBS program is an exception; its website provides information on the test addition process and the factors that are considered by the Advisory Council when considering addition of a

Table 9.1 NBS programs and testing facilities in Canada

Province/territory	*NBS program, testing facility, website*	*# of conditions included*[2]
British Columbia Yukon	BC Newborn Screening Program http://www.perinatalservicesbc.ca/our-services/screening-programs/newborn-screening-program	24
Alberta Nunavut (Kitimeot) Northwest Territories	Alberta Health Services, https://www.albertahealthservices.ca/info/page9014.aspx	22
Saskatchewan	Roy Romanow Provincial Laboratory https://www.saskhealthauthority.ca/facilities-locations/roy-romanov-provincial-laboratory/screening-and-reference-services	over 30
Manitoba	Cadham Provincial Laboratory https://www.gov.mb.ca/health/publichealth/cpl/baby.html#:~:text=For%20more%20information%20about%20newborn,at%20204%2D945%2D7458	around 40
Ontario Nunavut (Baffin)	Newborn Screening Ontario, Children's Hospital of Eastern Ontario https://www.newbornscreening.on.ca/	28
Québec	Québec Neonatal Blood and Urine Screening Program https://www.quebec.ca/en/health/advice-and-prevention/screening-and-carrier-testing-offer/blood-and-urine-screening-in-newborns	11 (by blood)

(continued)

[2] Information retrieved from the programs' websites on September 20th, 2022.

Table 9.1 (continued)

Province/territory	*NBS program, testing facility, website*	*# of conditions included*
New Brunswick Nova Scotia Prince Edward Island	Maritime Newborn Screening Program, IWK Health Centre https://www.iwk.nshealth.ca/newbornscreening	22
Newfoundland and Labrador	Provincial Medical Genetics Program, Health Sciences Centre St. Johns http://www.nlma.nl.ca/FileManager/Notices_and_Advisories/docs/2015/What_is_newborn_screening_brochure.pdf	19

Source Authors' own source

condition (Newborn Screening Ontario, n.d.). The lack of transparency also precludes any assessments of which decision processes engage patients and/or the public and consider their perspectives.

Governance Structures for Newborn Screening

For the most part, processes for adding new conditions to newborn screening panels are not guided by official government policies or regulations. But across all programs, governments have final authority over the decision to expand screening panels.

Most NBS programs have established standing Advisory Committees tasked with providing evidence-based scientific advice to governments and recommending changes to NBS panels. However, they possess no executive authority to make these changes to the panels themselves. In Ontario, for example, the Newborn Screening Ontario Advisory Council (NSO-AC) exists as a standing advisory committee that has within its mandate development of the process and review of proposals for potential new screening targets. In the Maritimes (i.e., New Brunswick, Nova Scotia, and Prince Edward Island), there is also a Diagnosis Committee that reviews submissions from medical professionals proposing additional conditions to the panel. Some Advisory Committees also strike specialized ad hoc working groups when evaluating whether or not to add new conditions to NBS screening panels.

These standing Advisory Committees consist largely of physicians with various specializations (pediatrics, public health, genetics, neonatology, endocrinology, etc.), laboratory staff, and occasionally economists (Québec) or government representatives (British Columbia, Ontario). Some Advisory Committees also require geographic representation among their members. For instance, Ontario's Advisory Committee includes members from across the province, and the Committee in the Maritimes includes representatives from all three participating provinces.

Most of the roles, responsibilities, and procedures of these Committees have evolved organically over time. In some provinces, according to some interviewees, issues as simple as meeting intervals are not formalised or do not proceed at a regular schedule. Some use terms of reference or similar guiding documents, but there are few external rules or pressures guiding their activities. One participant described their dissatisfaction with the informality of these arrangements:

> It is frustrating for clinicians and for the program and for the public, especially parents with children who are afflicted with these conditions not to have a clear or consistent process.

Decision Processes

With regard to decision-making processes for adding new conditions to NBS panels in Canadian programs, they tend to fall under two broad categories.

First, there are processes that are bound by legislation, regulation, directive, or other types of formalized guidance from a health agency or similar government body. Very few of the Committees' operations are covered by legislation. In most cases, the legislation mandates that an NBS program exists but is silent on their operations and decision-making processes for adding or removing conditions from the panel.

The exceptions are Saskatchewan and Alberta. Saskatchewan introduced *The Newborn Screening Regulations*, Chapter P-37.1 Reg 15, which formalized newborn screening procedures in 2014, including program administration, sample collection, testing and follow-up, disclosures, and adoption of guidelines. The regulations are, however, silent on the topics of risk governance and additions to the panel, but afford the minister of health tremendous latitude and flexibility on public engagement.

> The minister shall: (a) cause the [Newborn Screening Guidelines] to be made available to the public in any form or manner that the minister considers appropriate; and (b) take any steps that the minister considers appropriate to bring the guidelines, and the manner and form in which the guidelines are available, to the attention of the public.

Similarly, Alberta established a province-wide NBS program in 2009, alongside the formation of Alberta Health Services. This amalgamated and harmonized NBS program had previously been managed and administered by regional health authorities. However, both Saskatchewan and Alberta's regulations do not refer to the addition of new conditions to NBS panels.

Second, there are processes that are not bound by legal instruments or official policy. Participants within the interviews indicated that the decisions involving additions to NBS panels overwhelmingly fall into this category. While these activities are often guided by terms of reference or other internally developed procedures, there are no regulations guiding the development of terms of reference or their contents. Activities that fall into this category include internal deliberations by Advisory Committees, working groups and the government decision-makers who ultimately have the discretion to act or not act upon their advice. Within the Advisory Committees, working groups and other advisory bodies, the process of deciding if and when to make a recommendation is largely consensus-based. As one participant described it:

> No legislation, no. The Advisory Committee has terms of reference. It is done on a consensus basis… In terms of composition and structure of the committee, that was done by the [centre] when the program expanded. They consulted with the other provinces, tried to see what is done in other provinces and then established who should be on theirs.

Most participants were not aware of codified processes to guide or inform the decision-making of the Advisory Committees and similar bodies. In the absence of such guidance, Advisory Committees have developed processes and procedures internally as needed to provide recommendations to decision-makers. This includes decisions regarding adding new conditions to the NBS panel as well as the screening procedures themselves.

Factors Affecting Decisions About Panel Expansion

Advisory Committees draw on multiple information sources when looking for new conditions to add to the NBS panels. In addition to the scientific literature, participants across provinces also noted that the Advisory Committees' activities are heavily informed by the work of their counterparts across Canada, particularly Ontario, as well as the United States, the European Union, and the World Health Organization:

> Ontario does great research, and we just use their studies...quite often we rely on what other provinces do. That's how it works.

This is largely due to budgetary constraints, as Advisory Committees simply do not have the resources to exhaustively scan new scientific literature. This forces them to draw on knowledge generated and mobilized in other jurisdictions. Indeed, despite the provincial mandate of the programs, interviewees consistently referred to inter-provincial discussion and collegiality, albeit often unofficial.

Participants said that the final decision takes a range of factors into account, including scientific, economic, and political considerations. As noted earlier, occasionally, separate groups are struck to assess these considerations separately. For example, in British Columbia, a specialized working group reviews every new condition under consideration and iterates with the NBS Advisory Committee. One participant described the unique considerations required for adding a specific condition:

> We develop a working group that is specific to the condition that is being reviewed. So, for example, we just finished our review of spinal muscular atrophy. And so we invited a couple of neurologists with more expertise in neurogenetics to participate in that review.

For instance, if British Columbia's Advisory Committee endorses adding a condition to the panel, another specialized group develops a business case and cost-benefit analysis for presentation to the Ministry of Health for final decision. Interviewees indicated that government decision-makers are forced to weigh a number of considerations; as a result, the decision about the inclusion of a target or condition is not exclusively based on medical and scientific evidence.

> It took two years for the government to mandate the conditions be added to our panel. So there was quite a gap and during that time there were a lot of things happening. A change in government usually means a change in priorities and direction. [...]. There are lots of things to consider.

Indeed, all interviewees indicated that budgetary considerations are a key factor in the government's decision-making about panel additions. One participant stated that adding a single condition to the panel requires approximately $500,000 per year per condition but noted that adding a condition generates a "return on investment" of roughly 20 to one in medical costs. However, these savings accrue over decades and do not fit comfortably within budgeting or political cycles. The Advisory Committees appear to be well aware of these hurdles, and interviewees reported that they often make recommendations strategically with political considerations in mind. If budgetary constraints prevent the addition of a new condition to the panel, participants noted that Committees will often submit a rejected condition for reconsideration the following year, aware that the rejection may not have been made due to a lack of scientific merit.

As noted above, governments have the final say on whether or not a condition is ultimately added to an NBS panel, either directly or indirectly through budgeting decisions. It was not clear, however, from the interviews, how many steps removed the Advisory Committees are from these final decision-makers, be they the relevant minister or another senior official, or how their advice is weighed against other factors. The frequency and nature of interactions between the Advisory Committees and the decision-makers who take their advice was also unclear.

Public Involvement in the Screening Decision Process

Across all provinces examined, the Advisory Committees and the working groups consist of specialists and experts who are trained to evaluate the medical and technical rationales for adding new conditions to the NBS panel. There are no formal or structured opportunities for the public to become involved in this decision-making (although in Ontario members of the public can nominate conditions for review). The exception is in Nova Scotia, which at the time of our study was in the process of recruiting two parents to the panel, one from a city and another from a rural region of the province. In this case, the interviewee indicated that this was an initiative of the Advisory Committee, not a result of a government directive.

Although some participants noted that inclusion of parents is considered or planned, accessing the Advisory Committees generally requires parents to take initiative. Some participants noted resistance to parents and citizens sitting on expert panels, for a variety of reasons:

> It is difficult to find any one parent to speak for or represent the vast majority of parents. I am reluctant to have public representation on our advisory committee for this reason. Yet we must be in step with the wishes, values, and concerns of parents with respect to newborn screening. A more comprehensive and democratic way of doing this is through structured well-designed surveys and or group interviews. This can be resource intensive for any one province but could be coordinated at a national level.

Participants indicated that lack of expertise is the key obstacle to including the public in their decision-making. Some also mentioned a perceived lack of interest by the public, as demonstrated by the dearth of organized advocacy groups in this space. Advocacy groups that *are* organized and well-funded (one participant mentioned cystic fibrosis) are already included in the screening criteria, so there is no additional or incremental work that they can undertake with respect to the screening panel. Since new additions to the screening panel will generally be rare, public awareness may be low, and consequently advocacy may be limited.

Interviewees indicated that, in most provinces, there is minimal interaction between parents and citizens and the Advisory Committees and decision-makers who determine the details and composition of the NBS panels. Parents are largely passive participants in the screening process, their involvement limited to reading, conversations with medical professionals who are collecting samples, and providing the information necessary for informed consent. Some provinces do, however, seek post hoc patient feedback to improve the patient experience. For instance, in British Columbia, parent feedback is solicited after the fact to improve the overall NBS screening process.

Despite the noted reluctance by some to involve parents, other interviewees indicated a general openness toward greater public involvement, but none suggested it was a priority or an explicit part of the mandate. Participants noted that information on the NBS screening process, including new additions, is publicly accessible. However, interviewees did not suggest that promoting public awareness of NBS panels is a priority. Changes to the NBS screening panels are typically communicated to the public via websites, newsletters, ministerial press releases, and updates to the medical brochures and literature that medical professionals provide to new parents prior to the procedure.

Economic and Advisory Risk Management Tools Dominate

Decision-making about the composition of NBS panels is—whether implicitly or explicitly—an act of risk governance. Based on our documentary analysis and interviews and applying the REACT typology as a conceptual framework, we analyze the approaches taken to address the various risks related to NBS (Table 9.2).

Our findings indicate a relatively light use of **regulatory risk management approaches**. Certainly, governments are the final authority in decisions about panel composition, but the decision-making process and the composition and procedures of the advisory bodies are almost entirely unregulated in the jurisdictions examined (with the exceptions of Alberta

Table 9.2 Risk management in NBS panel decision-making in Canada

Risk management approach (REACT framework)	*Application in NBS panel decision-making*
Regulatory interventions	Governments have final decision authority, but the decision process itself is lightly regulated
Economic interventions	Cost-benefit considerations importantly drive governments' and advisory bodies' decisions
Advisory interventions	Advisory bodies assess potential panel additions and provide recommendations to governments
Community interventions	Overall, limited formalized opportunities for public involvement in decision making. Even information about the decision process is not publicly available in most programs
Technological interventions	Technological capability is a necessary condition for panel expansion, but limited insights from this study about how technology is being used to address risks associated with expansions

Source for REACT framework: Krewski et al. (2007) and Krewski et al. (2014). Authors' own source for application to NBS panel decision-making and Saskatchewan). Instead, the structures have evolved organically and occasionally function on an ad hoc basis.

In contrast, **economic approaches to risk management** seem paramount in the considerations of the advisory bodies and ultimately, governments' decisions about panel additions. Our findings indicate that governments use formal and informal economic analysis to measure and examine the societal impacts of adding conditions to the screening panels. What remains unclear are the parameters upon which these assessments are made: how are short-term budgetary burdens weighed against long-term health benefits and savings for the entire healthcare system?

Advisory risk management interventions are another key approach employed in the decision-making about NBS panel composition. All NBS programs have one or more advisory bodies, largely made up of healthcare professionals who are responsible for making recommendations regarding panel additions based on scientific evidence and their medical expertise. While these advisory bodies have no final decision authority, interviewees indicated that governments generally follow the committee's recommendations—if budgetary considerations allow. Some interviewees also indicated that advisory bodies sometimes anticipate such budgetary constraints when developing their recommendations.

Finally, before discussing community interventions, it should be noted that our results offer little insights into how **technological risk management tools** are applied in the decision-making process. Advisory committees for the screening programs involved in this study all include a representative from the testing laboratories, but we are not aware of more concrete mechanisms for considering technology as a risk mitigation tool. For example, one way of directly applying technological risk mitigation approaches would be to make an addition to the panel dependent on the use of a specific testing technology that reduced the risk of false positives. Our interviews do not provide evidence of such measures.

Great Potential to Increase Public Participation

According to our findings, **community-based risk management tools** currently receive limited attention. We see little evidence of formalized public involvement in decision-making processes. An exception is the opportunity for anyone in Ontario to suggest adding a condition to the screening panel. If engagement with parents takes place, it seems to be of an informal nature (for instance, advocacy with the government) or occur mostly after a new condition was added to the panel (for instance, in the BC program).

Table 9.3 categorizes the public participation activities identified in this study along the IAP2's spectrum (the *what*) and also considers these activities in reference to the four principles of democratization (the *how*). Based on the evidence considered, we find great potential for improvements across all four democratization principles: transparency, inclusiveness and representativeness, deliberative quality of the interaction, and accountability.

With respect to the levels of involvement and engagement, all interviewees indicated that NBS programs do inform the public about panel composition and decision outcomes regarding additions. These types of community risk management interventions—information and post hoc consultation—sit on the lower end of the spectrum of public participation, as defined by the International Association for Public Participation (International Association for Public Participation, 2018). However, with regard to transparency, information about the decision process is largely lacking. The exception to this is Newborn Screening Ontario, which offers publicly available information about the process and the criteria that the advisory committee considers when developing recommendations. However, what is clearly lacking for all programs is easily accessible and comprehensive public information about the decision process itself—who makes decisions and based on what information.

Few programs illustrated more active involvement of the public; outreach appears to be limited and tends to focus on consultation about implementation of the program, as opposed to the proposed decision to add a condition. While in theory any member of the public could nominate a condition using the process developed in Ontario, engagement practices generally appeared to focus on patients. This is despite the earlier proposition that there may be multiple stakeholder groups (including the general public not affected by a condition on the screening panel) that may be affected by or interested in NBS panel additions and that may hold different perspectives on the issue. Consequently, we suggest that consultations, where they exist, are generally not inclusive and representative of populations affected by changes in the NBS panel composition. Indeed, even when there were indications of improved engagement (e.g., the Maritimes), they tended to focus on parents.

We did not find evidence of public(s) having decision-making power; indeed, the final decision-making power appeared to lie with elected

Table 9.3 Public participation in decision-making for NBS panel additions

Spectrum of public participation (IAP2) Principles of democratization	*Inform publics about NBS and panel expansions*	*Consult publics for feedback on process and/or outcome*	*Involve publics in process to hear and consider concerns*	*Collaborate with publics at each step of decision process*	*Empower publics by giving final decision power*
Transparency: What information is available?	Communication by NBS programs largely about decision outcomes, not process (exception: ON)	n/a	The actual impact is not transparent because involvements mostly informal	n/a	n/a
Inclusiveness and representativeness: Who is invited/involved/heard? Who are the involved publics? (e.g., parents/advocacy groups/general public)	Existing information about NBS is accessible to the general public through program websites. Much of the information is targeted at parents though	Post hoc feedback from the public (BC)	Interviews indicated that there is a proposal for including parents in the advisory body of the Maritime NBS program NS Ontario enables the general public to suggest a panel addition	n/a	n/a
Deliberative quality: How are publics engaged?	Activities constitute one-way communication but websites typically include contacts for questions	Inconsistent and generally requires parents to take initiative; limited outreach on part of decision-makers (exception: QC, post hoc)	Not clear	n/a	n/a

Spectrum of public participation (IAP2) Principles of democratization	*Inform publics about NBS and panel expansions*	*Consult publics for feedback on process and/or outcome*	*Involve publics in process to hear and consider concerns*	*Collaborate with publics at each step of decision process*	*Empower publics by giving final decision power*
Accountability: How can citizens hold decision-makers accountable?	Elected governments have final authority and are democratically accountable to citizens; possibly problematic for jurisdictions without their own programs				n/a

Sources Spectrum of public participation (International Association for Public Participation, 2018), principles of democratization (Chapter 1), authors' own source for application to NBS screening government officials and the provincial public service. While governments are democratically accountable to their citizens, multiple jurisdictions in Canada do not have their own testing facilities, instead joining screening programs of other jurisdictions. In these jurisdictions, accountability relationships are weaker and the claim to government as a proxy for public involvement is reduced.

As we show in Table 9.1, Maritime provinces share a screening program and testing facilities. The Advisory Committee includes representatives from all three provinces, but final authority over panel additions lies with Nova Scotia government in Halifax, where the testing facility is physically located. For Nunavut and the Northwest Territories, which have joined Alberta's screening program and the Yukon, which has joined British Columbia, we have no indication from our interviews that representatives from the North are included in any of the involved bodies. As a result, there is a tension between democratic accountability and healthcare resourcing needs in the North.

Finally, while multiple interviewees identified a lack of expertise in engaging with the public, there was interest from some interviewees in greater public involvement. At the same time, others perceived lack of knowledge and interest on the side of the public. These comments are in line with what has become known as the 'knowledge-deficit model' of public engagement; the public is viewed as uninformed and thus unable to grasp the science upon which the decisions are based. While we have no evidence to indicate that public input would not be considered, the suggestion that a lack of understanding about NBS would preclude public involvement fails to consider the other societal risks outlined earlier, especially opportunity costs brought about by decisions made to fund certain healthcare interventions at the expense of others. This may indicate a lack of openness among some of the expert advisors involved in the process toward genuine democratization of decision-making or a perception that broader considerations are beyond their remit.

Where from Here? Avenues Forward for Decision-Makers and Scholars

This chapter discussed decisions about NBS panel additions as exercises of risk governance. While this study is exploratory in nature, we can identify multiple avenues forward for practitioners, including governments and NBS program leads, and future research directions for scholars.

A key takeaway from this study is that transparency about NBS panel decisions in Canada is generally low. While public information is available about decision outcomes (i.e., the list of conditions included in screening panels), even basic process information remains mostly hidden from the public, including who is involved in decision-making about panel expansions (e.g., the composition of the Advisory Committees) and how the decisions were made (e.g., explanations of why certain conditions were added or not). To improve transparency and accountability of decision-makers toward the public, this information should be as publicly accessible as possible.

Future research could examine options for diversifying the risk governance approaches used in decision-making about NBS panel composition. Such research should aim to better understand public perception of risks and benefits associated with NBS and help to examine the various publics affected by NBS—families, advocacy groups, and the public at large. The perceptions of risks and benefits associated with NBS may differ significantly across these groups. In particular for NBS programs serving diverse populations in multiple jurisdictions, such research may also identify how cultural differences may affect risk perceptions and preferences. Regarding the choice and design of mechanisms for public participation, the risk governance literature indicates that there is no one-size-fits-all approach. Therefore, research investigating risk perceptions and preferences for risk governance at a local level is crucial for strengthening the democratic character of decision-making across Canadian jurisdictions.

Resource limitations are at the heart of decision-making about NBS in Canada. Our work identified funding considerations as a key driver of decisions regarding panel additions. Moreover, they are likely a key driver in the design of the decision process itself. Specifically, community-based risk management approaches—currently used scarcely in Canada—can be resource-intensive in terms of time, expertise, and funding. Absent a champion for public involvement and engagement from within government, there will likely be little if any non-expert input into the process. Future research can help to inform arguments about why public engagement is worth the investment, as it can produce robust and acceptable decision outcomes and thus be worthy of investment.

There remains a great opportunity, and potential costs, in developing more transparent and engaging approaches to decision-making regarding the expansion of newborn screening panels. Indeed, a key question—much like that for newborn screening programs themselves—is whether such changes are worth the cost; a question that science cannot answer on its own.

Acknowledgements The authors thank Alexandra Guay-Charette for her excellent research assistance.

References

(1997). *Presidential/Congressional Commission on risk assessment and risk management. Risk assessment and risk management in regulatory decision-making* (Final Report. Vol. 2). United States Environmental Protection Agency Washington: https://cfpub.epa.gov/ncea/risk/recordisplay.cfm?deid=55006.

Bailey, D.B. Jr, Ackerman Porter, K., Andrews, S.M., Raspa, M., Gwaltney, A.Y., & Peay, H.L. (2021). Expert evaluation of strategies to modernize newborn screening in the United States. *JAMA Network Open*, 4(12), Article e2140998. https://doi.org/10.1001/jamanetworkopen.2021.40998.

Bailey, D.B. Jr, Beskow, L.M., Davis, A.M., & Skinner, D. (2006). Changing perspectives on the benefits of newborn screening. *Mental Retardation and Developmental Disabilities Research Reviews*, 12(4), 270–279.

Beck, U. (1992). *Risk society. Towards a new modernity.* Sage.

Bombard, Y., &, Miller F.A. (2012). Reply to Ross' commentary: Reproductive benefit through newborn screening: Preferences, policy and ethics. *European Journal of Human Genetics*, 20, 486–489.

Botkin, J.R., & Rothwell, E. (2016). Whole genome sequencing and newborn screening. *Current Genetic Medicine Reports*, 4(1), 1–6.

Buchbinder, M., & Timmermans, S. (2011). Newborn screening and maternal diagnosis: Rethinking family benefit. *Social Science and Medicine*, 73(7), 1014–1018.

Canadian Agency for Drugs and Technologies in Health. (2011, August 26). *Newborn screenings for disorders and abnormalities in Canada.* Retrieved September 22, 2022, from https://www.cadth.ca/sites/default/files/pdf/Newborn_Screening_es-26_e.pdf.

Corburn, J. (2005). Street science: Community knowledge and environmental health justice. The MIT Press.

Cornel, M.C., Rigter, T., Jansen, M.E., & Henneman, L. (2020). Neonatal and carrier screening for rare diseases: How innovation challenges screening criteria worldwide. *Journal of Community Genetics*, 12(2), 257–265.

Fiorino, D.J. (1990). Citizen participation and environmental risk: A survey of institutional mechanisms. *Science, Technology, & Human Values*, 15(2), 226–243.

International Association for Public Participation. (2018). *Public participation spectrum*. Retrieved September 22, 2022, from https://iap2canada.ca/foundations.

Jansen, M.E. (2017). Reply to a Mackie. *European Journal of Human Genetics*, 25(7), 791–792.

Jansen, M.E., Klein, A.W., Buitenhuis, E.C., Rodenburg, W., & Cornel, M.C. (2021). Expanded neonatal bloodspot screening programmes: An evaluation framework to discuss new conditions with stakeholders. *Frontiers in Pediatrics*, 9, 635353.

Jansen, M.E., Metternick-Jones, S.C., & Lister, K.J. (2016). International differences in the evaluation of conditions for newborn bloodspot screening: A review of scientific literature and policy documents. *European Journal of Human Genetics*, 25(1), 10–16.

Jardine, C., Turtiak, M., & Driedger, S.M. (2009). Public participation and risk governance: Opportunities and barriers. *International Journal of Risk Assessment and Management*, 13(3/4), 260–275.

Kahan, D.M. (2012). Cultural cognition as a conception of the cultural theory of risk. In S. Roeser, R. Hillerbrand, & M. Peterson (Eds.), *Handbook of risk theory: Epistemology, decision theory, ethics, and social implications of risk* (pp. 725–759). Springer.

Krewski, D., Hogan, V., Turner, M.C., Zeman, P.L., McDowell, I., Edwards, N., & Losos, J. (2007). An integrated framework for risk management and population health. *Human and Ecological Risk Assessment: An International Journal*, 13(6), 1288–1312.

Krewski, D., Turner, M.C., Lemyre, L., & Lee, J.E.C. (2012). Expert vs. public perception of population health risks in Canada. *Journal of Risk Research*, 15(6), 601–625.

Krewski, D., Westphal, M., Andersen, M.E., Paoli, G.M., Chiu, W.A., Al-Zoughool, M., Croteau, M.C., Burgoon, L.D., & Cote, I. (2014). A framework for the next generation of risk science. *Environmental Health Perspectives*, 122(8), 796.

Kuzma, J. (2016). Policy: Reboot the debate on genetic engineering. *Nature*, 531, 165–167.

Levy, H.L. (1998). Newborn screening by tandem mass spectrometry: A new era. *Clinical Chemistry*, 44(12), 2401–2402.

Newborn Screening Ontario (n.d.). https://www.newbornscreening.on.ca.

Nicholls, S.G., Newson, A.J., & Ashcroft, R.E. (2016). The need for ethics as well as evidence in evidence-based medicine. *Journal of Clinical Epidemiology*, 77, 7–10.

Nicholls, S.G., Wilson, B.J., Etchegary, H., Brehaut, J.C., Potter, B.K., Hayeems, R., Chakraborty, P., Milburn, J., Pullman, D., Turner, L., & Carroll, J.C.

(2014). Benefits and burdens of newborn screening: Public understanding and decision-making. *Personalized Medicine*, 11(6), 593–607.

Potter, B.K., Avard, D., Entwistle, V., Kennedy, C., Chakraborty, P., McGuire, M., & Wilson, B.J. (2009). Ethical, legal, and social issues in health technology assessment for prenatal/preconceptional and newborn screening: A workshop report. *Public Health Genomics*, 12(1), 4–10.

Potter, B.K., Avard, D., & Wilson, B.J. (2008). Newborn blood spot screening in four countries: Stakeholder involvement. *Journal of Public Health Policy*, 29(1), 121–142.

Renn, O. (2008). White paper on risk governance: Toward an integrative framework. In O. Renn & K.D. Walker (Eds.), *Global risk governance: Concept and practice using the IRGC framework* (pp. 3–73). Springer Netherlands.

Rogowski, W.H., Grosse, S.D., Schmidtke, J., & Marckmann, G. (2014). Criteria for fairly allocating scarce health-care resources to genetic tests: Which matter most? *European Journal of Human Genetics*, 22(1), 25–31.

Shih, S.T.F., Farrar, M.A., Wiley, V., & Chambers, G. (2021). Newborn screening for spinal muscular atrophy with disease-modifying therapies: A cost-effectiveness analysis. *Journal of Neurology, Neurosurgery and Psychiatry*, 92, 1296–1304.

Shone, S.M. (2019). Newborn screening policy decisions: Adding conditions. *North Carolina Medical Journal*, 80(1), 42–44.

Sims, E.J., Mugford, M., Clark, A., Aitken, D., McCormick, J., Mehta, G., Mehta, A., & UK Cystic Fibrosis Database Steering Committee. (2007). Economic implications of newborn screening for cystic fibrosis: A cost of illness retrospective cohort study. *The Lancet (British Edition)*, 369(9568), 1187–1195. https://www.sciencedirect.com/science/article/abs/pii/S0140673607605650.

Stern, P.C., & Fineberg, H.V. (1996). *Understanding risk: Informing decisions in a democratic society*. Committee on Risk Characterization, National Research Council. https://doi.org/10.17226/5138.

Stirling, A. (2007). Risk, precaution and science: Towards a more constructive policy debate. Talking point on the precautionary principle. *EMBO Reports*, 8(4), 309–315.

Therrell, B.L., Padilla, C.D., Loeber, J.G., Kneisser, I., Saadallah, A., Borrajo, G.J.C., & Adams, J. (2015). Current status of newborn screening worldwide: 2015. *Seminars in Perinatology*, 39(3), 171–187.

UK National Screening Committee. (2000). *Second report of the UK national screening committee*. London, UK, Department of Health.

Ulph, F., Wright, S., Dharni, N., Payne, K., Bennett, R., Roberts, S., Walshe, K., & Lavender, T. (2017). Provision of information about newborn screening antenatally: A sequential exploratory mixed-methods project. *Health Techno-ogyl Assessment Reports*, 21(55), 1–240.

van der Burg, S., & Oerlemans, A. (2018). Fostering caring relationships: Suggestions to rethink liberal perspectives on the ethics of newborn screening. *Bioethics*, 32(3), 171–183.

Watson, M.S., Lloyd-Puryear, M.A., & Howell, R.R. (2022). The progress and future of US newborn screening. *International Journal of Neonatal Screening*, 8(3), 41–66.

Webler, T., & Tuler, S. (2018). Four decades of public participation in risk decision making. *Risk Analysis*, 41(3), 503–518.

Wilson, J.M.G., & Jungner, G. (1968). Principles and practice of screening for disease (No. 24). World Health Organization. Retrieved September 22, 2022, from https://apps.who.int/iris/handle/10665/37650.

Zikmund-Fisher, B.J., Smith, D.M., Ubel, P.A., & Fagerlin, A. (2007). Validation of the subjective numeracy scale: Effects of low numeracy on comprehension of risk communications and utility elicitations. *Medical Decision Making*, 27(5), 663–671.

PART IV

Public Health

CHAPTER 10

Balancing Shared Decision-Making with Population-Based Recommendations: A Policy Perspective of PSA Testing and Mammography Screening

S. Michelle Driedger, Elizabeth J. Cooper, and Ryan Maier

Introduction

In any health system, policies and guidelines are necessary to ensure that scarce resources are managed optimally and in ways that reflect the best available evidence at the time. On the macro level, policies are often implemented on a system-wide basis, such as the case with mammography

S. M. Driedger (✉) · R. Maier
Department of Community Health Sciences, University of Manitoba, Winnipeg, MB, Canada
e-mail: Michelle.driedger@umanitoba.ca

R. Maier
e-mail: ryan.maier@umanitoba.ca

E. J. Cooper
University of Regina, Regina, SK, Canada
e-mail: Elizabeth.Cooper@uregina.ca

M. Gattinger (ed.), *Democratizing Risk Governance*,
https://doi.org/10.1007/978-3-031-24271-7_10

screening in Canada, where each province evaluates recommendations based on existing population-based evidence and sets their respective policy accordingly. On the micro level, policies are interpreted and implemented during the doctor-patient clinical encounter. Often, decisions are made in clinical contexts that contradict existing recommendations and policy. For example, a patient may insist on tests in the absence of clinically relevant symptoms, or a doctor may prescribe an otherwise unnecessary test simply because it has become part of their routine practice, as is the case of the prostate-specific antigen (PSA) test for men. In such cases, applying population-based guidelines at an individual level becomes challenging. We examine factors that can create tensions between these macro and micro levels, obstacles that can preclude their practical harmonization, and strategies promoted to bridge the divide—particularly shared decision-making within contexts of low democratization of risk. The contexts of this case study are mammography screening for breast cancer among women and PSA testing for prostate cancer among men. We draw insights from key informant interviews with policymakers responsible for screening programs in different jurisdictions in Canada as they discuss the challenges they face implementing population-based guidelines in clinical settings.

Background: Population-Based Cancer Screening Programs

A formalized population-based cancer screening program systematically invites otherwise asymptomatic people for testing. The program acts as a dragnet to identify undetected cancer before symptoms appear in order to treat it in its earliest phases of development and thereby prevent or delay its advancement (Morrison 1985). Screening should be clearly distinguished from cases where a doctor makes a clinical recommendation to order a diagnostic test for a patient that is presenting with symptoms or known risk factors. It should also be distinguished from opportunistic requests for a diagnostic test by a patient where no prior indication for it exists. With respect to our two cases in this chapter, mammography screening is a formally supported population-based program whereas PSA testing is not.

Uncertainty and Issues in Mammography Screening and PSA Testing

Mammography remains the standard clinical intervention to detect breast cancer in women and is the basis of existing screening programs. However, debates and uncertainty over mammography's utility in cancer detection persist. The efficacy of mammography screening (in terms of identifying incidence and mortality rates) often lacks consistency in study design and rigor, which can lead to uncertainty (Autier and Boniol 2018; Printz 2014). One study may conclude that mammography generally reduces mortality rates (Hirsch and Lyman 2011), while another may argue that there has been relatively little mortality benefit (Autier and Boniol 2018). Some argue that the risks of mammography screening, such as overdiagnosis or overtreatment, do not outweigh the benefits of detecting cancer and potentially saving younger women's lives (American College of Obstetricians-Gynecologists 2011). Overdiagnosis refers to diagnoses of conditions that may never have caused symptoms or death and overtreatment refers to treatment for conditions that if left untreated, were unlikely to cause symptoms or death (Bhatt and Klotz 2016). The debates can even pit medical disciplines against each other, with radiologists more commonly promoting mammography efficacy—especially for younger women, who may develop more aggressive cancers—against epidemiologists who may exhibit more caution toward mammography when interpreting trial results or retrospective analyses of data from screening programs in light of risks (Miller et al. 2014; Layne 2016; Ray et al. 2017). Diverse specialties may look at the same studies where the evidence seems clear, but weigh the evidence differently in their interpretations based on their underlying values. This has led to uncertainty in whether mammography screening is ultimately beneficial for women under 50 years of age (Autier and Boniol 2018).

Similar issues bedevil the use of the PSA test to detect prostate cancer. While the PSA test has been shown to assist in detecting potential prostate cancer and evaluating treatment strategies, its indications, the evidence, and its risk have provoked debate as to its own effectiveness as a potential screening tool. A high PSA score can be caused by numerous benign factors and prostate cancer can be present without a high PSA score (Obort et al. 2013). Such an unclear continuum of risk along the spectrum of PSA scores can lead to high rates of false-positives, overdiagnosis, and overtreatment. Reviews of studies have had contradictory results as to its efficacy as a means of testing for prostate cancer and whether such harms outweigh the relative reduction in cancer mortality (Croswell et al. 2011). The overall uncertainty as to the value of the PSA test has meant that there is no formalized prostate cancer screening program (Law et al. 2020).

Population-Based Health Policy Recommendations for Prostate and Breast Cancers in Canada

The Public Health Agency of Canada (PHAC) entrusts the Canadian Task Force on Preventive Health Care (CTFPHC) with developing clinical guidelines and recommendations (Canadian Task Force on Preventive Health Care 2019). The CTFPHC utilizes methods of evidence-based medicine wherein they conduct systematic reviews of clinical research in an effort to arrive at a consensus on which to base recommendations for patient management and care. Although these guidelines are intended to assist decision-making between doctors and individual patients, recommendations are largely drawn from population-level data. While population-based research has long since become a cornerstone in guiding evidence-based clinical practice, it has also provoked critique (Trinder and Reynolds 2000). Most notably, generalized conclusions at the population level can be difficult to translate into the individualized level of clinical practice, especially when these prove irreconcilable with the values of patients and doctors. Also, it has prompted accusations of promoting the rationing of health care (Kelly and Cronin 2011). Consequently, doctors may sometimes receive instructions to view practice recommendations not as templates of care for all patients, but as one factor among others to consider when making a decision on preventive care (Hoffman and Nguyen 2011).

The CTFPHC's current recommendations about prostate and breast cancer screening were published in 2014 and 2011, respectively. In each case, the CTFPHC concluded that relatively small reductions in prostate and breast cancer (for younger women) mortality resulting from screening were eclipsed by substantial harms, including false-positive test results, overdiagnosis, and overtreatment (Dunfield et al. 2014; Fitzpatrick-Lewis et al. 2011). The CTFPHC reaffirmed a recommendation held since 1994 that the PSA test should not be used to screen asymptomatic men of any age for prostate cancer (Bell et al. 2014), and recommended against routine mammograms to screen for breast cancer for average risk women aged 40 to 49 years (The Canadian Task Force on Preventive Health Care 2019). In 2018, as part of an updated review of the evidence, the CTFPHC reaffirmed these recommendations for mammography screening (Klarenbach et al. 2018).

Provincial/Territorial Implementation of PSA Testing and Mammography Screening Programs

While the CTFPHC makes screening recommendations, it is the policy-makers who are left to implement policies for their jurisdictions. Across Canada, there has been no formalized screening program for prostate cancer using the PSA test. Despite this, high rates of unorganized, opportunistic PSA testing remain in Canada even in the absence of formalized screening programs (Beaulac et al. 2006; Canadian Cancer Society's Advisory Committee on Cancer Statistics 2012). This means that although many asymptomatic men are being tested for prostate cancer, they are being exposed to potentially unnecessary harm. There is also variability over whether PSA testing is an insured health service across Canada. In some provinces (such as Manitoba), the cost of a PSA test is covered as a publicly financed health service, while in others (such as Ontario), the patient must bear the cost of the test (approximately \$30–50)[1] if the test is not ordered because of a clinical suspicion of prostate cancer (Ontario Government 2021). In Canada, which has a publicly funded health system, this variability in funding may reflect uncertainty around the value of the test.

For breast cancer, all Canadian provinces and territories (aside from Nunavut) have screening programs (as a fully insured health service) that include women aged 50–69. There is considerable variation between provinces for their practice guidelines for women under 50. Some begin at age 40 (such as Prince Edward Island), others permit self-referral of women 40–49 years old (such as British Columbia), and most remaining provinces (such as Alberta) only permit screening of women aged 40–49 if deemed high risk or have a physician referral (Canadian Partnership Against Cancer, 2021/2022). These current practice guidelines have undergone subtle shifts over the past several years and are also likely to

[1] At the time of our study during focus group data collection (data not included here), men reported paying \$30. Current lab rates list \$35. https://www.lifelabs.com/test/prostate-specific-antigen-psa-test/

continue evolving in the future (Canadian Partnership Against Cancer 2012). While the CTFPHC's recommendation is intended to prevent excessive harms due to false positives and overdiagnosis, the rates of false positives in Canada have been shown to be excessive, with younger women having a higher frequency than older women (Le et al. 2016).

The dissonance between recommendations and practice in prostate cancer screening, as well as the provincial variations in policy for breast cancer screening, demonstrate how the uncertainties of mammography screening and PSA testing efficacy can filter their way from the existing evidence, to policy recommendations, and down to clinical practice. In the clinical setting, these uncertainties may also be further complicated by people's political, economic, cultural, and personal values (all of which can inform the subjective weighting of risks versus benefits) or the practicalities of the clinical encounter (Atkins et al. 2005). Doctors may tend to rely more on their individual judgments and diagnostic routines (Goldman and Shih 2011), or patients may insist on asymptomatic testing and overestimate the benefits of mammography screening and PSA testing (Volk et al. 2003; Woloshin et al. 2000). To bridge the divide, Shared Decision-Making (SDM) has been promoted as a strategy that can be sensitive both to the nuances of the evidence (and perhaps reduce unnecessary testing) *and* the values of patients and doctors (who may be more inclined to test).

The Shared Decision-Making (SDM) Model

The shared decision-making (SDM) model has been lauded as a more equitable and empowering approach for guiding doctor-patient decision-making (Elwyn et al. 2012). SDM entails three stages: information exchange between the doctor and the patient about benefits and risks of a strategy or intervention as determined by existing evidence (generally initiated by the doctor) as well as the patient's values and preferences (generally solicited from the patient); deliberation on the options available; and finally the decision—with the ultimate decision being ideally consensus-based (Charles et al. 1997, 1999). SDM ideally leverages more egalitarian power distribution between patients and doctors (Goodyear-Smith and Buetow 2001), as well as fosters trust that most patients have in their healthcare provider (HCP) as an information source (Chawla and Arora 2013; Ipsos Reid 2012; Kraetschmer et al. 2004, Thom et al. 2004).

In situations where treatment or diagnostic options are clear, SDM may not be the most suitable approach for clinical decision-making (Schrager et al. 2017). However, when there may be more than one equally valid option based on available evidence and interpretations, SDM may be the best way for doctors and patients to discuss and weigh options that allow for the consideration of patient values and preferences. In this vein, SDM seems well-suited to discussions about mammography screening and PSA testing between doctors and their patients. The ambiguous nature of the evidence implores a discussion about risks and benefits of the tests, as well as cultural values, preferences, and beliefs with calls for greater use of SDM in cancer screening decisions for several years (Stefanek 2011). In fact, the CTFPHC guidelines for PSA testing (2014) and mammography (2011 and the 2018 update) specifically state that despite their recommendations against screening for specific populations, doctors should still actively discuss risks and benefits and facilitate decisions that respect the values and preferences of their patients.

Patient preference must be considered in the decision-making practice. Here too a spectrum exists: some patients want more autonomous control over decisions, others prefer to more passively leave decision-making to their doctor, compared to those who may want a blended collaborative role (Flynn et al. 2006; Levinson et al. 2005; Nies et al. 2017). Despite the heterogeneity in patient preferences, ostensibly SDM should be amenable to these kinds of individualized factors, and allow whatever preferences that exist to have their place in any decision-making process about screening (Elwyn et al. 2016). However, a number of studies on cancer screening decision-making have shown that SDM has not been well incorporated into practice (DuBenske et al. 2018; Feng et al. 2013; Katz et al. 2012; Hoffman et al. 2014).

Doctors may also have their own unique practices and preferences that can vary between individuals. This can be a barrier to SDM in terms of the application of best practice guidelines as well as informed decision-making. Physicians may be more likely to screen or test in excess of recommendations for a variety of reasons: they may defer to patient concerns/demands, disagree with guidelines, worry about missing potential cancer, or lack time to discuss risks and benefits (Haas et al. 2016). In addition, values around potential cost savings for the healthcare system versus individual rights to health care may play into decision-making practices. Preferences or constraints along these lines can lead to outcomes where people who may not necessarily require screening or testing receive it anyway (satisfying the desires of patients or doctors), which may not line up with clinical recommendations (Driedger et al. 2017).

SDM continues to be promoted as the preferred approach to making clinical decisions about cancer screening (Hoover et al. 2018; Lang et al 2018; Schrager et al 2017). We are currently left with a dilemma of how to reconcile evidence-based recommendations with the imperatives of SDM. On the one hand, prevailing guidelines posit that rates of mammography screening and PSA testing should be reduced to prevent unnecessary treatments. On the other hand, SDM promotes an understanding of risks and benefits, but it also invites individual preferences to inform the decision, which may be a demand for a test that could be unnecessary.

Methods: Key Informant Interviews with Senior-Level Policymakers

Senior-level policymakers ($n = 12$) were identified by members of the research team through purposive and snowball sampling and were located at provincial/territorial cancer agencies ($n = 6$), ministries of health in provinces/territories ($n = 3$), at national cancer advocacy organizations ($n = 2$) and expert panels/organizations that publish clinical guidelines ($n = 1$).

Ethics approval was obtained from the University of Manitoba Health Research Ethics Board (Reference number: H2010: 194). Following informed consent protocols, face-to-face and telephone interviews with persons responsible for cancer screening policy across Canada[2] were conducted between November 2012 and March 2014. All interviews were digitally audio-recorded, transcribed verbatim, verified for accuracy, and imported into NVivo9™. Detailed codebooks were established following an iterative process of reviewing transcript excerpts for all datasets and developing coding schemes following standard protocols (Richards 2014). After coding data, we identified key themes in the data using the constant comparative and concept-development approach (Strauss and Corbin 1998), searched for contradictory or contrasting perspectives, and used triangulation to identify areas of agreement/disagreement across the dataset. In the results presented

[2] Despite multiple attempts, no participants were recruited from the Maritimes.

below, selected quotes are included to represent key themes identified throughout the dataset and policymakers' names and positions are excluded to protect anonymity.

Results

Key informants shared a number of inter-related factors about the evolving nature of PSA testing and mammography screening, SDM and doctor-patient interaction, and potential influence for the integration of CTFPHC recommendations into clinical practice and policy. First, there has been shifting conceptions of the efficacy of mammography and PSA that being tested early and often is always beneficial. Second, participants described a shift that has occurred within the dynamics of clinical decision-making, with greater emphasis now being given to informed decision-making and patient autonomy. Third, these shifting paradigms underlying breast cancer screening and PSA testing require good science communication practice and health literacy skills. However, patients and doctors may be on differing levels of readiness to discuss new evidence and create space to allow for greater patient autonomy in informed decision-making. Fourth, individual patients can bring varying values and preferences that will influence their interpretation of evidence. Lastly, key informants frequently noted that doctors themselves may be one of the crucial obstacles that limit better integration of recommendations into practice.

Challenges in Shifting Entrenched Attitudes Among Different Stakeholders

Some of the participants asserted that for the past decades there had been a general consensus—even wholehearted enthusiasm—among the public, front-line medical staff, and policymakers as well, that asymptomatic screening and testing for cancers is generally a good thing to do. One key informant noted that the assumption that early screening or testing was always beneficial became a default "common sense" position. Any attempts to lower the rates of uptake for screening or testing could then easily be mischaracterized as being inherently wrong, biased in some way, or could leave health systems or doctors vulnerable to accusations that they are putting people's lives at risk just to save some money. When describing the massive exponential growth of PSA testing in their province over the previous decades, another key informant indicated that high rates of opportunistic testing had become so accepted that it had become "entrenched" and "out of control."

> So the clinicians writing up the requests for [tests], also the population, and the staff from the public health departments who have to administer these things ... they have the gung-ho attitude of let's do annual everything, screening everything, and if people are entitled to annual screening then cutting back must be a bad thing and we can't be seen to be cutting back for the sake of money.

Shifts to Individualized "Informed" Decision-Making and Communicating Risks and Benefits

Key informants held that clinical practice was not only shifting away from a default promotion of asymptomatic screening/testing, but also from paternalist dynamics of the doctor-patient encounter—where the doctor was expected to prescribe an intervention and the patient complies. Most agreed that these encounters are moving toward a model of shared (or 'informed'—the term that key informants used most often) decision-making.

> To whatever degree that the science is able to describe the benefits and risks [of screening], the decision-maker increasingly must be the individual [patient].

While most key informants agreed that individualized informed decision-making is the direction that practice is heading, many also maintained that this transition in clinical culture creates particular challenges. The overarching tension is that policymakers make decisions whether or not to set up screening programs based on population-based data, but that decisions about whether to screen or test for breast or prostate cancer are carried out at an individual patient-specific level. Reconciling the individual from the aggregate is fundamentally challenging because perceptions about patient benefits at the individual level may not be well aligned with population best-practice recommendations.

> This is sort of public health and a population approach to screening versus the individual. And we're [policymakers] not about the individual and

> that's often what the clinicians are focused on. We're talking about what's of benefit to a whole population target group [not the individual].

A challenge identified by many participants is that informed decision-making relies on the ability of doctors to adequately communicate health information and improve health literacy related to individual as well as population-based evidence. Most key informants agree that getting the messaging across is not simple or easy. Both patients and doctors may be new to discussions around decision-making and may struggle with either giving or receiving population-based risk/benefit information.

> I think that trying to communicate and talk about population level data and how that applies to the individual, I think that's quite new for patients and physicians too. It's a shift from how we used to function, and I think providers and patients are at different levels of readiness for that kind of role ... to move to an informed decision-making model, it's not that easy.

Key informants commonly noted that part of the challenge of communicating risk/benefit information is that the concepts conveyed may be confusing, and communicating the relative probabilities also means that the notion of uncertainty is very much part of the equation.

> We've moved from that sort of position of screening is always a good thing, to really then having to communicate the balance of benefit and harm. And how do you communicate uncertainty? Like, how do we get an individual to understand a number needed to screen to save a life, or what's the risk of overdiagnosis? Trying to make those concepts understandable to the public is a huge challenge.

Some key informants noted that patients may or may not even want that kind of information depending on the value that they assign it. Additionally, the evidence itself can be difficult to put into language that the lay public can understand and use to support their decision-making. The doctor-patient interaction is the assumed natural environment for having these conversations, but there may also be a need for a broader multi-sectoral push to communicate these concepts to the public in order to build up a more nuanced culture of understanding for uncertainty in evidence and policy.

> I think on the public side, at least as far as PSA goes, I don't think [educating] will be that bad. Because I think if you say to men, 'if you don't have any symptoms and you're under 50 you don't need this test,' they'll be okay with that. Because a lot of them, like I say, probably don't even know they're getting it anyway. But how do you educate an entire population? That's not easy. So we're going to have multiple avenues for public education, whether it's through media campaigns, one-on-one consultations, and mail outs, plus the physician piece. You've got to come at them from all angles.

On the other hand, some patients may prefer that the decision-making role should rest primarily with the doctor, and are more willing to eschew discussions of evidence and support whatever option their doctor proposes.

Communicating Patient Values and Preferences

While imparting information about screening and testing risks and benefits can be a challenging and novel task for doctors, policymakers acknowledged that each patient navigates their situation according to their own values, preferences, and priorities. For informed decision-making, the doctor needs to solicit this information from their patient and then balance it with communicating information about the risks and benefits of screening or testing in order to best support shared decision-making.

> The evidence [for cancer screening] may have the magnitude of benefits and people put different weights on the risks of the procedure. So that information may be there, but the value you put around the benefits or the risks may be different. And so it's not just a number kind of assessment, you know, a bigger number is better or a lower number is better. There is the human value that you attach to it, 'Does that matter to *me*?'

Other key informants agreed that a process of active information and values exchange was necessary to arrive at the optimal decision that balances fidelity to both prevailing recommendations as well as patient values. There are, as one participant noted, instances when the 'right' decision (as perceived by the patient—or the doctor for that matter) is not going to line up with existing recommendations given a particular patient's values and preferences. But, when good communication is practiced, the odds are increased that a doctor can be sensitive to patient values and that the evidence underpinning recommendations is understood and considered by the patient as well. Regardless of the decision, the important component is to preserve trust, amicability, and open communication between the doctor and patient.

> We don't even have a systematic approach to engage the public and individuals. Someone could be helped in the decision about getting a mammogram by just somebody who understands what their issues are. You know, so that a doctor who realizes the biggest issue for a woman is fear of breast cancer. It is such a big issue that she can't sleep at night because her aunt died of breast cancer. She should be screened if that's what she wants, but she should be told that there's risks of screening. We might get it wrong. You might have an unnecessary biopsy. And she says, 'I don't care, you can do 10 biopsies, I want to go to sleep at night knowing that the chance of me having cancer is as low as it can be. She should get screened. It'll save our health system millions of dollars in prescription drugs. You could look at it any way you want. And the person who says if I have breast cancer, that's God's way of telling me it's time to visit Him, I'll get it when I get it, I don't want to know until I have to. That person should not be screened. *Ever*. So, we're out of date in how we approach this stuff, but that's complicated. And governments hate complexity. They hate individual judgment—it's all got to be reduced to a one line sound bite. [It] drives me nuts.

Physician Practice, the Challenge of Change and Needing to Change the Path

Supporting shared decision-making practices often falls to doctors to implement better communication around health decision-making options. The rationale for shifting responsibility from the healthcare system and policymakers to doctors was raised by many of our key informants. Key informants regularly pointed to the catch-22 of implementing SDM. They discussed the need for personalized practice as doctors, established norms around screening acceptability, and the very real constraints of appointment length as barriers to soliciting patient preferences and improving health literacy around the benefits and harms of screening and/or testing. One key informant summed up many of these challenges.

> [PSA testing's] got the highest participation rate. And it's the program that's basically not recommended. So it's just bizarre. And when the clinical

> trials come out and showed that there really wasn't much benefit to PSA testing, okay you would hope that the physicians would take that and adjust their practice. But that's the other challenge, physicians' practice behavior. Oh boy, getting them to change is always really difficult. A lot of physicians will say, 'I *found* prostate cancer in some of these guys and the guy was asymptomatic. If I hadn't done this that might have been the one that would have been more aggressive. So isn't it easier for me to just test them all and not worry about it? Versus trying to be selective about who I should test?' We have programmed our population that breast screening, that finding cancer early is good, right? That just makes sense, doesn't it? It's pretty hard now to go out to the public and say, 'Well that doesn't really apply to prostate cancer. Finding it early might not make a difference.' It's a contradictory message ... I fully understand the General Practitioner in his or her position like, 'I don't have time to explain all this stuff to people and I know that we do find these cancers.' So that's where we're at.

Likewise, a few policymakers described the need to develop relatively conservative strategies for dealing with uncertainty about screening, such as needing to cultivate the support of physician groups before attempting to implement a new policy. By contrast, one cancer agency was contemplating a more disruptive response for change. In some provinces, the PSA test is included in the checklist of routine lab requisition forms. This one cancer agency was considering having the test removed from the routine forms, and implementing a different form specifically for the PSA test to stimulate physicians and patients to devote greater consideration before ordering the test for screening purposes. Instead of changing the mindset of physicians, that cancer agency wanted to change the pathway for how requisitions were ordered:

> Part of the problem with the PSA test is that it's too easy—it's just another check box on your lab blood requisition form. We want to have PSA taken off the form so that a physician would have to have a discussion with the patient about the test. A lot of guys are having a PSA test and they don't even know it, because the doctor takes blood and checks off cholesterol and this and that, and the guy doesn't know everything that's being checked.

Discussion

While numerous other studies have focused on the roles that doctors and patients play in making decisions about cancer screening, the perspectives of policymakers shed fresh light on those responsible for interpreting emerging evidence and recommendations and implementing guidelines for clinical practice. Upstream interventions that use population-based evidence and a systems perspective to determine efficacy and feasibility influencing downstream outcomes are central to the democratization of risk. SDM is essential at the micro level to unpack what constitutes an informed decision when faced with scenarios where uncertainty is central to the situation.

Some participants may have perceived SDM to be more embedded in practice than it is due to generalized discourses around the need for SDM and patient-centered care, rather than as a change that is still very much required and in its infancy. Many stakeholders may be new to SDM and there may be challenges involved in shifting values and preferences in a meaningful way toward a conversation that focuses on ensuring people comprehend risks and benefits. The assumption that SDM and its egalitarian emphasis was 'the norm' in screening decisions may reflect tensions between the macro level where policymakers reside and the micro level where clinician and patients interact. An additional clue that some mischaracterization of SDM dynamics existed in the comments of key informants was their common use of the term 'informed' decision-making, rather than 'shared.' Although their description of the shifts that were occurring were apt to capture the dynamics of SDM (e.g., a bi-directional flow of information between doctors and patients and collaborative decisions), 'informed' is not synonymous with 'shared.' 'Informed' decision-making is unidirectional with information flowing from the doctor and the patient making the decision (Schrager et al. 2017). True (and as participants noted) 'shared' and 'informed' decision-making places the patient in more of a decision-making role than paternalistic care, but participants' use of 'informed' suggests that there may be lingering confusion on the policy level of what is being promoted or realized by SDM for breast and prostate cancer screening discussions. Key informants were keen to pinpoint the challenges involved in SDM's potential application which mirrored challenges identified in existing literature. While at times patients might be happy to function within an informed decision-making context, there remains dissonance between policy and practice expectations. A lack of dialogue and reciprocal information exchange is a long-standing gap reported by patients about their discussions with doctors about cancer screening decisions (Hoffman et al. 2014; Katz et al. 2012). Yet patient preferences for cancer screening and testing are crucial in situations where uncertainty is present and patient-based considerations of risk/benefit trade-offs can be a key part of the decision-making equation (Gunn et al. 2021; Howard et al. 2015; Nguyen et al. 2021). This fact likely underpins the continued promotion of SDM in cancer screening decisions and the tensions between cultural values and stakeholder preferences.

Study participants identified a lack of risk/benefit information about mammography or PSA testing from doctors to their patients as a challenge to SDM adoption. Participants agreed that sharing population-based information about uncertainties and probabilities can be a difficult task, given the conflicts, subtleties, and nuances of the evidence (Bell et al. 2017; Keating and Pace 2018; Lang et al. 2018). Some participants believed that conveying risk/benefit and uncertainty information about testing or screening is something to which doctors may not yet be that accustomed. Doctors may avoid SDM conversations and ignore prevailing guidelines, especially if they have failed to previously catch cases of asymptomatic cancer or they simply lack the time for sharing risk/benefit information (Haas et al. 2016). A lack of discussion has consequences, as several studies have shown that men are less likely to opt for PSA testing if they are given the benefits and risks of testing and made aware of the existing uncertainties (Flood et al. 1996; Volk et al. 2003). Another study has also shown that if physicians are given a primer on PSA testing, during follow-up visits with men, there is increased likelihood for SDM and greater potential for reduced PSA testing (Feng et al. 2013).

Some participants identified a core challenge—or contradiction—of applying population-based evidence into the clinical encounter and were wholly aware of the limitations and strictures of their perspective as policymakers. From their point of view, they see population-level trends whereas physicians and patients are often looking downstream at individual-level outcomes. As we found, participants spent considerable time discussing the micro-dynamics of doctor-patient interaction—an unavoidable nod toward the bridge that population-based evidence must cross. This seeming dissonance between the macro and micro levels has indeed formed the basis of previous criticism of clinical guidelines based on population-level evidence (Trinder and Reynolds 2000). The natural extension of this dilemma is that (as noted above) even within the parameters of SDM—or especially so, given the prominence it accords to patient preference—there will be many instances where the decision that is made will not align with evidence-based recommendations. In many cases, the odds of unnecessary testing taking place can be reduced if the patient knows which way the evidence is leaning and can negotiate their own values in light of it (Klarenbach et al. 2018). Thus, for mammography and PSA testing, doctors need to also allow the time for values clarification (if a patient so desires) alongside providing the relevant evidence (despite its relative complexity). Our study findings support the continued call for doctors to better integrate SDM into their discussions with patients (Bell et al. 2017; Dasarathy and Rajesh 2020). If asymptomatic mammography screening and PSA testing is to be better negotiated with SDM—with its greater attentiveness to being preference-sensitive health care—perhaps some degree of what is considered unnecessary testing (by the standards of population-based evidence) will be unavoidable. Even if the decision does not align with existing recommendations, we can hope that the discussions around it were equitable and informed, the preferences of the parties were recognized, the potential for a decision that matches evidence-based guidelines was advanced, and that the relationship between decision-making parties was fostered.

While patients may require their own help in comprehending the evidence through a variety of means (i.e., public service announcements, advertising campaigns, clinical discussions), a valuable insight provided by some participants was that doctors themselves may also need some continuing education on the newer and nuanced styles of interpreting and communicating the "gray areas" of evidence for breast and prostate cancer screening. As we illustrate through our results, we may still be quite early on in our understanding of the paradigm shifts that have taken place and how those who are faced with responding to them may be struggling to do so.

Conclusion

This chapter focused on a real tension concerning public involvement in health risk decisions. At the macro risk governance level, there is little engagement with the public around population-based recommendations or how health systems implement national recommendations in their jurisdictions. At the micro doctor-patient level, there are several opportunities through SDM for individual values and preferences to influence decisions. This dual tension at the macro and micro level is not wholly inappropriate in a health system that must adhere to using public resources in an equitable and defensible manner. Policymaker participants provided a unique perspective on the challenges facing SDM adoption as well as tensions inherent in establishing clinical policy for diverse populations from the more lofty heights of the policy realm. Fundamentally, more research into when 'less care' is in fact 'better care' is needed not only about how to navigate these conversations at the micro level between the doctor and patient, but also at the systems level when making policies based on effective use of scarce resources to best serve the population.

Acknowledgements The authors wish to thank the study participants for sharing their time, experiences, and insights regarding decision-making about breast and prostate cancer screening. This research was supported by a grant from the Canadian Cancer Society Research Institute held by author SMD entitled "Advancing quality in cancer control and cancer system performance in the face of uncertainty" (grant #700589). The funder had no role in study design, data collection and analysis, decision to publish, or preparation of the manuscript. We gratefully acknowledge the contributions of Gary Annable (University of Manitoba), who was a full-time researcher with the project, and project co-lead, Dr. Melissa Browers (University of Ottawa), for their helpful comments on a much earlier draft of this chapter.

References

American College of Obstetricians-Gynecologists. (2011). Practice bulletin no. 122: Breast cancer screening. *Obstetrics and Gynecology*, 118(2 Pt 1), 372–382.

Atkins, D., Siegel, J & Slutsky, J. (2005). Making policy when the evidence is in dispute. *Health Affairs*, 24(1), 102–113.

Autier, P. & Boniol, M. (2018). Mammography screening: A major issue in medicine. *European Journal of Cancer*, 90, 34–62.

Beaulac, J. A., Fry, N.R., & Onysko, J. (2006). Lifetime and recent prostate specific antigen (PSA) screening of men for prostate cancer in Canada. *Canadian Journal of Public Health*, 97(3), 171–176.

Bell, N., Connor, S., Gorber, A., Joffres, M., Singh, H., Dickinson, J., Shaw, E., Dunfield, L., & Tonelli, M. (2014). Recommendations on screening for prostate cancer with the prostate-specific antigen test. *Canadian Medical Association Journal,* 186(16), 1225–1234. https://doi.org/10.1503/cmaj.140703

Bell, N., Grad, R., Dickinson, J., Singh, H., Moore, A., Kasperavicius, D., & Kretschmer, K. (2017). Better decision making in preventive health screening: Balancing benefits and harms. *Canadian Family Physician/Medecin de Famille Canadien,* 63(7), 521–524.

Bhatt, J., & Klotz, L. (2016). Overtreatment in cancer—is it a problem? *Expert Opinion on Pharmacotherapy,* 17(1), 1–5. https://doi.org/10.1517/14656566.2016.1115481

Canadian Cancer Society's Advisory Committee on Cancer Statistics. (2012). Canadian cancer statistics 2012. Toronto, Canadian Cancer Society. Retrieved April 28, 2021, from https://www.cancer.ca/~/media/cancer.ca/CW/cancer%20information/cancer%20101/Canadian%20cancer%20statistics/Canadian-Cancer-Statistics-2012-EN.pdf?la=en#:~:text=An%20estimated%20186%2C400%20new%20cases%20of%20cancer%20and%2075%2C700%20cancer,48%25%20in%20women

Canadian Partnership Against Cancer. (2012). Breast control in Canada: A system performance focus report. Toronto, Canadian Partnership Against Cancer.

Canadian Partnership Against Cancer. (2021/2022). Breast Cancer Screening in Canada: 2021/2022. Retrieved September 16, 2022 from https://www.partnershipagainstcancer.ca/topics/breast-cancer-screening-in-canada-2021-2022/summary/

Canadian Task Force on Preventive Health Care. (2019). About Us. Retrieved January 13, 2021, from http://canadiantaskforce.ca/about/.

Charles, C., Gafni, A., & Whelan, T. (1997). Shared decision-making in the medical encounter: What does it mean? (Or it takes at least two to tango). *Social Science and Medicine,* 44(5), 681–692.

Charles, C., Gafni, A., & Whelan, T. (1999). Decision-making in the physician–patient encounter: revisiting the shared treatment decision-making model. *Social Science and Medicine,* 49(5), 651–661.

Chawla, N., & Arora, N. K. (2013). Why do some patients prefer to leave decisions up to the doctor: Lack of self-efficacy or a matter of trust? *Journal of Cancer Survivorship,* 7(4), 592–601.

Croswell, J. M., Kramer, B. S., & Crawford, E. D. (2011). Screening for prostate cancer with PSA testing: current status and future directions. *Oncology (Williston Park),* 25(6), 452–460, 463.

Dasarathy, J., & Rajesh, R. (2020). PSA cancer screening: a case for shared decision-making. *Journal of Family Practice,* 69(1), 26;28; 30;32; 46. PMID: 32017831.

Driedger, S. M., Annable, G., Turner, D., Brouwers, M., & Maier, R. (2017). Can you un-ring the bell? A qualitative study examination of the role of affect in cancer screening decision making. *BMC Cancer,* 17(647). DOI https://doi.org/10.1186/s12885-017-3596-7.

DuBenske, L. L., Schrager, S. B., Hitchcock, M. E., Kane, A. K., Little, T. A., McDowell, H. E., & Burnside, E. S. (2018). Key elements of mammography shared decision-making: a scoping review of the literature. *Journal of General Internal Medicine,* 33(10):1805–1814. https://doi.org/10.1007/s11606-018-4576-6. Epub 2018 Jul 20. PMID: 30030738; PMCID: PMC6153221.

Dunfield, L., Usman, A., Fitzpatrick-Lewis, D., & Shane, A. (2014). Screening for prostate cancer with prostate specific antigen and treatment of early-stage or screen-detected prostate cancer: A systematic review of the clinical benefits and harms. Retrieved March 15, 2021, from https://canadiantaskforce.ca/wp-content/uploads/2014/11/2014-prostate-cancer-systematic-review-en.pdf

Elwyn, G., Frosch, D., Thomson, R., Joseph-Williams, N., Lloyd, A., Kinnersley, P., Cording, E., Tomson, D., Dodd, C., Rollnick, S., Edwards, A., & Barry, M. (2012). Shared decision making: a model for clinical practice. *Journal of General Internal Medicine,* 27(10), 1361–1367.

Elwyn, G., Frosch, D. L., & Kobrin, S. (2016). Implementing shared decision-making: consider all the consequences. *Implementation Science,* IS(11), 114–114.

Feng, B., Srinivasan, M., Hoffman, J. R., Rainwater, J. A., Griffin, E., Dragojevic, M., Day, F. C., & Wilkes, M. S. (2013). Physician communication regarding prostate cancer screening: Analysis of unannounced standardized patient visits. *Annals of Family Medicine,* 11(4), 315–323.

Fitzpatrick-Lewis, D., Hodgson, N., Ciliska, D., Pierson, L., Gauld, M., & Liu, Y. Y. (2011). Breast Cancer Screening. Hamilton, Ontario, McMaster University.

Flood, A. B., Wennberg, J. E., Nease, Jr., R. F., Fowler, Jr., F. J., Ding, J., & Hynes, L.M. (1996). The importance of patient preference in the decision to screen for prostate cancer. Prostate patient outcomes research team. *Journal of General Internal Medicine,* 1(6),342–349.

Flynn, K. E., Smith, M. A., & Vanness, D. (2006). A typology of preferences for participation in healthcare decision making. *Social Science and Medicine,* 63(5), 1158–1169.

Goldman, J. J., & Shih, T. L. (2011). The limitations of evidence-based medicine--applying population-based recommendations to individual patients. *Virtual Mentor,* 13(1), 26–30.

Goodyear-Smith, F., & Buetow, S. (2001). Power issues in the doctor-patient relationship. *Health Care Analysis,* 9(4), 449–462.

Gunn, C. M., Maschke, A., Paasche-Orlow, M. K., Kressin, N. R., Schonberg, M. A., & Battaglia, T. A. (2021). Engaging women with limited health literacy in mammography decision-making: Perspectives of patients and primary care providers. *Journal of General Internal Medicine,* 36(4),938–945. https://doi.org/10.1007/s11606-020-06213-2. Epub 2020 Sep 15. PMID: 32935318; PMCID: PMC8042081.

Haas, J. S., Sprague, B. L., Klabunde, C. N., Tosteson, A, N., Chen, J. S., Bitton, A., Beaber, E. F., Onega, T., Kim, J. J., MacLean, C. D., Harris, K., Yamartino, P., Howe, K., Pearson, L., Feldman, S., Brawarsky, P., & Schapira, M. M. (2016). Provider attitudes and screening practices following changes in breast and cervical cancer screening guidelines. *Journal of General Internal Medicine,* 31(1), 52–59.

Hirsch, B. R., & Lyman, G. H. (2011). Breast cancer screening with mammography. *Current Oncology Report,* 13(1), 63–70.

Hoffman, K. E., & Nguyen, P. L. (2011). The debate over prostate cancer screening guidelines. *Virtual Mentor,* 13(1), 10–15.

Hoffman, R. M., Elmore, J. G., Fairfield, K. M., Gerstein, B. S., Levin, C. A., & Pignone, M. P. (2014). Lack of shared decision making in cancer screening discussions: results from a national survey. *American Journal of Preventative Medicine,* 47(3), 251–259.

Hoover, D. S., Pappadis, M. R., Housten, A. J., Krishnan, S., Weller, S. C., Giordano, S. H., Bevers, T. B., Goodwin, J. S., & Volk, R. J. (2018). Preferences for communicating about breast cancer screening among racially/ethnically diverse older women. *Health Communication,* 1–5.

Howard, K., Salkeld, G. P. Patel, M. I., Mann, G. J., & Pignone, M. P. (2015). Men's preferences and trade-offs for prostate cancer screening: A discrete choice experiment. *Health Expectations,* 18(6), 3123–3135.

Ipsos Reid. (2012). Life-savers, medical professionals top the list of most trusted professionals. Toronto, Ipsos Reid.

Katz, M. L., Broder-Oldach, B., Fisher, J. L., King, J., Eubanks, K., Fleming, K., & Paskett, E. D. (2012). Patient-provider discussions about colorectal cancer screening: who initiates elements of informed decision making? *Journal of General Internal Medicine,* 27(9), 1135–1141.

Keating, N. L. & Pace, L. E. (2018). Breast cancer screening in 2018: Time for shared decision making. *Journal of the American Medical Association,* 319(17), 1814–1815.

Kelly, A. M. & Cronin, P. (2011). Rationing and health care reform: Not a question of if, but when. *Journal of the American College of Radiology,* 8(12), 830–837.

Klarenbach, S., Sims-Jones, N., Lewin, G., Singh, H., Thériault, G., Tonelli, M., Doull, M., Courage, S., Garcia, A. J., & Thombs, B. D. (2018). Recommendations on screening for breast cancer in women aged 40–74 years who are not at increased risk for breast cancer. *Canadian Medical Association Journal,* 190(49), E1441.

Kraetschmer, N., Sharpe, N., Urowitz, S., & Deber, R. B. (2004). How does trust affect patient preferences for participation in decision-making? *Health Expectations,* 7(4), 317–326.

Lang, E., Bell, N. R., Dickinson, J. A., Grad, R., Kasperavicius, D., Moore, A. E., Singh, H., Theriault, G., Wilson, B. J., & Stacey, D. (2018). Eliciting patient values and preferences to inform shared decision making in preventive screening. *Canadian Family Physician,* 64(1), 28–31.

Layne, G. P. (2016). Screening mammography: Controversy and recommendations. *West Virginia Medical Journal,* 112(6), 26–29.

Law, K. W., Nguyen, D. D., Barkin, J., & Zorn, K. C., (2020). Diagnosis of prostate cancer: the implications and proper utilization of PSA and its variants; indications and use of MRI and biomarkers. *Canadian Journal of Urology,* 27(27 Suppl. 1), 3–10. PMID: 32101694.

Le, M. T., Mothersill, C. E., Seymour, C. B., & McNeill F. E. (2016, September). Is the false-positive rate in mammography in North America too high? *British Journal of Radiology,* 89(1065), 20160045.

Levinson, W., Kao, A., Kuby, A., & Thisted, R. A. (2005). Not all patients want to participate in decision making: A national study of public preferences. *Journal of General Internal Medicine,* 20(6), 531–535.

Miller, A. B., Wall, C., Baines, C.J., Sun, P., To, T., & Narod, S. A. (2014). Twenty five year follow-up for breast cancer incidence and mortality of the Canadian National Breast Screening Study: Randomised screening trial. *British Medical Journal,* 348, g366.

Morrison, A. (1985). *Screening in Chronic Disease.* New York, Oxford University Press.

Nies, Y. H., Islahudin, F., Chong, W. W., Abdullah, N., Ismail, F., Ahmad Bustamam, R. S., Wong, Y. F., Saladina, J. J., & Mohamed Shah, N. (2017). Treatment decision-making among breast cancer patients in Malaysia. *Patient Prefer Adherence,* 11, 1767–1777.

Nguyen, D. D., Trinh, Q. D., Cole, A. P., Kilbridge, K. L., Mahal B. A., Hayn, M., Hansen, M., Han, P. K. J., & Sammon, J. D. (2021). Impact of health literacy on shared decision making for prostate-specific antigen screening in the United States. *Cancer,* 27(2), 249–256. https://doi.org/10.1002/cncr.33239. Epub 2020 Nov 9. PMID: 33165954.

Obort, A. S., Ajadi, M. B., & Akinloye, O. (2013). Prostate-specific antigen: Any successor in sight? *Reviews in Urology,* 15(3), 97–107.

Ontario Government. (2021). *Prostate cancer screening.* Retrieved April 28, 2021 from https://www.ontario.ca/page/prostate-cancer-screening

Printz, C. (2014). Mammogram debate flares up: Latest breast cancer screening study fuels controversy. *Cancer,* 120(12), 1755–1756.

Ray, K. M., Joe, B. N., Freimanis, R. I., Sickles, E. A., & Hendrick, R. E. (2017). Screening mammography in women 40–49 years old: Current evidence. *American Journal of Roentgenology,* 210(2), 264–270.
Richards, L. (2014). *Handling qualitative data* (3rd ed.). Sage Publishing.
Schrager, S. B., Phillips, G., & Burnside, E. (2017). A simple approach to shared decision making in cancer screening. *Family Practice Management,* 24(3), 5–10.
Stefanek, M. E. (2011). Uninformed compliance or informed choice? A needed shift in our approach to cancer screening. *Journal of the National Cancer Institute,* 103(24), 1821–1826.
Strauss, A. L., & Corbin, J. (1998). *Basics of qualitative research: Techniques and procedures for developing grounded theory.* Sage Publications Inc.
The Canadian Task Force on Preventive Health Care. (2011). Recommendations on screening for breast cancer in average-risk women aged 40–74 years. *Canadian Medical Association Journal,* 183(17), 1991–2001.
Thom, D. H., Hall, M. A., & Pawlson, L. G. (2004). Measuring patients' trust in physicians when assessing quality of care. *Health Affairs,* 23(4), 124–132.
Trinder, L. & Reynolds, S. (Eds.). (2000). *Evidence-based practice: A critical appraisal.* Oxford.
Volk, R. J., Spann, S. J., Cass, A. R., & Hawley, S. T. (2003). Patient education for informed decision making about prostate cancer screening: A randomized controlled trial with 1-year follow-up. *Annals of Family Medicine,* 1(1), 22–28.
Woloshin, S., Schwartz, L. M., Byram, S. J., Sox, H. C., Fischhoff, B., & Welch, H. G. (2000). Women's understanding of the mammography screening debate. *Archives of Internal Medicine,* 160(10), 1434–1440.

CHAPTER 11

Public Engagement on Childhood Vaccination: Democratizing Policy Decision-Making Through Public Deliberation

Kim H. Chuong, Amanda Rotella, Elizabeth J. Cooper, and Kieran C. O'Doherty

Introduction[1]

Immunization is considered one of the most successful and cost-effective public health interventions by the World Health Organization, preventing

[1] Since writing this chapter, the introduction of vaccines for COVID-19 has dramatically affected public discourse on vaccination. Because the main focus of this chapter is on democratization of risk in the context of childhood vaccination, the COVID-19 vaccine is strictly speaking not within the scope of the chapter. However, much public discourse on COVID-19 vaccines has focused on the vaccination of children, and controversy has surrounded questions of whether vaccination should be mandatory. We therefore include an appendix at the end of the chapter that outlines key developments since the COVID-19 vaccine became available.

K. H. Chuong · K. C. O'Doherty (✉)
Department of Psychology, College of Social and Applied Human Sciences, University of Guelph, Guelph, ON, Canada
e-mail: odohertk@uoguelph.ca

M. Gattinger (ed.), *Democratizing Risk Governance*,
https://doi.org/10.1007/978-3-031-24271-7_11

an estimated 2 to 3 million deaths per year globally (WHO 2018). Yet, there is a growing concern that vaccination rates are insufficient to effectively control the spread of infectious diseases. According to the 2018 assessment report by the WHO Strategic Advisory Group of Experts on Immunization, major outbreaks of measles and diphtheria had recently occurred in several regions across the world attributable to low national vaccination coverage or pockets of low coverage within a region. When a sufficiently high proportion of a population becomes immune to a disease, either through vaccination or from having previously been infected, "herd immunity" is achieved, which inhibits the spread of the infectious disease. Herd immunity protects susceptible individuals who are not vaccinated or under-vaccinated[2] for a variety of reasons (e.g., allergies, medical complications, accessibility of vaccines), by reducing the risk of infection (Fine et al. 2011; WHO Strategic Advisory Group of Experts on Immunization 2018). The percentage of a population required to be vaccinated to achieve herd immunity varies according to the disease, where herd immunity threshold estimates range from 75% for mumps to 94% for pertussis and measles (Plans-Rubió 2012). These high thresholds create challenges from a public health perspective as it is difficult to vaccinate large proportions of populations to achieve herd immunity. It is important to note that vaccination is generally preferred as a mechanism for achieving herd immunity (as opposed to substantial portions of the population being exposed to the illness) as vaccines typically are safer, with fewer long-term health implications than the disease itself.

As with most health interventions, there are benefits and risks associated with vaccination. Vaccination confers individual benefits of gaining immunity as well as collective benefits in the form of herd immunity. But vaccination does not fully prevent infection and there is always a possibility of contracting the disease during an outbreak, even among

A. Rotella
Department of Psychology, Northumbria University, Newcastle Upon Tyne, UK
e-mail: amanda.rotella@northumbria.ac.uk

E. J. Cooper
University of Regina, Regina, SK, Canada
e-mail: Elizabeth.Cooper@uregina.ca

[2] Under-vaccinated here refers to selective, delayed, or partial vaccination.

people who are vaccinated (Law Reform of Commission of Saskatchewan 2009). In addition, some individuals experience adverse events when being vaccinated, even among vaccines deemed safe and effective and which meet regulatory standards (Keelan and Wilson 2011; Looker and Kelly 2011). Most of the adverse events are minor and short term, such as localized soreness or swelling at the injection site. The government approval process is such that approved vaccines have low risk of serious or life-threatening injuries. In Ontario, the annual report on vaccine safety for 2017 documented 696 reports of adverse events following immunization (Public Health Ontario 2017). Of these, 558 were classified as mild (sore arms, rashes, allergic skin reactions), and 26 were classified as serious (see p. 44 of the Report for a detailed description of the adverse events), equivalent to 3 per 1 million doses distributed. Given the benefits and risks associated with vaccination, there is considerable diversity in public opinion about when and how to vaccinate children and formal recommendations can vary by country or region (Song et al. 2014).

There has been some controversy relating to claims about vaccine safety and efficacy. Understandably, public health officials tend to be wary of any claims that undermine belief in the safety and efficacy of vaccines since this translates into lower vaccination rates, which in turn compromises the goals of achieving herd immunity. Indeed, many claims about lack of vaccine safety and efficacy have been shown to be unfounded (Asturias et al. 2016; Conklin et al. 2021). While these considerations are both important and prominent in scholarly literature, our focus here is somewhat different. In the context of the theme of this book on democratization of risk decision-making, our attention is on the role of public voices in making decisions about vaccination policy. Our argument is that key policy decisions rely not only on expert knowledge and scientific data, but also on values and consideration of how to make trade-offs between competing values. For example, if the safety and efficacy of vaccines are accepted, policy decisions still need to be made about whether vaccination is compulsory, mandatory, or simply recommended. Making this decision requires that a trade-off is made between values that, in this situation, are in conflict with each other. Favouring values of protection and public health may indicate a policy decision of mandatory or even compulsory vaccination. On the other hand, favouring values of autonomy and diversity may indicate a policy of recommendation only. In line with principles of deliberative democracy, our argument is that such decisions therefore require meaningful democratic consideration and public input.

To date, there has been little opportunity for the Canadian public to engage and provide input into public policy decisions regarding childhood vaccination. Developing mechanisms for public dialogue and input is important to inform policy decisions that are reflective of the needs, interests, and values of Canadian publics. Public trust is crucial for the successful implementation of any public health program. Chafe et al. (2011) asserted that a loss of public confidence is the biggest threat to the Canadian healthcare system and increased public engagement is a means to restore confidence and rejuvenate healthcare reform. However, "despite extensive stakeholder engagement frameworks, few strategies for public input have been implemented at major agencies that award funds for research, recommend expensive drugs and vaccines, or set major health priorities" (Chafe et al. 2011, p. 165). Active and meaningful public engagement in health service delivery and health research is considered essential to quality improvement of health services, greater responsiveness to public needs, and more legitimate, transparent, and accountable decision-making (Esmail et al 2015; Kovacs Burns et al. 2014; Manafo et al. 2018).

Public engagement through deliberative processes has garnered increasing interest from public health researchers and policy makers on a number of health-related topics, including priority setting, planning and governance of health services, and health technology assessment (Degeling et al. 2015). Calls for deliberative approaches into vaccine-related discussions and policy decisions have also been made (Hendrix et al. 2016) and implemented (Marshall et al. 2014; Parrella et al. 2016; The Keystone Center 2009). In this chapter, we describe an academic-led public deliberation event on childhood vaccination in Ontario, Canada (O'Doherty et al. 2021), and examine this public engagement on childhood vaccination for its role in the democratization of science and policy. To our knowledge, there has not been any official, government-sponsored public engagement event for members of the public in Ontario to deliberate on the topic of childhood vaccination. Broadly speaking, public deliberation is based on deliberative democratic ideals that lay citizens should have a voice in the discussions and debates about important social issues and their potential decision options. Public deliberation is useful for policy issues that involve competing ethical values about a public good, high controversy, and considerations of both technical/expert and cultural/real-world knowledge (Solomon and Abelson 2012). It seeks to create conditions that allow for respectful dialogue among citizens and between policy makers and citizens. The process of deliberation fosters the exchange of opinions and reasons for or against particular propositions. Public deliberation aims to arrive at public decisions that all parties view as legitimate (Abelson et al. 2003; Blacksher 2013; Chambers 2003). In this sense, public deliberation is distinct from other participatory approaches in health and health care. Whereas other participatory approaches may have the primary goal of advancing health equity and social justice through redistribution of resources and empowerment of marginalized and minority groups, the purpose of public deliberation is to create spaces for public dialogue on an issue and to develop policy solutions that are broadly perceived as legitimate (Blacksher 2013).

We begin by providing a brief overview of the current regulatory context in Ontario with regard to childhood vaccination. We then outline the Ontario Vaccine Deliberation and the recommendations that were generated and endorsed by the participants through small and large group discussions during the deliberation. We draw on the Ontario Vaccine Deliberation, as well as other research, to illustrate the importance of engaging publics in childhood vaccination decision-making. Our discussion focuses most heavily on the three main issues that were raised by members of the public during the deliberation: (i) mandatory vaccination and non-medical exemptions, (ii) communication about vaccination, and (iii) compensation for individuals and families who experience serious adverse events following immunization. Further, we discuss the collective recommendations made by the Ontario Vaccine Deliberation group in the context of the broader literature on vaccination and deliberative processes.

Current Regulatory Context in Ontario

In Canada, the federal, provincial, and territorial governments share responsibility in the delivery of health care, including immunization (Public Health Agency of Canada 2014). At the national level, the Biologics and Genetic Therapies Directorate (BGTD) of Health Canada (HC) is the federal authority responsible for regulating the quality, safety, and efficacy of all biologic drugs, including vaccines. The National Advisory Committee on Immunizations (NACI) makes recommendations to the Public Health Agency of Canada (PHAC) for the use of vaccines that are currently or newly approved for use in humans. NACI is an independent advisory body of experts in the fields of pediatrics, infectious diseases, immunology, medical microbiology, internal medicine, and public health. Additionally, the National Immunization Strategy provides a framework for interjurisdictional collaboration on immunization planning and programming.

The provincial and territorial governments are responsible for the administration and delivery of immunization-related programs within their jurisdiction. Provincial/territorial immunization policies and schedules are developed in consultation with an expert advisory committee based on identified priorities, resource availability, and recommendations from expert sources such as NACI. In Ontario, the Provincial Infectious Diseases Advisory Committee on Immunization (PIDAC-I) advises Public Health Ontario (PHO) and the Ontario Ministry of Health and Long-Term Care (MHLTC) on matters related to vaccination.

Childhood vaccination is not compulsory under federal, provincial, or territorial laws in Canada. Currently, Ontario and News Brunswick are the only two provinces that require vaccination for children for school attendance. It should be noted that there is no standard approach to mandatory vaccination as vaccination programs that are described as "mandatory" vary widely across different countries (MacDonald et al. 2018). Mandatory vaccination has been distinguished from compulsory vaccination in legal aspects: parents are legally free to not vaccinate their children in the former, whereas vaccine refusal is treated as a crime and comes with legal penalties in the latter (Giubilini 2019; Navin and Largent 2017). However, a mandatory vaccination program may withhold access to valuable social goods or services (e.g., enrollment in public school or daycare) if parents refuse to vaccinate their children. In jurisdictions with mandatory vaccination legislation, exemptions are granted for medical clauses and may be granted for religious and conscientious clauses. Ontario's *Immunization of School Pupils Act* (ISPA) requires that children and adolescents attending primary and secondary schools be vaccinated against designated diseases, unless they have a valid exemption (Government of Ontario 2017). Parents keep a record of immunization for their children, which will be submitted and updated with a local public health unit upon school entry. Proof of immunization is required for nine infectious diseases: diphtheria, tetanus, polio, measles, mumps, rubella, meningococcal disease, pertussis (whooping cough), and varicella (chicken pox) for children born in 2010 or later (Ontario Ministry of Health and Long-Term Care 2018).

Valid exemptions in Ontario include (1) medical exemptions and (2) non-medical exemptions based on reasons of religion or conscience. Medical exemptions are granted upon receipt of a statement signed by a physician or registered nurse that vaccination may be detrimental to the health of the person (ISPA 1990), such as if the child has a medical condition that prevents receiving the vaccine. A medical exemption is also granted if the vaccine is deemed unnecessary for reason of past infection or laboratory evidence of immunity (ISPA 1990). As of September 2017, legislation requires that parents must complete an immunization educational session at a local public health unit prior to submitting a non-medical exemption for their children (Government of Ontario 2019). Parents receive a certificate at the end of the session, which they submit along with a statement of religious or conscientious belief signed by a commissioner for taking affidavits if they choose to proceed with the application.

Information on the rationale for childhood vaccination, vaccine schedules, disease risks, and risks of adverse reactions are publicly available via official government websites, distributed by healthcare centres and community resource centres. For example, the Public Health Agency of Canada provides *A Parents' Guide to Vaccination* which can be downloaded freely, ordered in print, and is distributed by some healthcare providers (Public Health Agency of Canada 2018). Free resources are also available online from various national institutions and organizations, such as the Canadian Paediatric Society (2019) and CanImmunize (2021). Communication about possible risks in a transparent manner is certainly critical to promote public trust and support for vaccination programs. It is also necessary to enable people to make informed choices regarding immunizations. However, survey research with Canadian parents has found that safety concerns remain, even among parents who are supportive and vaccinate their children according to recommended vaccination schedules (Greenberg et al. 2017).

The term, "vaccine hesitancy," has been used to describe a broad continuum of public attitudes and associated behaviors toward vaccination, ranging from total acceptance to complete refusal (Bedford et al. 2018; Dubé et al. 2013; Larson et al. 2014). Vaccine hesitancy is defined by the WHO Strategic Advisory Group of Experts (SAGE) Working Group on Vaccine Hesitancy as a "delay in acceptance or refusal of vaccination despite availability of vaccination services" (MacDonald 2015, p. 4163). Individuals who are seen to be vaccine-hesitant may refuse all vaccines, refuse some vaccines but agree to others, delay and seek alternative vaccination schedules, or accept vaccines but are unsure of doing so (Larson et al. 2014).

Parental decision-making about childhood vaccination is increasingly recognized as being nuanced and complex. The knowledge deficit model, which presumes that laypeople are resistant to vaccination due to misunderstanding or ignorance of science, has been increasingly criticized (e.g., Goldenberg 2016, 2021; Hobson-West 2003). Goldenberg (2016) argued that vaccine hesitancy is better framed as a problem of public mistrust of scientific institutions and experts, including the values underlying expert consensus in support of vaccination and their ties to the pharmaceutical industry. Moreover, many parents approach vaccine safety with consideration of the personal needs of their children and view the potential of rare but serious adverse events as a safety priority, rather than viewing it as a reasonable risk (Goldenberg 2016). From a public health perspective, the risks associated with serious adverse events are balanced by the benefits of wide-scale immunization in the population. However, individual and public perceptions of what constitute reasonable risks can be evaluated quite differently compared to those viewing health decision-making through a population lens. Parental decisions to accept or refuse a vaccine for their children are also grounded in factors other than risk information based on scientific data (Hobson-West 2003), such as past experiences with other vaccines or health services, religious and moral convictions, alternative health discourses (e.g., homeopathy), and access to conflicting information and controversies about vaccination in the media, among others (Dubé et al. 2013; Hobson-West 2003; MacDonald 2015).

From a public health perspective, it is thus important to understand publics' responses to vaccination and the reasons underlying them. In the context of living in a democracy, an additional requirement is to create spaces in which there can be meaningful dialogue about specifically those issues that are potentially divisive when making public policy decisions. In the sections below, we demonstrate how public deliberation can be a practical approach for dialogue and learning about public perspectives, values, and concerns that need to be addressed and taken into account in policy decision-making.

Overview of the Ontario Vaccine Deliberation

The Ontario Vaccine Deliberation was a public deliberation event on childhood vaccination in Ontario, hosted and organized by a research team from the University of Guelph (O'Doherty et al. 2021). The deliberation was held in Waterloo, Ontario, over two weekends in October 2017. Twenty-five participants, over the age of 18, took part in a 4-day deliberation event about childhood vaccination. Participants were provided with an information booklet prior to the event to ensure that everyone had the same base of knowledge of the topic (for a copy of the information booklet, see https://osf.io/t54e2/). During the deliberation, expert speakers provided information that reflected various key societal positions and interests, as well as critical technical information. Expert speakers were chosen to present on issues related to public health, complementary and alternative medicine, vaccine safety, parental perspectives, and philosophical and historical approaches to vaccination. Participants then formed small groups to discuss and deliberate on key questions with trained facilitators. Following small group discussions, participants convened as a large group to further explore issues identified, work toward a collective position on those issues, and vote on statements that represented the group's collective position. The last day of the deliberation focused on summarizing the recommendations derived from those statements and a final ratification vote to ensure all recommendations were captured accurately (see O'Doherty et al. 2021, for more details on process and outcomes).

The Ontario Vaccine Deliberation resulted in 20 recommendations in three key areas:

1. Mandatory vaccination and exemptions
2. Communication about vaccines and vaccination
3. Adverse event following immunization (AEFI) reporting and compensation.

These recommendations were the deliberative conclusions generated and recognized by the participants as theirs through intense small and large group discussions and a voting process.

Mandatory Vaccination and Exemptions

All participants of the deliberation expressed support for mandatory childhood vaccination for all children in Ontario. There was also strong agreement that parents, legal guardians, and/or custodians have a responsibility to the health of the community by vaccinating their children. These recommendations appear to reflect broader public value placed on mandatory vaccination in Ontario, or at least on the necessity of vaccinating children. Survey research with nationally representative samples had reported that a majority of Canadian parents with children aged 5 and younger were supportive of recommended vaccines for children (Dubé et al. 2018; Greenberg et al. 2017). Despite their support, a considerable percentage of parents had concerns about potential adverse reactions, including the link between vaccines and autism (Greenberg et al. 2017). Many parents also agreed that vaccination should remain a parental choice (Greenberg et al. 2017).

Mandatory vaccination programs are often controversial as they reduce individual choice and autonomy to some degree, leading to ethical concerns of coercion (MacDonald et al. 2018). Evidence for the validity of this concern is evident in controversies that followed the implementation of mandatory vaccination for COVID-19 in Canada and elsewhere (Bardosh et al. 2022; Flood et al. 2021; McLaren 2022; Smith 2022).

The main triggers for a shift to enact, strengthen, or contemplate mandatory vaccination in several countries include outbreaks of vaccine-preventable diseases (including in Italy, France, Serbia, and California), a failure of less coercive methods to increase vaccination rates, and the global goal of eliminating polio (MacDonald et al. 2018). Vaccination legislation and regulations may be changed as context or political will changes (MacDonald et al. 2018). In different political, economic, and sociocultural contexts, mandatory vaccination for children may or may not be met with majority support. For instance, amidst a recent measles outbreak in Vancouver, British Columbia, news headlines featured that a majority of Canadians supported mandatory vaccination for children entering school (Azpiri 2019; Young 2019). In Manitoba, vaccination against measles used to be required for school attendance, although this is no longer the case. In March 2018, there was an effort to reintroduce mandatory vaccination for school children in Manitoba for measles, as well as other common childhood vaccinations; however, the majority of school board delegates from across the province voted against a motion[3] (Malone 2018). The vote demonstrated a rare instance of democratic engagement of an important group of stakeholders in decisions regarding childhood vaccination, albeit at the school board level. Arguably, the successful implementation of mandatory vaccination depends on public trust and support for such a program.

While methods such as ballots and survey polling can gauge the extent of public support for a mandatory vaccination program, these methods often do not provide a means to understand and incorporate the perspectives and values of the publics into public health decision-making. How vaccination is conceptualized can bring forth emphases on particular ethical values while masking others (O'Doherty et al. 2017). Many of the participants in the Ontario Vaccine Deliberation supported the notion that mandatory vaccination means children who are not vaccinated and without valid exemptions should be excluded from school or organized activities. A key reason included a responsibility to protect oneself and the whole society through vaccination. However, several participants were against the idea. The reasons for those opposed included the right of everyone to go to school in Canada according to the *Educational Act* and that it would not be fair to exclude children from these activities. In terms of exemptions, all participants were supportive of medical exemptions. More controversial was the issue of non-medical exemptions for religious or conscientious beliefs. For many participants, religious or conscientious beliefs should not be grounds for exemptions. The reasons provided were diverse, including that only science and medicine should provide guidelines for exemptions, and that personal and religious beliefs have no role in societal decisions. A few participants supported allowing for non-medical exemptions or abstained from voting during the engagement session. Some of these participants expressed that religious beliefs should be respected and they invoked the national image of Canada as a diverse and tolerant country that should not exclude people because of their religious choices. The persistent disagreement among the participants even after intense deliberation highlights the difficulties associated with societal decisions about mandatory vaccination, and (in the case of mandatory vaccination) consequences for non-compliance and whether non-medical exemptions should be allowed. These difficulties reflect the

[3] A key reason for the vote against the motion was that the right to education would be jeopardized by the measure.

diverse values that participants might have drawn on in their reasoning, including the values of personal choice and social responsibility to protect, or at least not harm, the community.

Academic debate surrounding childhood vaccination and non-medical exemptions has implicated various competing ethical values. These include the values of individual autonomy, fairness, beneficence, and distributive justice of benefits and burdens between those who vaccinate and those who do not (Hendrix et al. 2016). In the context of childhood vaccination, it is parental autonomy in the choices they make on behalf of their children that is central to discussions. The principle of the least restrictive alternative has also been invoked in that policies that are less intrusive on individual/parental autonomy should generally be preferred over more intrusive alternatives such as mandatory vaccination with heavy penalties for lack of compliance (MacDonald et al. 2018; Verweij and Dawson 2004). Kass's (2001) public health ethics framework holds that public health programs are required to minimize burdens and "if 2 options exist to address a public health problem, we are required, ethically to choose the approach that poses fewer risks to other moral claims, such as liberty, privacy, opportunity, and justice, assuming benefits are not significantly reduced" (p. 1780). A key ethical consideration is then under what circumstances the state is justified in intervening and constraining parental autonomy to decide whether to vaccinate or not vaccinate their children.

Claims for more intrusive vaccination programs (i.e., mandatory or compulsory vaccination) have been made on the grounds that vaccine refusal imposes unjust harm, or risk of harm, upon other people (e.g., Brennan 2018; Flanigan 2014). Verweij and Dawson (2004) proposed an ethical principle for vaccination programs whereby "participation should, generally, be voluntary unless compulsory vaccination is essential to prevent a concrete and serious harm" (p. 3123), which can be a risk to the health of a child at an individual level or a risk to the realization of herd immunity as a public good at a societal level. Under this principle, parental autonomy is important but not absolute (Verweij and Dawson 2004). Similarly, Isaacs et al. (2009) asserted that parental objections to vaccination should be respected unless they would impede public health measures to realize herd immunity. Giubilini (2019) argued for compulsory vaccination and against allowing non-medical exemptions on the basis that vaccination policies should not aim only for the pragmatic realization of herd immunity, but for the fair distribution of the burdens of its realization. With that aim, fairness is conceived to be a value that need not and should not be compromised by the principles of individual autonomy and least restrictive alternative. Fairness, according to this perspective, requires that everyone who can be vaccinated makes their contribution to realize herd immunity as a public good; there is an individual and a collective responsibility to get vaccinated as well as an institutional responsibility to implement policies that can guarantee the realization of herd immunity (Giubilini 2019). In contrast, Navin and Largent (2017) claimed that allowing non-medical exemptions is ethically and socio-politically preferred over their elimination, and the best way to keep exemption rates low is to make the application process more burdensome. Arguably, this is the route that Ontario has taken with the requirements of notarization of application and attendance of an educational session.

While there are rich discussions about ethical values among academics and vaccine experts, much less attention has been given to the values that members of the public may draw on in their justifications for or against childhood vaccination. Members of the public may or may not invoke ethical values in the same way as academic debates have. Therefore, public deliberations like the Ontario Vaccine Deliberation provide an opportunity for members of the public to articulate and deliberate on the different values underlying vaccination and exemption policies in their own terms. Vaccination policy decisions may lead to unintended social and political consequences, particularly if they are implemented without public input. For example, the State of California passed legislation in 2016 to eliminate non-medical exemptions after a serious measles outbreak. The ethical aspects of this measure have attracted considerable debate in academic and public forums over moral obligations to respect individual/parental rights. Concerns over negative social and political consequences, including parents removing their children from public schools or daycare centers and increased political polarization, have also been raised (Navin and Largent 2017). Furthermore, there is evidence of an increase in medical exemptions in some counties that previously had the highest rates of personal belief exemptions (Delamater et al. 2017).

Communication About Vaccines and Vaccination

During the Ontario Vaccine Deliberation, there was strong support among participants for communicating about childhood vaccination through multiple channels. Suggestions included providing information to parents and prospective parents during pregnancy and after birth, as well as providing education through the school curriculum. The latter was considered important as children could let parents know about their experiences in school. Thus, both children and parents would be educated about childhood vaccination. Participants also suggested that public health agencies need to consider alternative communication mechanisms, such as social media, advertising in health professionals' offices, and advertising in public spaces. There was strong support that information should be scientific and unbiased, addressing the risks, benefits, and concerns of childhood vaccination. Most participants agreed that information about vaccine safety and effectiveness that comes from peer-reviewed scientific studies is better than information from other sources. Two participants either opposed or abstained from voting on this recommendation. The reasons included consideration that other information, such as personal experience, can be just as valuable, and sometimes personal experience, tradition, and religious beliefs are more powerful than scientific information.

Vaccine Injury Compensation

Among participants in the Ontario Vaccine Deliberation, there was strong support for tracking and mandatory reporting of adverse events following immunization (AEFI) by health professionals to Public Health Units. There was also strong support among the participants for compensation of serious life-altering adverse events leading to diminished capacity, although participants acknowledged the difficulty of defining "serious life-altering." At the time of the Ontario Vaccine Deliberation, Québec was the only province in Canada with a public program to compensate for serious injuries or death caused by voluntary vaccination (Gouvernement du Québec 2019). The lack of a national scheme to compensate for vaccine-related injuries in Canada had been identified as a policy gap (Keelan and Wilson 2011), even though the issue had long been debated by vaccine experts and policy makers. One of the earliest statements in support of a national compensation program for government-approved vaccines was published in 1986 by the Canadian Paediatric Society (Canadian Paediatric Society 1986; Keelan and Wilson 2011). A public petition had also been created by Bob Martin, who developed Guillain–Barré syndrome after receiving an influenza vaccine in 2010 (https://www.thepetitionsite.com/882/711/468/ceate-a-no-fault-vaccine-compensation-program-for-candians/). In December 2020, Health Canada announced a Vaccine Injury Support Program (VISP) for all provinces and territories, with the exception of Québec, which continues to be covered under its own program. "The purpose of the VISP is to ensure that all people in Canada who have experienced a serious and permanent injury as a result of receiving a Health Canada authorized vaccine, administered in Canada on or after December 8, 2020, have fair and timely access to financial support" (https://vaccineinjurysupport.ca/en). The program is open to all individuals, including children. The VISP began to accept claims in June 2021. Claims are first subjected to an administrative review by a case manager. Those that are deemed admissible will go through a process of collecting medical records and then assessment by a Medical Review Board.

The VISP is premised on the notion of no-fault compensation. A global survey of WHO Member States identified 25 jurisdictions with no-fault vaccine injury compensation programs in 2018, with most jurisdictions categorized as high-income countries (WHO 2019). No-fault compensation is based on the premise that there are unavoidable or unintended risks associated with vaccination even for vaccines that are properly designed, manufactured, and delivered (Keelan and Wilson 2011; Law Reform Commission of Saskatchewan 2009; Looker and Kelly 2011; Manitoba Law Reform Commission 2000). As such, parties injured by vaccines or their legal representatives are entitled to compensation without the need to prove negligence or fault by vaccine manufacturers, vaccine providers, or the health care system. Without a formal no-fault compensation scheme, the only recourse is through civil litigation under the tort law. A tort lawsuit requires the plaintiff to prove that another party is liable for the injury due to negligence or failure to adequately warn of the risks and secure informed consent. The burden of proof is on the plaintiff to establish a causal link between the injury and the negligent act. This process is considered problematic as it is often difficult to establish negligence and causation in the case of immunization. Moreover, civil litigation is costly and time consuming, making it an inaccessible course of action for many vaccine recipients. No-fault compensation schemes are believed to provide a fairer and more efficient process for people to seek redress for vaccine-related injuries. While a causal link still needs to be established based on a balance of probabilities, the standard of proof is less strict compared to the legal standard (Keelan and Wilson 2011;

Law Reform Commission of Saskatchewan 2009; Looker & Kelly 2011; Manitoba Law Reform Commission 2000).

A lack of impetus to create a no-fault compensation program might be due to an absence of political, economic, and social pressures (Keelan and Wilson 2011). For example, the US National Vaccine Injury Compensation Program (VICP) was established in 1986 after a prolonged liability crisis when a large number of vaccine manufacturers stopped producing the Diphtheria-Pertussis-Tetanus (DPT) vaccine for children in response to civil litigations over injuries (Keelan and Wilson 2011). According to Cook and Evans (2011), the VICP has been a "key component in stabilizing the US vaccine market through liability protection to both vaccine companies and health care providers" since then (p. S74). In terms of immunization rates, there is no published evidence on whether the availability of a no-fault vaccine injury compensation program has an impact on immunization uptake (Keelan and Wilson 2011). According to Browne (2016), the Canadian Medical Association previously rejected a motion to implement a national vaccine injury compensation program due to concerns that such a program could constitute an admission to the dangers of immunization, undermine public confidence, and deter people from vaccinating themselves or their children. To the contrary, arguments have been made that no-fault compensation could increase public confidence and support in childhood vaccination (Keelan and Wilson 2011; Law Reform Commission of Saskatchewan 2009; Manitoba Law Reform Commission 2000).

On the whole, there is much agreement that there is an ethical imperative of providing a no-fault vaccine compensation program. Childhood vaccination is a preventive measure that greatly reduces the burden of many infectious diseases and benefits the community through herd immunity. However, parents do expose their healthy children to risks when they vaccinate their children. Arguments have been made that there is a community responsibility to compensate those who are injured (Looker and Kelly 2011; Manitoba Law Reform Commission 2000). Mello (2008) asserts that the ethical principles of fairness and solidarity mean members of the community should not have to bear the burdens of realizing the public good of herd immunity alone. Among individuals who are vaccinated, the injured and the uninjured bear unequal shares of the burdens and, thus, mechanisms are needed to provide "a safety net" for those whose sacrifice is large (Mello 2008). As the Manitoba Law Reform Commission observed in its 2000 report, "although vaccination is not compulsory, there is considerable governmental and social pressure to participate in the immunization process" (p. 15). The arguments for vaccine injury compensation are even more compelling in the context of mandatory vaccination, where individual/parental choice and autonomy are relatively reduced by government mandates.

No-fault vaccine injury compensation programs vary across jurisdictions that have them. Looker and Kelly (2011) identified six common elements to these programs: administration and funding, eligibility, process and decision-making, standard of proof, elements of compensation, and litigation rights. Most of the programs are administered and funded by state or national governments, or a mixture of both as is the case in Japan. Finland, Norway, and Sweden use a manufacturers' levy, while Taiwan (China) and USA retain centralized government control by imposing a vaccine tax on manufacturers (Looker and Kelly 2011). Among the participants of the Ontario Vaccine Deliberation, there was strong support that a fund should be established with contributions from both the pharmaceutical industry and the government to compensate individuals on a case-by-case basis. We do not contend that this is necessarily the most effective mechanism to establish a sustainable source of funding for a vaccine injury compensation program. Rather, we highlight the importance of soliciting public input among other considerations in the design of vaccine-related programs and establishing potential mechanisms for public input on a longer-term basis. For example, the Manitoba Law Reform Commission (2000) recommended that the Health Minister appoint a medical expert as the Director of Childhood Vaccination Injury Compensation, who would handle vaccine injury claims. If a claim was rejected by the Director, the claimant could appeal to an administrative tribunal known as the Childhood Vaccination Injury Compensation Appeal Board. Members of the Board would consist of immunology experts, lawyers, and lay persons. We cannot claim that the recently implemented VISP is a consequence of recommendations from members of the public who participated in the Ontario Vaccine Deliberation. However, it is striking to observe that these public recommendations presaged a decision by Health Canada and are in line with international ethical and legal scholarship that rejects paternalistic arguments that a national vaccine injury compensation program would constitute an admission to the dangers of immunization and thereby deter people from vaccinating themselves or their children (see Browne, 2016).

Conclusion

In this chapter, we have argued that childhood vaccination is an issue that stands to gain from the input of public deliberation processes. The argument for involving lay publics in policy decisions about childhood vaccination is based on the observation that such decisions require not only scientific expertise, but also consideration of how competing values are to be balanced in policies that affect all members of a society (e.g., how to balance autonomy in decisions affecting one's children and ensuring community protection from infectious diseases). To be effective, such policies also require relationships of trust between government decision-makers, scientists, vaccine producers, health professionals, and publics. In this context, we have discussed how public deliberation can be a practical and meaningful approach to engage members of the public in a formal process of knowledge exchange and dialogue about childhood vaccination. Importantly, the purpose of public deliberation is not to influence people's opinions in a particular direction or to convince them to take a certain position. Rather, the purpose is to broaden the considerations that are brought to bear on a policy issue and to expand the range of voices that are part of the decision-making process. Engaging publics to deliberate on the topic of childhood vaccination can yield a better understanding of public perspectives; more importantly, public deliberation provides an avenue toward more democratic, legitimate, and accountable policy decisions.

Appendix: Applying Democratization Processes Through Public Deliberation to the COVID-19 Pandemic

As of September 26, 2022, there have been 620,413,942 cases of coronavirus disease (COVID-19) worldwide, with 6,540,871 reported deaths (Worldometer 2022). This pandemic was caused by the SARS-CoV-2 virus, a novel infectious disease affecting respiratory illnesses. Given its rapid spread, with often serious complications, including debilitation and death, this spurred the need to rapidly develop a vaccine and vaccination program for populations across the world. Many of the debates relating to COVID-19 are not strictly within the scope of this chapter since they are neither about childhood vaccination nor about democratization of risk decision-making. However, the COVID-19 pandemic has shown in detail how the effects of infectious disease are mediated by societal factors, and controversy surrounding COVID-19 vaccines has been particularly pronounced for children. In this appendix, we therefore discuss some implications of COVID-19 vaccinations for broader debates about public responses to vaccination and for democratizing policy decision-making.

How the COVID-19 Pandemic Relates to Childhood Vaccination

As stated in the chapter, immunization is one of the most successful public health interventions worldwide, preventing millions of deaths per year, with its greatest impact on childhood diseases (WHO 2018). Routine childhood vaccination in Canada covers many illnesses, including diphtheria, hepatitis B, human papillomavirus (HPV), influenza, measles, meningococcus, mumps, pertussis (whooping cough), polio, rotavirus, rubella, tetanus, and varicella (chicken pox). This is a broad array of diseases, including both bacteria and viruses, each having different characteristics when it comes to spread, prognosis, and treatment. Due to the differences in characteristics, achieving herd immunity varies by disease, with estimates ranging from 75% for mumps to 94% for pertussis and measles (Plans-Rubió 2012).

Whereas most of these infections are primarily described as childhood diseases or are more fatal for children, this is not true for COVID-19, where risk of hospitalization and mortality increases with age (Starke et al. 2021). Moreover, it has been noted that in the case of SARS-CoV-2, the classical concept of herd immunity may not apply (Morens et al. 2022). These differences can influence decisions to vaccinate children, where parents' decisions to vaccinate for COVID-19 are related to perceived susceptibility and benefit of the vaccines, low trust in the safety of the vaccines, or if they viewed them as unnecessary (Du et al. 2021; Humble et al. 2021; Qin et al. 2022). Indeed, routine childhood vaccinations have been found to be viewed as more essential, safe, and effective compared to the COVID-19 vaccines (Temsah et al. 2021).

Although COVID-19 vaccines are approved in Canada for children as young as 6 months old, vaccines for COVID-19 were not available to children under the age of 5 until July 2022 (Shakil 2022) and were only approved for children 5 to 11 years old in November 2021 (Health Canada 2021). For these reasons, COVID-19 vaccination is somewhat distinct from other routine childhood vaccinations and so presents a somewhat separate case from that discussed throughout the chapter.

Vaccination Against COVID-19

SARS-CoV-2 is a rapidly mutating virus, with over 5 variants of concern[4] having been reported to date (September, 2022) since its discovery in early 2020 (World Health Organization 2022). To reduce and prevent illness, many vaccines have been developed worldwide. Of these (at the time of writing), six COVID-19 vaccines have been approved for administration in Canada, which use four types of technologies: mRNA vaccines (Moderna, Pfizer-BioNTech), viral vector-based vaccines (AstraZeneca, Johnson & Johnson), protein subunit vaccines (Novavax), and plant-based vaccines (Medicago) (Government of Canada 2022a, b). Notably, there are substantial differences between the vaccines administered for COVID-19 and those for childhood vaccination; vaccines for COVID-19 were developed rapidly as an emergency response to the pandemic, with several types employing new vaccine technologies, and can have lower effectiveness due to the prevalence of multiple strains of COVID-19 (Vasireddy et al. 2021). Public knowledge and trust of the vaccines have remained low, and they are viewed less favorably than routine childhood vaccinations (Temsah et al. 2021).

How the COVID-19 Pandemic Relates to Democratization

(Lack of) Public Consultation in Policy Decisions About COVID-19 Vaccination

Since the beginning of the pandemic, the regulatory context of vaccines has shifted. Although the approach to childhood vaccination is largely unchanged across Canada, a new vaccine injury compensation fund has been introduced (Vaccine Injury Support Program, VISP), and policies regarding the COVID-19 vaccination program have changed and adapted over time. Because the responsibilities for the administration and delivery of immunization programs are within the jurisdiction of the provincial and territorial governments, there has also been substantial variation in

[4] Most variants have little to no impact on disease spread or prognosis. Variants of concern are those that the World Health Organization has identified as variants that affect the virus properties, such as how it spreads, disease severity, vaccine performance, or other public health and social measures.

COVID-19 vaccination policies. For example, mandatory vaccine requirements to enter public spaces and for workers employed in certain fields (e.g., healthcare workers) varied by province and territory.

In parallel with policies on childhood vaccination, there have been rich discussions about ethics and policies about COVID-19 vaccination among academics, vaccine experts, politicians, and public health workers. Notably, and similar to the discussions on childhood vaccination, much less attention has been given to involving broader publics in discussions and decisions about how competing social values are to be traded off in policies relating to COVID-19 vaccination programs (e.g., should vaccination be mandatory or just recommended). Having limited avenues for public input, combined with the top-down policy directives, there have been disconnects between scientists, politicians, and the public. In fact, lower public trust in governments has been related to lower willingness to accept COVID-19 vaccination (Wouters et al. 2021), and it has been argued that the governmental response to COVID-19 is a key influence on public confidence in vaccination (Sabahelzain et al. 2021).

Mandatory Vaccination

As explained in the main chapter, the term "mandatory vaccination" is generally seen to encompass policies that exclude children from school or organized activities without vaccination or valid exemptions. Notably, no such policy has been implemented in schools for children with regard to COVID-19 vaccination in Canada. Despite this, there are parallels with restrictions on adults. At the federal level, the Government of Canada required all federal public servants and employees in the federally regulated air, rail, and marine transport sectors to be vaccinated in fall 2021 (Wong et al. 2021). Similarly, Ontario required hospitals, home, and community care personnel to provide proof of full vaccination or of a medical exception, while British Columbia required long-term care and assisted living workers to be vaccinated (Wong et al. 2021). Mandatory vaccination was announced and enforced in different regions, at different time points, across Canada.

A notable change with regard to vaccination in Canada was the unprecedented degree of enforcement of mandatory COVID-19 vaccination. Noncompliance without a medical exemption in several provinces and industries could have led to job loss or forced leave without pay. For example, all federal public servants in the Core Public Administration

(CPA) were required to be vaccinated, whether they were teleworking, remote working, or working on-site (Government of Canada 2021). Although this requirement was suspended in June 2022, employees who did not comply at the time were placed on administrative leave without pay (Government of Canada 2022b).

Indeed, the Ontario Vaccine Deliberation indicated that mandatory vaccination is a central and divisive issue in arriving at policy decisions. Although the COVID-19 context is quite different from the routine childhood vaccinations that were the topic of the Ontario Vaccine Deliberation, debates relating to the values of personal choice, fairness, and social responsibility to protect the community featured quite similarly in both the confines of the formal setting of the Ontario Vaccine Deliberation and broader public discourse relating to the COVID-19 vaccines.

Democratization

Despite there being many differences between routine childhood vaccination and COVID-19 vaccination, discussions in both contexts highlight the need for increased public participation and deliberation in determining vaccination policies. To date, policy decisions have been determined by public health agencies, scientists, and politicians, with limited avenues for public input. Depending on public confidence in leadership, this top-down strategy can reduce public confidence and vaccine uptake (Sabahelzain et al. 2021; Wouters et al. 2021).

As we explored in the main chapter, the case for direct involvement of broader publics in policy decisions about vaccination rests on the argument that these decisions involve multiple values that are often competing (e.g., allowing personal choice and aiming for broad vaccination coverage), that they affect all sectors of society, and that they require a certain degree of trust in the vaccines and those producing and administering them (O'Doherty et al. 2021). This argument pertains equally, or possibly even more, to the case of COVID-19. Unfortunately, there seems to have been very little opportunity for meaningful public input in high-level decisions about vaccination policy decisions for COVID-19. While public health responses and communication might have gone beyond providing more information about risks to the public to a broader array of communication strategies as a product of the pandemic (see Goldenberg 2021), overall strategies were still largely top-down, without public input. That is, communication relating to COVID-19 and COVID-19 vaccines was typically one-way, driven by experts and government authorities. While this might be appropriate for technical aspects of the vaccine and information about the virus and its effects, it does not address the problem of how competing societal values and interests are to be represented in policy responses.

At the same time, the COVID-19 pandemic has been characterized by the spread of misinformation about COVID-19 itself and the vaccines, arguably to a larger degree than has been the case for childhood vaccination, although the dialogue around these has changed in several ways (e.g., discussion of COVID-19 vaccines is less focused on autism and childhood illnesses). Given that misinformation is spread beyond geographical borders online, it is difficult to determine how public input at a local level would influence the spread of misinformation about vaccination. However, because public deliberation processes have the potential to increase trust in policy decisions (O'Doherty et al. 2021; Solomon and Abelson 2012), we can speculate that people residing in an area that incorporates public deliberation in their decision-making processes would have more trust in information from public sources, and therefore be more likely to look for information from government sources, and possibly reduce their consumption of misinformation related to vaccination.

To our knowledge, there has not been any official, government-sponsored public engagement event for members of the public in Canada to deliberate on COVID-19 vaccination, and very little internationally (see, for example, Scherer et al. 2022). Creating mechanisms for public involvement such as public deliberations would increase citizen participation in important social issues based on democratic ideals and would allow for a more nuanced and respectful approach in understanding public perceptions and concerns regarding vaccination, with particular relevance to mandatory vaccination and communication about vaccines. Ultimately, governmental responses to public health risks such as those posed by COVID-19 stand to gain in both legitimacy and effectiveness by incorporating a wider range of perspectives and values through public dialogue.

References

Abelson, J., Forest, P. G., Eyles, J., Smith, P., Martin, E., & Gauvin, F. P. (2003). Deliberations about deliberative methods: Issues in the design and evaluation of public participation processes. *Social Science & Medicine, 57*(2), 239–251. https://doi.org/10.1016/S0277-9536(02)00343-X

Asturias, E., Wharton, M., Pless, R., MacDonald, N., Chen, R., Andrews, N., Salisbury, D., Dodoo, A., Hartigan-Go, K., Zuber, P. (2016). *Vaccine, 17*(34): 3342–3349. https://doi.org/10.1016/j.vaccine.2016.05.018

Azpiri, J. (2019, February 21). Strong majority of Canadians support mandatory vaccinations for children entering school: Poll. *Global News*. Retrieved from https://globalnews.ca/news/4979899/strong-majority-of-canadians-support-mandatory-vaccinations-for-children-entering-school-poll/

Bardosh, K., de Figueriedo, A., Gur-Arie, R., Jamrozik, E., Doidge, J., Lemmens, T., Keshavjee, S., Graham, J., Baral, S. (2022). The unintended consequences of COVID-19 vaccine policy: Why mandates, passports and restrictions may cause more harm than good. *BMJ Global Health, 7*(5):3008684.

Bedford, H., Attwell, K., Danchin, M., Marshall, H., Corben, P., & Leask, J. (2018). Vaccine hesitancy, refusal, and access barriers: The need for clarity in terminology. *Vaccine, 36*, 6556–6558.

Blacksher, E. (2013). Participatory and deliberative practices in health: Meanings, distinctions, and implications for health equity. *Journal of Public Deliberation, 9*(1), Article 6.

Brennan, J. (2018). A libertarian case for mandatory vaccination. *BMJ Medical Ethics, 44*, 37–43. https://doi.org/10.1136/medethics-2016-103486

Browne, K. (2016, May 13). *Vaccine injury compensation and the common good* (Impact Ethics post) Retrieved from https://impactethics.ca/2016/05/13/vaccine-injury-compensation-and-the-common-good/

Canadian Paediatric Society. (1986). In support of a compensation plan for vaccine-associated injuries. *Canadian Medical Association Journal, 135*(7), 747–749.

Canadian Paediatric Society. (2019). *Immunization and vaccines*. Retrieved from https://www.cps.ca/en/issues-questions/immunization

CanImmunize. (2021). *CanImmunize Knowledge Centre*. Retrieved from https://learn.canimmunize.ca/en/knowledge-centre?_ga=2.268101384.760924602.1620866089-243904744.1620866089.

Chafe, R., Levinson, W., & Hébert, P. C. (2011). The need for public engagement in choosing health priorities. *CAMJ, 183*(2), 165. https://doi.org/10.1503/cmaj.101517

Chambers, S. (2003). Deliberative democratic theory. *Annual Review of Political Science, 6*, 307–326. https://doi.org/10.1146/annurev.polisci.6.121901.085538

Cook, K. M., & Evans, G. (2011). The National Vaccine Injury Compensation Program. *Pediatrics, 127*, S74–S77. https://doi.org/10.1542/peds.2010-1722K

Conklin, L., Hviid, A., Orenstein, W., Pollard, A., Wharton, M., Zuber, P. (2021). *BMJ Global Health, 6*(Suppl 2), e004898. https://doi.org/10.1136/bmjgh-2020-004898.

Degeling, C., Carter, S. M., & Rychetnik, L. (2015). Which public and why deliberate? A scoping review of public deliberation in public health and health policy research. *Social Science and Medicine, 131,* 114–121. https://doi.org/10.1016/j.socscimed.2015.03.009

Delamater, P. L., Leslie, T. F., & Yang, Y. T. (2017). Change in medical exemptions from immunization in California after elimination of personal belief exemptions. *JAMA, 318*(9), 863–864.

Du, M., Tao, L., & Liu, J. (2021). The association between risk perception and COVID-19 vaccine hesitancy for children among reproductive women in China: An online survey. *Frontiers in medicine*, 1494.

Dubé, E., Laberge, C., Guay, M., Bramadat, P., Roy, R., & Bettinger, J. (2013). Vaccine hesitancy: An overview. *Human Vaccines and Immunotherapeutics, 9*(8), 1763–1773. https://doi.org/10.4161/hv.24657

Dubé, E., Gagnon, D., Ouakki, M., Bettinger, J. A., Witteman, H. O., MacDonald, S., ... Greyson, D, on behalf of the Canadian Immunization Research Network. (2018). Measuring vaccine acceptance among Canadian parents: A survey of the Canadian Immunization Research Network. *Vaccine, 36,* 545–552. https://doi.org/10.1016/j.vaccine.2017.12.005

Esmail, L., Moore, E. & Rein, A. (2015). Evaluating patient and stakeholder engagement in research: Moving from theory to practice. *Journal of Comparative Effectiveness Research, 4*(2), 133–145. https://doi.org/10.2217/cer.14.79

Fine, P., Eames, K., & Heymann, D. L. (2011). "Herd immunity": A rough guide. *Clinical Infectious Diseases, 52*(7), 911–916. https://doi.org/10.1093/cid/cir007

Flanigan, J. (2014). A defense for compulsory vaccination. *HEC Forum, 26*(1), 5–25. https://doi.org/10.1007/s10730-013-9221-5

Flood, C., Thomas, B., & Wilson, K. (2021). Mandatory vaccination for health care workers: An analysis of law and policy. *Canadian Medical Association Journal, 193*(6). E217–220. https://doi.org/10.1503/cmaj.202755

Giubilini, A. (2019). *The ethics of vaccination*. Cham, Switzerland: Palgrave Pivot. https://doi.org/10.1007/978-3-030-02068-2

Goldenberg, M. J. (2016). Public misunderstanding of science? Reframing the problem of vaccine hesitancy. *Perspectives on Science, 24*(5), 552-581. https://doi.org/10.1162/POSC_a_00223

Goldenberg, M. (2021). *Vaccine hesitancy: Public trust, expertise, and the war on science*. University of Pittsburgh Press.

Government of Canada (2021, December 7). Government of Canada will require employees in all federally regulated workplaces to be vaccinated against COVID-19. *Government of Canada*. https://www.canada.ca/en/employment-social-development/news/2021/12/government-of-canada-will-require-employees-in-all-federally-regulated-workplaces-to-be-vaccinated-against-covid-19.html

Government of Canada (2022a, August 5). Approved COVID-19 Vaccines. *Government of Canada*. https://www.canada.ca/en/health-canada/services/drugs-health-products/covid19-industry/drugs-vaccines-treatments/vaccines.html

Government of Canada (2022b, June 23). COVID-19 vaccination requirement for federal public servants. *Government of Canada*. https://www.canada.ca/en/government/publicservice/covid-19/vaccination-public-service.html

Government of Ontario. (2017). *Immunization of School Pupils Act, RSO, 1990*. Retrieved from https://www.ontario.ca/laws/statute/90i01

Government of Ontario. (2019). *Vaccines for children at school*. Retrieved from https://www.ontario.ca/page/vaccines-children-school

Gouvernement du Québec. (2019). *Vaccine Injury Compensation Program*. Retrieved from https://www.quebec.ca/en/health/advice-and-prevention/vaccination/vaccine-injury-compensation-program/

Greenberg, J., Dubé, E., & Driedger, M. (2017). Vaccine hesitancy: In search of the risk communication comfort zone. *PLoS Currents*. https://doi.org/10.1371/currents.outbreaks.0561a011117a1d1f9596e24949e8690b

Health Canada. (2021, November 19). Health Canada authorizes use of Comirnaty (the Pfizer-BioNTech COVID-19 vaccine) in children 5 to 11 years of age Retrieved from https://www.canada.ca/en/health-canada/news/2021/11/health-canada-authorizes-use-of-comirnaty-the-pfizer-biontech-covid-19-vaccine-in-children-5-to-11-years-of-age.html

Hendrix, K. S., Sturm, L. A., Zimet, G. D., & Meslin, E. M. (2016). Ethics and childhood vaccination policy in the United States. *American Journal of Public Health, 106*(2), 273–278. https://doi.org/10.2105/AJPH.2015.302952

Hobson-West, P. (2003). Understanding vaccine resistance: Moving beyond risk. *Health, Risk & Society, 5*(3), 273–283. https://doi.org/10.1080/13698570310001606978

Humble, R. M., Sell, H., Dubé, E., MacDonald, N. E., Robinson, J., Driedger, S. M., ... & MacDonald, S. E. (2021). Canadian parents' perceptions of COVID-19 vaccination and intention to vaccinate their children: Results from a cross-sectional national survey. *Vaccine, 39*(52), 7669–7676.

Isaacs, D., Kilham, H., Leask, J., & Tobin, B. (2009). Ethical issues in immunisation. *Vaccine, 27*, 615–618. https://doi.org/10.1016/j.vaccine.2008.11.002

Kass, N. E. (2001). An ethics framework for public health. *American Journal of Public Health, 91*(11), 1776–1782.

Keelan, J., & Wilson, K. (2011). Designing a no-fault vaccine-injury compensation programme for Canada: Lessons learned from an international analysis of programmes. Toronto, ON: The University of Toronto's Munk School of Global Affairs. Retrieved from https://munkschool.utoronto.ca/research-articles/designing-a-no-fault-vaccine-injury-compensation-programme-for-canada-lessons-learned-from-an-international-analysis-of-programmes/

Kovacs Burns, K., Bellows, M., Eigenseher, C., & Gallivan, J. (2014). 'Practical' resources to support patient and family engagement in healthcare decisions: A scoping review. *BMC Health Services Research, 14*, 175. https://doi.org/10.1186/1472-6963-14-175

Larson, H. J., Jarrett, C., Eckersberger, E., Smith, D. M. D., & Paterson, P. (2014). Understanding vaccine hesitancy around vaccines and vaccination from a global perspective: A systematic review of published literature, 2007–2012. *Vaccine, 32*(19), 2150-2159. https://doi.org/10.1016/j.vaccine.2014.01.081

Law Reform Commission of Saskatchewan. (2009). *Vaccination and the law: Report to the Minister of Justice*. Law Reform Commission of Saskatchewan. Retrieved from https://lawreformcommission.sk.ca/vaccinationfinal.pdf

Looker, C., & Kelly, H. (2011). No-fault compensation following adverse events attributed to vaccination: A review of international programmes. *Bulletin of World Health Organization, 89*(5), 371–378. https://doi.org/10.2471/BLT.10.081901

MacDonald, N. E. (2015). Vaccine hesitancy: Definition, scope and determinants. *Vaccine, 33*(34), 4161-4164. https://doi.org/10.1016/j.vaccine.2015.04.036

MacDonald, N. E., Harmon, S., Dube, E., Steenbeck, A., Crowcroft, N., Opel, D. J., ... Butler, R. (2018). Mandatory infant and childhood immunization: Rationales, issues and knowledge gaps. *Vaccine, 36*, 5811–5818. https://doi.org/10.1016/j.vaccine.2018.08.042

Malone, K. L. (2018, March 19). Manitoba school boards voted against push to have all school children vaccinated. *CTV News*. Retrieved from https://www.ctvnews.ca/health/manitoba-school-boards-vote-against-push-to-have-all-school-children-vaccinated-1.3848675

Manafo, E., Petermann, L., Mason-Lai, P. & Vandall-Walker, V. (2018). Patient engagement in Canada: A scoping review of the 'how' and 'what' of patient engagement in health research. *Health Research Policy and Systems, 16*, 5. https://doi.org/10.1186/s12961-018-0282-4

Manitoba Law Reform Commission. (2000). *Compensation of vaccine-damaged children* (report # 104). Winnipeg, Manitoba: Manitoba Law Reform Commission. Retrieved from http://www.manitobalawreform.ca/pubs/pdf/archives/104-full_report.pdf

Marshall, H. S., C. Proeve, J. Collins, R. Tooher, M. O'Keefe, T. Burgess, R. Skinner, M. Watson, H. Ashmeade, & Braunack-Mayer, A. (2014). Eliciting youth and adult recommendations through citizens' juries to improve school based adolescent immunisation programs. *Vaccine, 32*(21), 2434–2440. https://doi.org/10.1016/j.vaccine.2014.02.098

McLaren, P. (2022). Some thoughts on Canada's 'Freedom Convy' and the settler colonial state. *Educational Philosophy and Theory, 54*(7), 867–870. https://doi.org/10.1080/00131857.2022.2051478

Mello, M. M. (2008). Rationalizing vaccine injury compensation. *Bioethics, 22*(1), 32–42. https://doi.org/10.1111/j.1467-8519.2007.00590.x

Morens, D. M., Folkers, G. K., & Fauci, A. S. (2022). The concept of classical herd immunity may not apply to COVID-19. *The Journal of Infectious Diseases*. https://doi.org/10.1093/infdis/jiac109

Navin, M. C., & Largent, M. A. (2017). Improving nonmedical vaccine exemption policies: Three case studies. *Public Health Ethics, 10*(3), 225–234. https://doi.org/10.1093/phe/phw047

O'Doherty, K. C., Smith, C., & McMurtry, C. M. (2017). Vaccine hesitancy: Ethical considerations from multiple perspectives In P. Bramadat, M. Guay, J. A. Bettinger. & R. Roy (Eds.), *Public health in the age of anxiety* (pp. 56–79). Toronto, ON: University of Toronto Press.

O'Doherty, K. C., Crann, S., Bucci, L. M., Burgess, M. M., Chauhan, A., Goldenberg, M., McMurtry, C. M., White, J., & Willison, D. (2021). Deliberation on childhood vaccination in Canada: Public input on ethical trade-offs in vaccination policy. *AJOB Empirical Bioethics, 12*(4), 253–265. https://doi.org/10.1080/23294515.2021.1941416

Ontario Ministry of Health and Long-Term Care (2018). *Immunization*. http://www.health.gov.on.ca/en/pro/programs/immunization/ispa.aspx

Parrella, A., Braunack-Mayer, A., Collins, J., Clarke, M., Tooher, R., Ratcliffe, J., & Marshall, H. (2016). Prioritizing government funding of adolescent vaccinations: Recommendations from young people on a citizen's jury. *Vaccine, 34*(31), 3592–3597. https://doi.org/10.1016/j.vaccine.2016.05.019

Plans-Rubió, P. (2012). Evaluation of the establishment of herd immunity in the population by means of serological surveys and vaccination coverage. *Human Vaccines and Immunotherapeutics, 8*(2), 184–188.

Public Health Ontario. (2017). Annual Report on Vaccine Safety in Ontario, 2017. Retrieved from https://www.publichealthontario.ca/-/media/documents/A/2018/annual-vaccine-safety-report-2017.pdf

Public Health Agency of Canada. (2014). *Canada Immunization Guide*. Retrieved from https://www.canada.ca/en/public-health/services/canadian-immunization-guide.html

Public Health Agency of Canada. (2018). *A Parents' Guide to Vaccination*. Retrieved from https://www.canada.ca/en/public-health/services/publications/healthy-living/parent-guide-vaccination.html

Qin, C., Wang, R., Tao, L., Liu, M., & Liu, J. (2022). Association Between Risk Perception and Acceptance for a Booster Dose of COVID-19 Vaccine to Children Among Child Caregivers in China. *Frontiers in public health, 10*, 834572.

Sabahelzain, M. M., Hartigan-Go, K., & Larson, H. J. (2021). The politics of COVID-19 vaccine confidence. *Current Opinion in Immunology, 71*, 92–96.

Scherer, M., Kamler, A., Weiss, L., Blacksher, E., Jeavons, J., & Gold, M. R. (2022). Using Public Deliberation to Set Priorities: The Case of COVID-19 Vaccine Access in New York City. *Journal of Public Health Management and Practice*, *28*(1), 86–94.

Shakil, I. (2022, July 14). Canada clears Moderna's COVID vaccine for children under 5. *Reuters.* https://www.reuters.com/world/americas/canada-approves-modernas-covid-vaccine-children-aged-6-months-older-2022-07-14/

Smith, M. (2022). Evaluating potential unintended consequences of COVID-19 vaccine mandates and passports. *BMJ Global Health, 7*(7). https://doi.org/10.1136/bmjgh-2022-009759

Solomon, S., & Abelson, J. (2012). Why and when should we use public deliberation? *Hastings Center Report, 42*(2), 17–20. https://doi.org/10.1002/hast.27

Song, G., Silva, C. L., & Jenkins Smith, H. C. (2014). Cultural worldview and preference for childhood vaccination policy. *Policy Studies Journal, 42*(4), 528–554.

Starke, K. R., Reissig, D., Petereit-Haack, G., Schmauder, S., Nienhaus, A., & Seidler, A. (2021). The isolated effect of age on the risk of COVID-19 severe outcomes: A systematic review with meta-analysis. *BMJ global health*, *6*(12), e006434.

Temsah, M. H., Alhuzaimi, A. N., Aljamaan, F., Bahkali, F., Al-Eyadhy, A., Alrabiaah, A., ... & Alhasan, K. (2021). Parental attitudes and hesitancy about COVID-19 vs. routine childhood vaccinations: A national survey. *Frontiers in Public Health*, 1513.

The Keystone Center. (2009). *The Public Engagement Project on the H1N1 Pandemic Influenza Vaccination Program.* https://www.keystone.org/wp-content/uploads/2015/08/093009-H1N1-Pandemic-Influenza-Vaccination-Prog-Engagement-Report.pdf

Vasireddy, D., Vanaparthy, R., Mohan, G., Malayala, S. V., & Atluri, P. (2021). Review of COVID-19 variants and COVID-19 vaccine efficacy: What the clinician should know? *Journal of Clinical Medicine Research*, *13*(6), 317.

Verweij, M., & Dawson, A. (2004). Ethical principles for collective immunisation programmes. *Vaccine, 22*, 3122–3126. https://doi.org/10.1016/j.vaccine.2004.01.062

WHO Strategic Advisory Group of Experts on Immunization. (2018). *2018 Assessment Report of the Global Vaccine Action Plan*. Geneva, Switzerland: World Health Organization.

Wong, S., O'Hearn Davies, T., & Giguère, A. (2021, September 23). Shift towards mandatory vaccine policies in Canada. *Norton Rose Fulbright*. https://www.nortonrosefulbright.com/en-il/knowledge/publications/5ac54903/shift-towards-mandatory-vaccine-policies-in-canada

World Health Organization. (2018). *Immunization*. Retrieved from https://data.unicef.org/topic/child-health/immunization/

World Health Organization. (2019). Global Advisory Committee on Vaccine Safety, 5–6 December 2018. *Weekly epidemiological record, 94*(4), 45–52. Retrieved from https://www.who.int/wer/2019/wer9404/en/

World Health Organization (2022, September 30). Tracking SARS-CoV-2 variants. *World Health Organization*. https://www.who.int/activities/tracking-SARS-CoV-2-variants

Wouters, O. J., Shadlen, K. C., Salcher-Konrad, M., Pollard, A. J., Larson, H. J., Teerawattananon, Y., & Jit, M. (2021). Challenges in ensuring global access to COVID-19 vaccines: Production, affordability, allocation, and deployment. *The Lancet, 397*(10278), 1023–1034.

Young, L. (2019, April 3). 9 in 10 Canadians believe vaccines should be mandatory for school: Ipsos poll. Global News. Retrieved from https://globalnews.ca/news/5117520/mandatory-vaccination-ipsos-poll/

CHAPTER 12

Narratives and the Water Fluoridation Controversy

Andrea M. L. Perrella, Simon J. Kiss, and Ketan Shankardass

Introduction

Water fluoridation is a simple, effective, inexpensive, and safe public health intervention for preventing tooth decay. It is not a new practice. The first Canadian experiment with water fluoridation occurred in 1942 when a health board recommended the practice in Brantford, Ontario. Soon after, many other municipalities embarked on a similar public health plan. By 1953, more than 1000 communities, large and small, had either

A. M. L. Perrella (✉)
Department of Political Science, Wilfrid Laurier University, Waterloo, ON, Canada
e-mail: aperrella@wlu.ca

S. J. Kiss
Digital Media and Journalism, Wilfrid Laurier University, Brantford, ON, Canada

K. Shankardass
Department of Health Sciences, Wilfrid Laurier University, Waterloo, ON, Canada

M. Gattinger (ed.), *Democratizing Risk Governance*,
https://doi.org/10.1007/978-3-031-24271-7_12

begun a fluoridation practice or had approved such a plan (Stadt 1954). Now, fluoridated water is fairly widespread, particularly in Canada and the United States, where fluoridated water is available for about half the population. In Canada, fluoridated water was available to about 14 million people in 2017 (PHAC 2017).[1]

At its inception, water fluoridation was readily embraced. It was a major public health achievement, highly economical (Ran et al. 2016), and also not necessarily artificial. Fluoride can be found naturally in water supplies, and its varying quantities are associated with different effects. Early twentieth-century research into tooth mottling identified high levels of fluoride in water as the cause, but also found that even individuals with mottled teeth appear resistant to tooth decay (McKay 1925). It was later shown that smaller amounts of fluoride in water, about 0.7 mg/L, can still provide a good level of prevention against cavities while avoiding the mottling problem (see also Arnold, Jr. 1943). Such early research led to municipal governments intervening to bring fluoride levels to that optimal level, either by adding fluoride to water, or by removing fluoride in cases where natural levels are higher than necessary.

But since its early history, there have been moments of resistance to fluoridation, either by elected officials in municipal councils or in the population, expressed in referendums. And while one might be able to comprehend risk-averse attitudes among populations in the 1950s who faced a rapidly changing world with technology taking over even the most mundane of human experiences such as drinking water, water fluoridation remains controversial to this day. Relatively recently in Canada, several cities have held elections that resulted in voters choosing to stop water fluoridation (Perrella and Kiss 2015; Kiss et al. 2018). Between 2012 and 2017, almost 40 communities across Canada have ceased water fluoridation (PHAC 2017). More have considered the issue, and there appears to be little abatement in the pace of such anti-fluoridation opposition. There are anti-fluoridation movements all over Canada. This leads to the main question of inquiry: What does it take to move a public against water fluoridation?

[1] Levels vary considerably across sub-regions. Manitoba and the Northwest Territories rank high with 68 to 73% of their populations having access to fluoridated water (naturally occurring or added), while at the other extremes are New Brunswick (9.62%), Newfoundland and Labrador (1.59%), Yukon (no water fluoridation), British Columbia (1.26%), and Québec (2.49%) (PHAC 2017).

This is an important question given the current lack of universal public insurance for dental care in Canada[2] and the potential for a lack of fluoridation to exacerbate existing inequalities in oral health (Chari et al. 2022). But there is a more general concern relating to public health interventions, which can be expensive and sometimes risky. If the public can get riled up to oppose water fluoridation, which is relatively inexpensive, effective, and safe, then what can we expect when members of the public face more significant and imperative health threats? Consider, for instance, the anti-vaccination movement, or socially conservative parents who oppose their daughters receiving the human papillomavirus vaccine. In a far more contemporary sense, consider those who doubt the efficacy and safety of any of the COVID-19 vaccines approved or any health mandates (e.g., masking) that governments and organizations impose on their publics. The list goes on. Few other areas of health science—and public policy—get people more impassioned and politicized than public health, and among them, water fluoridation holds a special place.

What is remarkable in all this is that support for water fluoridation normally begins favorably. Over time, the few surveys that are available show that when people are simply asked about their views of water fluoridation, there is often clear support. The following are examples spanning nearly 50 years that show the percentage of respondents in favour of water fluoridation:

- National U.S. survey, 1973, 67% (Nunn et al. 1973).
- United Kingdom, 1977 to 2007, a review of 11 surveys shows support ranging from 56 to 79% (Catleugh, Delves and Bellaby, n.d.).
- Yates County, NY, 1997, 84% (Health Counts in Yates! n.d.).
- New Zealand, 2009, 58% (Whyman et al. 2016).
- Canada, 2009, 62% (Quiñonez and Locker 2009).
- West Midlands, UK, 2010, 67% (National Health Service, West Midlands 2010).
- Michigan, 2013, 70% (Marketing Resource Group 2013).
- Waterloo Region, 2013, 60% (Perrella and Kiss 2015).
- Calgary, 2021, 68% (Kanygin 2021).

[2] It should be noted that there is significant movement in this policy area. A 2022 agreement between the minority Liberal government and the smaller New Democratic Party included expanded access to dental care as a key element (Thompson 2022).

Consistently, the majority of the public seems to support water fluoridation, often at high levels. But when the same public is agitated, it can turn skeptical or force a water authority to cease fluoridation. How does this opinion reversal happen? What does it take for a favourable position to flip? That is the key question being explored here.

The context in which public opinion can shift in a consequential way is the main vehicle used to oppose—or cease—water fluoridation: a plebiscite. The public is often asked in the form of a vote to take an unambiguous side: To fluoridate or not to fluoridate? This is important to point out as the context of the discourse within an elected council is qualitatively different than one held more widely among the electorate. Elected officials can debate and interact with experts to decide such matters by weighing the evidence. Elected officials can change their mind in such a context, too. And while debates within official chambers can be charged with emotions, for sure, they are not always just about facts; ideology and interpretations matter. But the dynamic of a popular election on the topic elevates more prominently psychological forces, as most citizens are not required to be as engaged in any one political topic. They are not required to witness the questioning of experts. They are not required to review reports. Their general lack of sophistication and knowledge about public affairs (Delli Carpini and Keeter 1996; Fournier 2002) ill-equips them to engage with matters of a more technical nature, such as water treatment.

Partly for this reason, people are susceptible to heuristics and narratives. Through mental shortcuts and gripping images, people can be persuaded to hold an opinion independent of any factual basis (Hochschild and Einstein 2015). They can also be swayed to hold some doubt over a matter about which they had not fully considered. As it pertains to our study, people who would hold favourable views of fluoridation could then become more doubtful after having been exposed to opposing narratives, or when narratives prime certain features that place fluoridation in a more negative light. More technically, the public can be moved if faced with a narrative that follows "emphasis framing" (see Chapter 1; see also Cacciatore et al. 2016). They may not be fully convinced of opposing arguments, but particular emphasis frames may raise the level of doubt to make them more risk averse. Consequently, when asked in a referendum, they feel more comfortable to err on the side of caution: Better to vote against water fluoridation. It is this possibility that we explore.

Fluoridation Narratives

Much of the debate about water fluoridation revolves around its "safety" or its "toxicity." However, this debate is often confused. What does it mean for a substance to be "safe?" In reality, any substance, natural or synthetic, can be extremely unsafe at some level. But toxicity is less about the substance than about its dose (Gardner 2009; Kiss 2015). This is true of many commonly consumed products. Here are some examples of common substances and their median lethal dose (LD_{50}) per kilogram:

- Caffeine, 192 mg (Boyd 1959);
- Nicotine, 6.5-to-13 mg (Mayer 2014);
- Vitamin C, 11,900 mg (Nelson 2018).

When toxicologists say something is "safe," they mean it is safe *at that dose*. In the case of water fluoridation, the optimal level of fluoride for people to consume is 0.7 mg/L on a daily basis. This provides fluoride's well-documented preventative properties while avoiding any of its potential risks. This amount is well below an unsafe dose. The LD_{50} of sodium fluoride (the most common form of fluoride) is 52 mg/kg (Environmental Health and Safety 2001). And fluoride toxicity has happened! Normally, this occurs from groundwater wells containing fluoride flowing from granite. In some cases, such as in India, fluoride concentrations can reach as high as 70 mg/L, but "crippling" skeletal fluorosis has occurred in areas with fluoride concentrations at much lower levels of 2.8 mg/L (Gupta and Ayoob 2016: 15). Water treatment facilities can extract excess fluoride from water to reach the optimal level of 0.7 mg/L, far below known levels of toxicity.[3] The foregoing is not meant just to be a technical discussion of the nature of toxicology or to highlight that water fluoridation can help protect the public from toxic levels of the substance; instead, it is to highlight the way in which technical discussions of "risk" and "safety" depend on a nuanced understanding of dose and exposure, which are often lost in a loaded, emotionally charged, and often simple public debate.

[3] There are other potential harmful effects at lower doses. For instance, risk of fluorosis (bone disease) can occur when ingesting at least 6 mg of fluoride a day (WHO 2017), which is still difficult to reach as this requires drinking at least eight litres of water.

How do people determine what is risky or safe? Often, science plays a minimal role (Kahan et al. 2011). Instead, perception of risk is shaped by how an issue is discussed and how that interacts with personal worldviews. When an issue involves the deliberation of wide publics, there are opportunities to articulate a wide range of perspectives, some factual, some exaggerated, and some invoking various frames or normative points of view. The wide accessibility of online resources can even render more prominent marginal perspectives. As noted by Shanahan et al. (2011), the contemporary context over policy discussion and change has fewer gatekeepers, with members of the public able to access online information and interpretations. This can create a chaotic context that renders policy narratives as effective summaries of different arguments. The question asked here is: Which narratives are more able to move public opinion on water fluoridation?

Some narratives are common in fluoridation debates. To begin, there is a more strictly technical, or scientific, argument. When public health professionals or dentists are asked about the safety and effectiveness of fluoride, few go much beyond just stating that, yes, it is effective, and, yes, at low doses, it is safe. Enough said. But this narrative competes with others.

A common factual counter-argument is that the substance has been linked to bone disease (skeletal fluorosis) and even bone cancer (Clemmesen 1983). While correct, this requires a much higher dose. Bone disease is more likely to occur if one consumes over a long period (about 10 years) water containing fluoride above the recommended amount.[4] It is also factually correct to say fluoride is safe at low doses. But debates rarely engage in the finer technical points. Instead, the public is faced with two competing and equally convincing arguments about the toxicity of fluoride.

Furthermore, when discussing fluoride as a chemical, opponents often use the term "hydrofluorosilicic acid," a compound commonly used for water fluoridation. Certainly, it is natural for individuals, especially those without a chemistry background, to react with some shock to hear that

[4] There are different studies on what constitutes a high dose. Li et al. (2001) find bone fractures more likely to occur with long-term (minimum 25 years) consumption of water that contains 4.32 ppm of fluoride. Pratusha et al. (2011) raise that threshold as high as 10 ppm.

an "acid" is added to their drinking water. And rightfully so. Hydrofluorosilicic acid, with a pH level of 1.2, is corrosive (Heneke and Carson 2001). But lost in the discussion is that when this acid is added to drinking water, it dissolves and breaks apart, leaving behind the fluoride ions needed for dental protection (Urbansky 2002). This is an important point as one of the largest challenges in this debate is to explain its complex chemistry.

The above factual argument and counter-arguments must also compete with normative (or moral) frames. The debate on whether to fluoridate sometimes takes a turn on whether it is the right thing to do, independent of its safety. One common pro-fluoridation narrative views society as a community in which individuals look after each other, even if that requires some state intervention (Beauchamp 1985; Kaul v. City of Chehalis 1954; Perrella and Kiss 2015). To protect individuals—especially children—against tooth decay, adding a small amount of fluoride to water seems like a small price to pay, in terms of both the degree of liberty that is infringed and also the actual cost to taxpayers.

The counter argument to this favourable frame invokes liberty (Carstairs and Elder 2008). There have been several attempts to bring this argument to court challenges, although in most cases, courts rule against those who argue fluoridation infringes on constitutionally protected individual rights (Block 1986; Pratt et al. 2002). Does the state have the right to add chemicals to our water, without our consent? Constitutional law aside, people may question the legitimacy of a state that adds "chemicals" or "drugs" to water. If people want to consume such chemicals, it should be a personal choice, they argue. They add that since fluoride is available in toothpaste, it is up to individuals—and parents—to control their own personal exposure and that of their children. This more libertarian frame is often used as a non-technical argument about the morality of water fluoridation.

Combined, these diverse narratives sort along two dimensions. One dimension is based on fact, where on the one hand fluoride can be said to be safe, albeit at low doses, and on the other hand fluoride can be said to be toxic, albeit at high doses. The second dimension is based on a moral or normative frame, where one pole represents libertarian views that the state must not add chemicals to water without our consent, while the opposing pole represents a communitarian frame, whereby society, and by extension, the state, sometimes needs to impose standards for our collective benefit.

Together, these dimensions form a four-cell typology, displayed in Fig. 12.1. The top-right quadrant represents those who view water fluoridation as safe and believe the practice provides a public good. The bottom-right quadrant can be described as libertarians, as they regard water fluoridation as safe, but oppose it on the principle that it infringes on personal rights. The top-left quadrant represents those, such as "eco-warriors," who doubt the safety of water fluoridation, but still believe in the principle of a public good and the necessity of state intervention. They just think fluoride is hazardous. These same individuals also generally question the safety of genetically modified food and believe in an organic diet. Finally, the bottom-left quadrant describes those who doubt the safety of fluoride, and also question the legitimacy of state intervention.

The bulk of the opposition to water fluoridation is a coalition of three of the quadrants, all but the top-right. Anti-fluoridation campaigns have invoked the fluoride-as-hazard and fluoridation-as-infringement arguments, often together. Keep in mind: Water fluoridation is widely practiced, and there usually is support for it. But public deliberation can agitate this support and cause it to oppose the practice. This suggests that support is not deeply rooted; people can be persuaded to doubt their

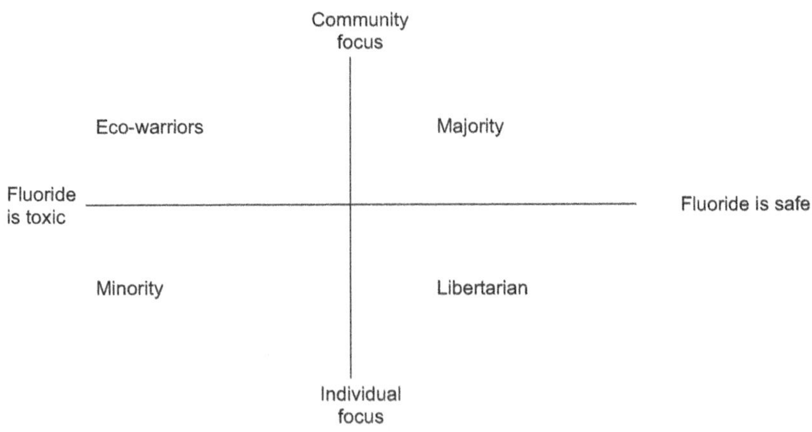

Fig. 12.1 Typology of factual and moral narratives (*Source* Authors' own source)

support for fluoridation, and with enough people changing their mind, opposition can grow significantly.

But not all narratives can be assumed to weigh equally. It is plausible that some are more effective. But which ones? Is it the factually based argument that suggests fluoride is hazardous? Is it the moral argument that emphasizes individual rights and questions the role of the state? Is it both combined? What happens when people are faced with competing narratives, favourable and unfavourable? Do they cancel each other out, or is one more persuasive? In sum, which narratives, or combination of narratives, are more likely to shift opinion?

In addition, are some people more susceptible to particular narratives? This possibility is explored by considering whether certain cognitive characteristics are more likely to yield greater susceptibility. In particular, the following is considered: (1) knowledge, specifically about science and fluoride; (2) trust (or skepticism) in science; and (3) political attitudes, specifically ideology. Let's briefly consider each of these three main dimensions.

First, as noted in much of the literature, citizens have low levels of knowledge about politics (Delli Carpini and Keeter 1996; Fournier 2002). Their grasp of science is no better (Durant et al. 1989). Consequently, when faced to deliberate over a policy matter that requires some knowledge of science, and specifically, over some knowledge of fluoride as a chemical and its properties, citizens are either prone to not understand or are susceptible to inaccurate and sometimes false perspectives. As noted by Martin (1991), "... claims do not have to be scientifically correct in order to be persuasive. The claims about fluoridation and cancer were effective politically, even though many scientific refutations were published" (141).

Second, it is evident people are mobilizing against scientific reason. Anti-vaccination movements, alternative health fads and climate-change denial are three current examples of skepticism. These, along with opposition to water fluoridation, frustrate efforts to develop effective science-based health policy, regardless of sound and irrefutable scientific research. As noted by Camargo and Grant (2015), many such movements grow out of individuals listening to each other, not to science, as the scientists often are ill-equipped to communicate complex matters to the general public (see also, Collins and Evans 2008, 2014). Such belief systems sometimes can be fairly constrained, resisting opposing views, as was demonstrated by some during the COVID-19 pandemic (Miller

2020; see also Wood et al. 2012). Therefore, one would expect that when confronted by a variety of narratives, skeptics are less likely to be swayed by factual evidence.

Third, it is evident that much of the anti-science movement, particularly as it relates to vaccinations, climate change, and (more recently) anti-masking, shows an ideological bent. It seems much of the skepticism toward many public health measures comes from those of the right, and those who support—or identify with—parties (and leaders) of the right. This has been confirmed by some research (Dillard et al. 2021; McCright et al. 2013; Kirst et al. 2017). Perhaps the most prominent representative of such science skepticism is none other than former U.S. president Donald Trump, who has spoken out against masks during the COVID-19 pandemic, not to mention climate change.[5] However, it should be noted that opposition to public health practices does not always have a distinctive ideological flavor. Certainly, some orientations can explain such opposition. For instance, school programs to have teenage girls vaccinated against the human papillomavirus often face resistance from parents who feel this program challenges their authority, or that the practice presses against traditional beliefs about gender norms (Perrella and Kiss 2015; see also Kahan et al. 2015; Kiss et al. 2020). But opposition to water fluoridation, in particular, can stem from both traditional, right-wing views and those who situate themselves on the left. As noted by Carstairs and Elder (2008), libertarians, environmentalists, "naturalists," and those who oppose corporate capitalism may see each other as allies in opposing fluoridation.

Data and Methods

In order to determine how narratives can affect support for fluoridation, experimental questions were administered through an online survey, conducted in April of 2017, to respondents in both Canada and the

[5] Trump has since changed his mind in one respect by encouraging parents to have their children vaccinated, although many of his supporters continue to spread an anti-vaccination perspective (D'Antonio 2019).

United States.[6] A total of 3854 respondents were gathered,[7] and each was asked to indicate their support for fluoridation on a scale of 1 ("extremely opposed") to 7 ("extremely supportive"). Prior to registering their support, we presented respondents with one of 10 experimental treatments. These were structured to prime a variety of four considerations, or prompts, each drawn from the same dimensions illustrated in Fig. 12.1. In addition, one group of respondents served as a control, where they were simply asked to score their view of water fluoridation without having been exposed to either a fact- or normative-based prompt. The goal is to determine if any prompt, i.e., narrative, is able to shift public opinion away from some default position.

In the first dimension, "fact," respondents were given two prompts: (1) "Health Canada[8] recommends 0.7 mg/L of fluoride in water to prevent tooth decay"; (2) "The risk of skeletal fluorosis (bone disease) rises when people consume 10 mg of fluoride a day for about 10 years." The second dimension featured two moral-based prompts: (1) "Low income children are often at risk of getting cavities because their family cannot afford dental care"; (2) "Governments sometimes add medicines or chemicals in drinking water without people's consent."

Four questions contained only one of the four prompts listed above. Six other questions exposed respondents to a blend of narratives. Some received conflicting prompts: one favourable to fluoridation (or positive) and one opposing (or negative). Some received two reinforcing prompts: one factual and the other a frame. The idea was to cover as much variety as possible in order to simulate the competitive and informational diversity that would characterize an actual fluoridation plebiscite campaign. The total list of questions appears in Table 12.1.

Of these prompts, it is expected that those that prime facts are less effective than those that emphasize moral frames (Druckman and Bolsen 2011). As previously noted, arguments based on science or fact compete against more normative frames. This will be displayed by comparing mean scores of each experimental group against the control.

[6] The survey, administered by Dynata, was made possible through financial support from the Laurier Institute for the Study of Public Opinion and Policy.

[7] The total number of respondents reached was higher, 4241, but 387 were excluded for not providing responses. Among the valid entries, 2,105 respondents were from the United States and 1749 from Canada (1240 in English, 509 in French).

[8] The American version referenced the Centers for Disease Control.

Table 12.1 List of question groups

Experimental group	*Question wording*
1: Factual, positive	Health Canada recommends 0.7 mg/L of fluoride in water to prevent tooth decay
2: Moral, positive	Low income children are often at risk of getting cavities because their family cannot afford dental care
3: Factual, negative	The risk of mild forms of skeletal fluorosis (bone disease) rises when people consume 10 mg of fluoride a day for about 10 years
4: Moral, negative	Governments sometimes add medicines or chemicals in drinking water without people's consent
5: Factual and moral, both positive	Health Canada recommends 0.7 mg/L of fluoride in water to prevent tooth decay. Also, low income children are often at risk of getting cavities because their family cannot afford dental care
6: Factual and moral, both negative	The risk of mild forms of skeletal fluorosis (bone disease) rises when people consume 10 mg of fluoride a day for about 10 years. Also, governments sometimes add medicines or chemicals in drinking water without people's consent
7: Moral, positive, and negative	Low income children are often at risk of getting cavities because their family cannot afford dental care. Also, governments sometimes add medicines or chemicals in drinking water without people's consent
8: Factual, positive, and negative	Health Canada recommends 0.7 mg/L of fluoride in water to prevent tooth decay. Also, the risk of mild forms of skeletal fluorosis (bone disease) rises when people consume 10 mg of fluoride a day for about 10 years
9: Moral, positive, and factual, negative	Low income children are often at risk of getting cavities because their family cannot afford dental care. Also, the risk of mild forms of skeletal fluorosis (bone disease) rises when people consume 10 mg of fluoride a day for about 10 years

(continued)

Table 12.1 (continued)

Experimental group	*Question wording*
10: Factual, positive, and moral, negative	Health Canada recommends 0.7 mg/L of fluoride in water to prevent tooth decay. Also, governments sometimes add medicines or chemicals in drinking water without people's consent

Source Authors' own source

Following these questions, respondents were randomly assigned to one of two questions that specifically focuses on ways to describe the actual chemical commonly used to add fluoride to water. As noted, critics of fluoridation often refer to hydrofluorosilicic acid as the ingredient added to water, hoping people would become scared at the thought of their drinking water containing any sort of acid. Also noted is the fairly complicated chemical process involved when hydrofluorosilicic acid is added to water. If this process is left insufficiently explained, it is likely people will oppose fluoridation. To test this, the first question simply stated: "One of the ways that water utilities add fluoride to the local water supply is to add hydrofluorosilicic acid to the water supply." The second question contained more information: "One of the ways that water utilities add fluoride to the local water supply is to add hydrofluorosilicic acid to the water supply. This chemical dissolves and separates into two parts, fluoride and water."[9] Of the two, the second question appears more benign, and respondents are expected to be less opposed to fluoridation when given the more factually complete description. This will be evident by noting changes in mean scores of support for water fluoridation.

Questions from the first iteration are examined further to determine whether their persuasiveness, or susceptibility, depends on any of the three key attributes discussed earlier, namely knowledge of science, skepticism toward science, and ideology.

Regarding knowledge of science, an argument can be made that people who are knowledgeable are expected to be more swayed by

[9] The chemical reaction is a bit more involved (see, for example, Haneke and Carson 2001). But the main point is not to explain fully the chemical process, but to leave respondents with a sense that the chemical (and the resultant reaction) is far more benign than the "acid" name suggests.

factual accounts, while those who are less informed are expected to be more affected by normative frames. However, and as noted in this volume, there is evidence to suggest otherwise, whereby even among the highly educated we can see the presence of motivated reasoning that impedes what would otherwise be labeled rational thought (Hochschild and Sen 2015; Kahan et al. 2012; Kraft et al. 2015). To test this hypothesis, respondents were disaggregated into groups based on their scientific literacy. Four questions were used. The first was a "True or False" question that asked respondents whether "Lasers work by focusing sound waves" (correct answer is *false*). The second asked respondents to indicate whether sound or light travels faster (correct answer is *light*). Another two questions measured knowledge about fluoride, specifically. The first asked: "Is it true or false that fluoride occurs naturally in many water sources like rivers and lakes?" (correct answer is *true*). The second asked: "Is it true or is it false that fluoride only prevents cavities by being applied directly to the teeth?" (correct answer is *false*).[10] The four responses were combined into a 0-to-1 index, where those who answered all questions correctly scored a 1, while those unable to provide a correct answer to any question scored a 0.

Our second susceptibility test considered skepticism, or trust, toward science. It is expected that those more skeptical of science—particularly science related to public health—to be less supportive of fluoridation, but also more likely to be swayed by negative normative prompts. The survey contained a question that asked: "Does science solve or cause problems?" Respondents were offered a seven-point scale ranging from 1, to indicate "science replaces older problems with new ones," to 7 to indicate "science overcomes problems." This measure was rescaled to 0-to-1 to render it consistent with the others.

The third test of susceptibility considered ideology. The survey contained a question that asked respondents to situate themselves on an ideological spectrum. Canadian respondents were asked to assign themselves a score from 1, to indicate the extreme left, to 7, to indicate the extreme right. American respondents were given the same scale, but

[10] In both cases, when processing responses to the two science-knowledge and the two fluoride-knowledge questions, respondents who indicated that they either "Did not know" or "Refused" to answer were assumed to not know the correct answer, thereby assigned a score of 0.

the extremities were labeled "extremely liberal" and "extremely conservative," respectively. In both cases, responses were rescaled to a 0-to-1 scale. Here, it is expected that those who are more to the right to be less supportive of water fluoridation, and their support more likely to be further bolstered by normative (particularly libertarian) frames. However, as suggested by McCright et al. (2013), those on the left are more likely to support the science behind public health. They are expected to be more likely swayed by the communitarian moral frame. It is therefore plausible that those in the centre will be more supportive than those close to either extreme.

The effect of our different experimental narratives and their potential to move opinion when one takes into account scientific knowledge, trust in science, and ideology is analyzed by comparing mean scores of each of the 10 experimental groups against the control. Regression models are also generated whereby the 10 experimental groups are entered as dummy variables with the control group acting as the reference. Also included in the regression models are control variables for country, gender, income, education, and age.[11]

Results

As noted at the onset, support for water fluoridation is generally positive. But when agitated, support can drop to the point where public health authorities may be forced to either cease water fluoridation programs, or pull back from even considering it. Our sample also shows a baseline level of support. Respondents indicated their support for water fluoridation using a score of 1 (extremely opposed) to 7 (extremely supportive), with 4 designated for a neutral, "not sure," response. Overall, support among those in the placebo, or control, group leans on the positive side, with a mean score of 4.5 out of 7. A third of respondents chose the neutral "not sure" (see Fig. 12.2). Even if that neutral category is discounted, far more

[11] Country is a dummy variable whereby 1 for Canada and 0 for the United States. Gender is also a dummy variable whereby women are scored a 1. Income is measured with a question that asks respondents to select one of nine different incomes, with the lowest category selected by those with an income of less than $25,000, and the highest category selected by those with an income of more than $500,000. The education variable has 10 categories, from "some elementary school" all the way up to "completed graduate school." In both cases, the variables were recoded to a 0-to-1 scale. Respondent ages in the sample run from 18 to 81, which was also rescaled to 0-to-1.

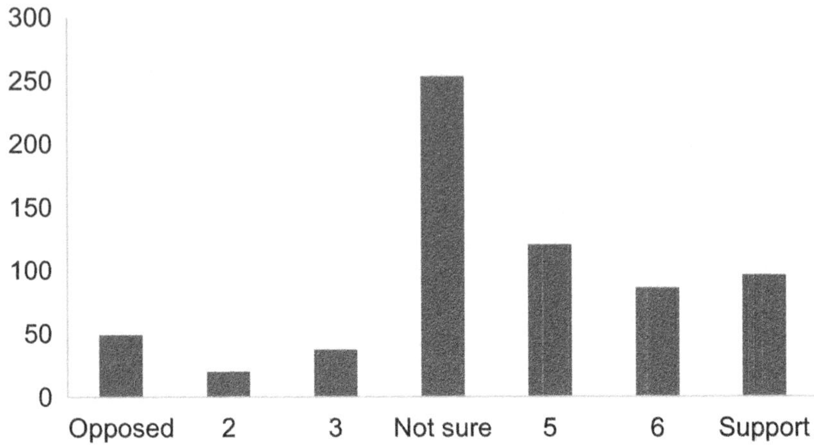

Fig. 12.2 Support for water fluoridation (control group) (*Source* Authors' own source)

respondents land on the positive end ($n = 302$) than on the opposed end ($n = 106$). Clearly, we see here what was seen elsewhere: Opinion toward water fluoridation is not, by default, negative. It is decidedly positive. From this baseline of support, it is possible to examine whether certain narratives are more likely to push or pull support one way or another.

Effect of Narratives

Figure 12.3 displays three series, all showing difference in level of support from the control group. The first series is for the entirety of the dataset. The other two separate results for Canada and the United States. The 10 different narratives are sorted based in descending order from those that move opinion in the "total" series more toward supporting fluoridation, down to those that move opinion in the opposite direction.

One evident pattern is the higher number of negatives than positives. Six out of ten scenarios show lower support compared to the control group. A second notable finding, and an unsurprising one, is the 0.77-point decline in support occurring when two negative prompts are paired together. In contrast, a pairing of two positive prompts improves support by 0.17 points, a difference that is only marginally significant at $p < 0.10$

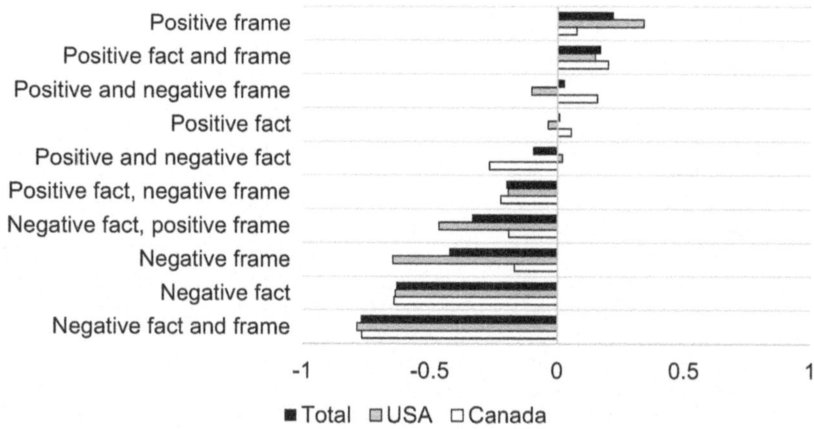

Fig. 12.3 Distance from control group (*Source* Authors' own source)

(see Table 12.2). A third key observation is the ability for a negative prompt, be it fact or frame, to pull down support, even when paired against a positive prompt. However, in Table 12.2, when the experiments are tested with a battery of statistical controls, only two such contrasted pairings yield significant results: Group 9 (negative fact, positive frame) and Group 10 (positive fact, negative frame). These patterns hold generally for the two countries when examined separately. A fourth key finding is the strength of the bone-disease prompt. Unlike what was expected, the negative-fact narrative may be even more effective than the negative-normative narrative. The largest coefficients in Table 12.2 are for Group 3 (negative fact, $B = -0.637$) and Group 6 (negative fact and frame, $B = -0.757$), while Group 4, which was exposed only to the negative moral frame, produces a coefficient of $B = -0.441$.

It does appear, however, that there are some differences between Canada and the United States. A combined positive and negative frame (Group 7) strengthens support for water fluoridation in Canada, but has the opposite effect in the United States. However, this difference fails to reach significance.

Overall, one general conclusion is that support for water fluoridation is not bolstered by reassuring factual arguments. Even when a positive fact is paired with supportive moral arguments, public opinion seems unaffected. What seems effective, however, are negative narratives, be they

Table 12.2 Effect of 10 experimental narratives

	B	*Std. Err*	*Beta*	*Sig*
Experimental group				
Positive fact	0.075	0.122	0.012	0.539
Positive frame	0.204	0.118	0.033	0.084
Negative fact	−0.637	0.122	−0.098	0.000
Negative frame	−0.441	0.122	−0.068	0.000
Positive fact and frame	0.215	0.120	0.034	0.072
Negative fact and frame	−0.757	0.121	−0.118	0.000
Positive and negative frame	0.050	0.120	0.008	0.677
Positive and negative fact	−0.063	0.120	−0.010	0.598
Negative fact, positive frame	−0.304	0.121	−0.048	0.012
Positive fact, negative frame	−0.246	0.122	−0.038	0.044
Country	−0.128	0.060	−0.036	0.033
Gender	−0.147	0.059	−0.042	0.013
Income	0.647	0.152	0.079	0.000
Education	0.971	0.126	0.135	0.000
Age	0.882	0.127	0.114	0.000
Intercept	3.642	0.121		0.000
Adj. R-sq	0.083			
S.E.E	1.682			
N	3485			

Source Authors' own source

about the toxicity of fluoride or about the state's legitimacy to impose fluoridation. As suggested earlier, public support for fluoride may not be solidly grounded and seems easily budged. This becomes even more important when considering the second iteration of experiments.

In the second round, respondents were assigned to one of two groups. The idea here is to determine if their opinion on water fluoridation changes based on the contentious reference to the chemical compound hydrofluorosilicic acid. Opponents to water fluoridation often refer to this compound in their public statements with the hope that members of the public will turn skeptical of fluoridation because of the fear that this involves adding an acid to water. But, as noted earlier, what is not always discussed is how hydrofluorosilicic acid dissolves and breaks down when added to water. It is expected respondents will become less skeptical by this more benign (and lengthy) explanation of a common water fluoridation process.

It should be noted that this second experiment appears in the survey immediately after the first. Therefore, most respondents reach this point after having already been primed by whatever prompts they had received in the prior question. What this next iteration simulates is a context of ongoing debates: They receive one narrative, and then they receive another, potentially more complex one. What happens next? Does it shift opinion at all?

Overall, the two experimental groups differ. Those who were given the simpler "hydrofluorosilicic acid" prompt scored slightly lower than those given the more benign and longer prompt, 4.13 and 4.28 out of 7, respectively. This difference is statistically significant, but overall the gap is very small. This likely reflects the very diverse sample, with most respondents having been exposed to different narratives in the first experiment. When compared to results from the first iteration, the less benign prompt appears to pull down support, pretty much across all the groups. Figures 12.4a and b show differences in support levels from the first iteration to the second. Negative values suggest respondents become less supportive of fluoridation after this second round of prompts. Here, a decline in support is visible, even among respondents who in the first iteration were in the control group. Among those who received the simpler prompt, their support of fluoridation declined by 0.31 points (see Fig. 12.4a). The longer and more informative prompt does little to change opinion; among respondents who were exposed to the more benign prompt about how hydrofluorosilicic acid reacts in water, changes are more subtle (see Fig. 12.4b). These results challenge any notion that citizens would grow to support fluoridation if they were given more information about the chemical process. It appears they may not.

Scientific Literacy

Perhaps factual narratives fail because of a lack of scientific knowledge. There is no expectation for ordinary citizens to understand the chemistry of water fluoridation, just as they need not know the chemistry of dish soap. But it might help. Knowledge of science, maybe even a little, may make one more likely to accept, or at least listen to, a scientific argument for fluoridation. To test this, the sample is disaggregated into groups based on their level of scientific literacy.

The survey included questions that measure how much people know about science, in general, and about water fluoridation, more particularly.

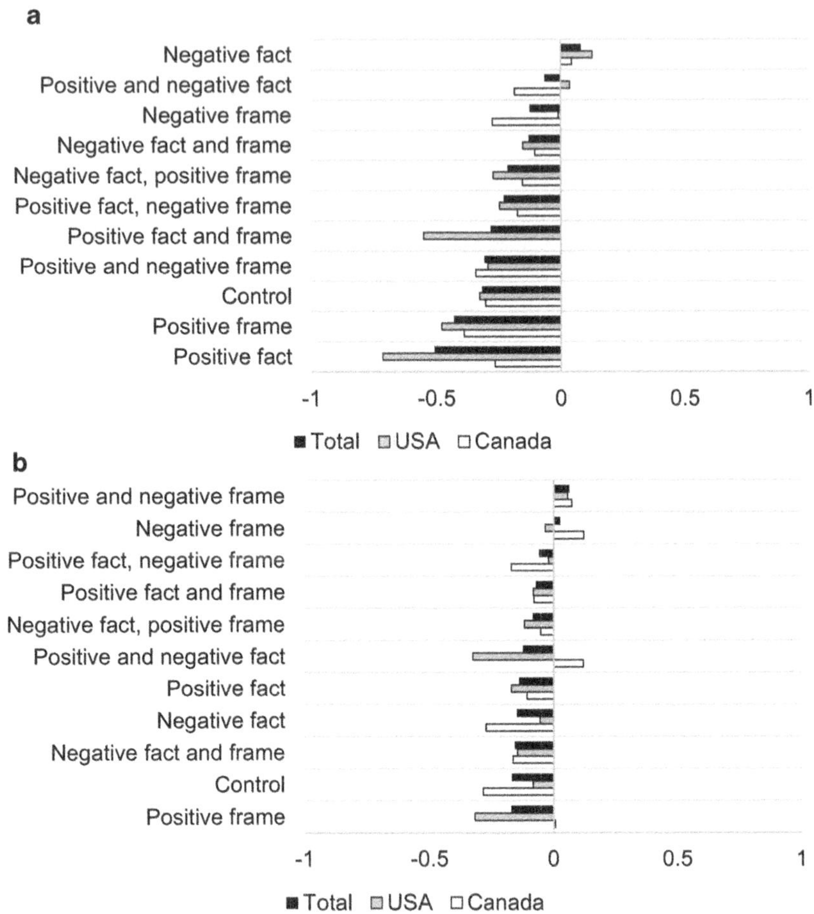

Fig. 12.4 (**a**) Simpler hydrofluorosilicic acid prompt; (**b**) Complex hydrofluorosilicic acid prompt (*Source* Authors' own source)

Overall, on a range of 0 to 1, the sample scored an average of 0.46, with standard deviation of 0.26. It should be noted that about 40% of the sample scored no better than 0.25, and a little over one-third scored 0.50, which produces a slightly positively skewed distribution. Hence, this distribution was collapsed into three categories. Scores from 0 to 0.25 are the "low" category; those who scored 0.5 are the "middle" category;

and those who scored higher are in the "high" category. The mean scores across the 10 experimental groups are examined for each of these three subsets.

The first subset considers respondents with the lowest level of scientific knowledge. Again, as seen in previous scenarios, the presence of the negative frames, for the most part, outweighs the effect of any other positive prompts (see Fig. 12.5a). But the effects of the narratives are more limited. In Table 12.3 (Model 1), only Group 3 (negative fact) and Group 6 (negative fact and frame) show significant results. Group 2 (positive frame) and Group 10 (positive fact, negative frame) are marginally significant at $p < 0.10$. Overall, the negative frame matters more than any factual account for this low-knowledge group. This holds overall, but in separate country-specific regression models (not shown here), Americans' views of fluoridation improve if offered a narrative that features only the positive frame, but among Canadian respondents, views sour when presented either with a negative fact alone, or if combined with a negative frame.

The next subset examines those with mid-level knowledge. Results, shown in Fig. 12.5b, are similar to what appears in Fig. 12.5a, repeating the same pattern seen elsewhere, with narratives more likely to pull respondents away from supporting fluoridation than toward supporting it. The largest changes, however, are the two negative narratives (Groups 3 and 6). There seems to be some potential for positive narratives to increase support for fluoridation, but according to Model 2 of Table 12.3, these two groups show no significant effects.

Among the high-knowledge subset, remarkably, almost all experimental groups show weakening support for water fluoridation, regardless of narrative, especially among Americans (see Fig. 12.5c). Not all of these effects are statistically significant, however. In Model 3 of Table 12.3, support for fluoridation weakens in the presence of any negative prompt, be it factual or normative, alone or in combination. Those who are more scientifically literate do not grow more supportive of fluoridation when provided factual accounts about the chemical's safety. Instead, the mention of any negative perspective, be it factual or normative, is enough to raise doubts among them. Why that occurs is unclear. Perhaps our measure of science literacy does not go far enough, and just reflects those with enough fluency to be attentive to different perspectives of the water fluoridation discourse, but not scientifically sophisticated enough to weigh the evidence, and thus, may be prompted toward being cautious.

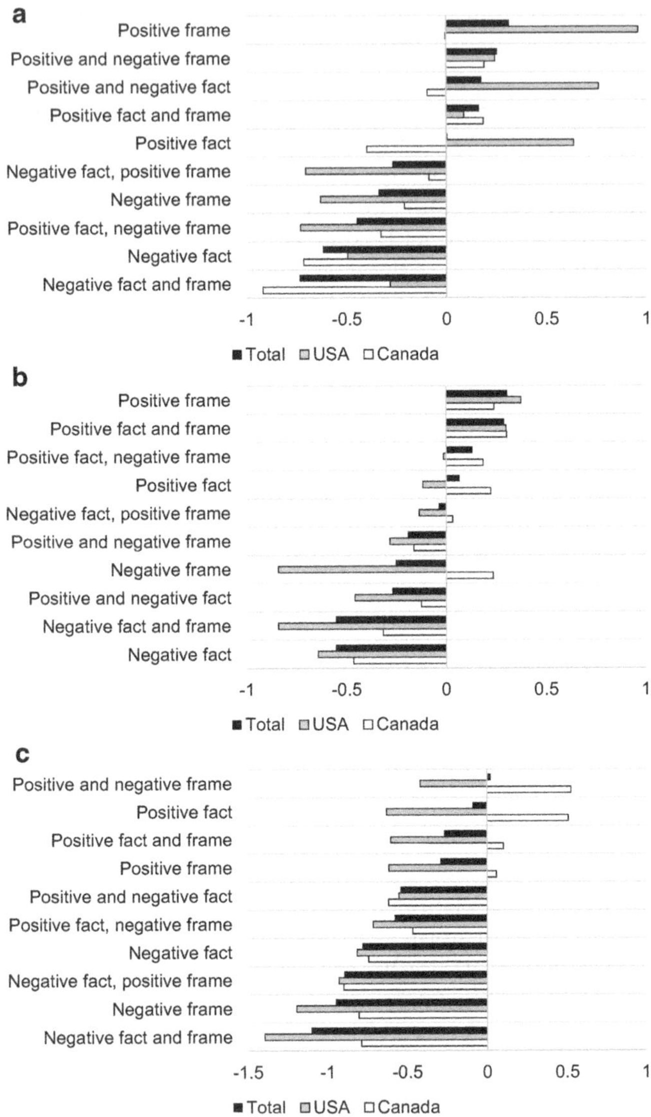

Fig. 12.5 (**a**) Low knowledge; (**b**) Mid-level knowledge; (**c**) High knowledge (*Source* Authors' own source)

Table 12.3 Effect of narratives considering scientific knowledge

	Model 1: Low knowledge				*Model 2: Mid-level knowledge*				*Model 3: High knowledge*			
	B	*Std. Err*	*Beta*	*Sig*	*B*	*Std. Err*	*Beta*	*Sig*	*B*	*Std. Err*	*Beta*	*Sig*
Experimental group												
Positive fact	0.124	0.220	0.019	0.573	0.003	0.233	0.001	0.989	−0.003	0.272	0.000	0.990
Positive frame	0.386	0.207	0.065	0.063	0.244	0.229	0.040	0.287	−0.351	0.268	−0.052	0.191
Negative fact	−0.536	0.212	−0.088	0.012	−0.594	0.242	−0.090	0.014	−0.648	0.279	−0.092	0.021
Negative frame	−0.278	0.214	−0.045	0.193	−0.235	0.253	−0.034	0.353	−1.049	0.283	−0.146	0.000
Positive fact and frame	0.308	0.221	0.048	0.163	0.262	0.227	0.043	0.248	−0.371	0.252	−0.060	0.141
Negative fact and frame	−0.692	0.212	−0.113	0.001	−0.627	0.241	−0.095	0.009	−1.053	0.285	−0.145	0.000
Positive and negative frame	0.313	0.209	0.052	0.136	−0.230	0.243	−0.035	0.344	0.144	0.262	0.022	0.582
Positive and negative fact	0.266	0.216	0.042	0.218	−0.284	0.243	−0.043	0.242	−0.484	0.250	−0.079	0.054
Negative fact, positive frame	−0.020	0.226	−0.003	0.930	−0.011	0.237	−0.002	0.964	−0.992	0.253	−0.160	0.000
Positive fact, negative frame	−0.367	0.212	−0.060	0.083	0.076	0.253	0.011	0.763	−0.763	0.262	−0.118	0.004
Country	−0.348	0.112	−0.097	0.002	−0.149	0.119	−0.041	0.211	−0.266	0.133	−0.073	0.046
Gender	−0.261	0.108	−0.074	0.016	0.002	0.116	0.001	0.988	−0.088	0.135	−0.024	0.513
Income	0.948	0.294	0.107	0.001	0.825	0.308	0.096	0.008	0.804	0.321	0.100	0.012
Education	0.462	0.228	0.065	0.043	1.032	0.245	0.142	0.000	0.986	0.281	0.131	0.000
Age	0.898	0.236	0.114	0.000	0.564	0.250	0.072	0.024	1.594	0.277	0.202	0.000
Intercept	3.917	0.225		0.000	3.666	0.243		0.000	3.852	0.276		0.000
Adj. R-sq	0.088				0.058				0.122			
S.E.E	1.63				1.748				1.707			
N	1032				968				744			

Source Authors' own source

Science Skepticism

Our next test examines the effect of skepticism toward science. Overall, the sample seems to have a great deal of trust in science. On the 0-to-1 scale, the mean score is 0.74, with more than half of the respondents scoring above 0.65. Unlike our measure of scientific literacy, our measure for skepticism is far less symmetrical, making it impractical to divide up the sample into three segments. Therefore, the sample is divided into two groups: lower trust (scores of 0 to 0.50) and higher trust (higher than 0.5).

First, comparing the two groups shows that, indeed, those who are more skeptical of science show a lower level of support for water fluoridation, yielding mean scores of 3.96 to 4.61 ($t = -10.151$, $p < 0.001$). This confirms our expectation. Whether each group displays differences in susceptibility is our next focus.

Among those with low levels of trust toward science, negative narratives—alone or in combination with positive narratives—appear more likely to weaken support than positive narratives are able to raise it (Fig. 12.6a). This is confirmed in Model 1 of Table 12.4, where the only significant effects are found for Groups 3, 4, and 6. Unlike our expectations, narratives need not be a frame. Any negative narrative can make this group more opposed.

Similar results are obtained among those with a higher level of trust for science. However, there is some indication that positive narratives can strengthen support for fluoridation (see Fig. 12.6b). Model 2 of Table 12.4, however, suggests that the most promising narrative is the communitarian one (Group 2), but here, results are only marginally significant ($p < 0.10$). Overall, even among this more trusting group, negative narratives outweigh the positives.

Ideology

The final test looks at ideology, which is divided into three general subsets: left, centre and right. The expectation is that those in the center should show higher levels of support compared to those further away, either on the right or the left. When each group's mean level of support for water fluoridation is examined, expectations appear to be met, but not convincingly. Those on the left scored a mean of 4.3; those on the right scored 4.7; and those in the centre scored 4.2. Those in the centre appear

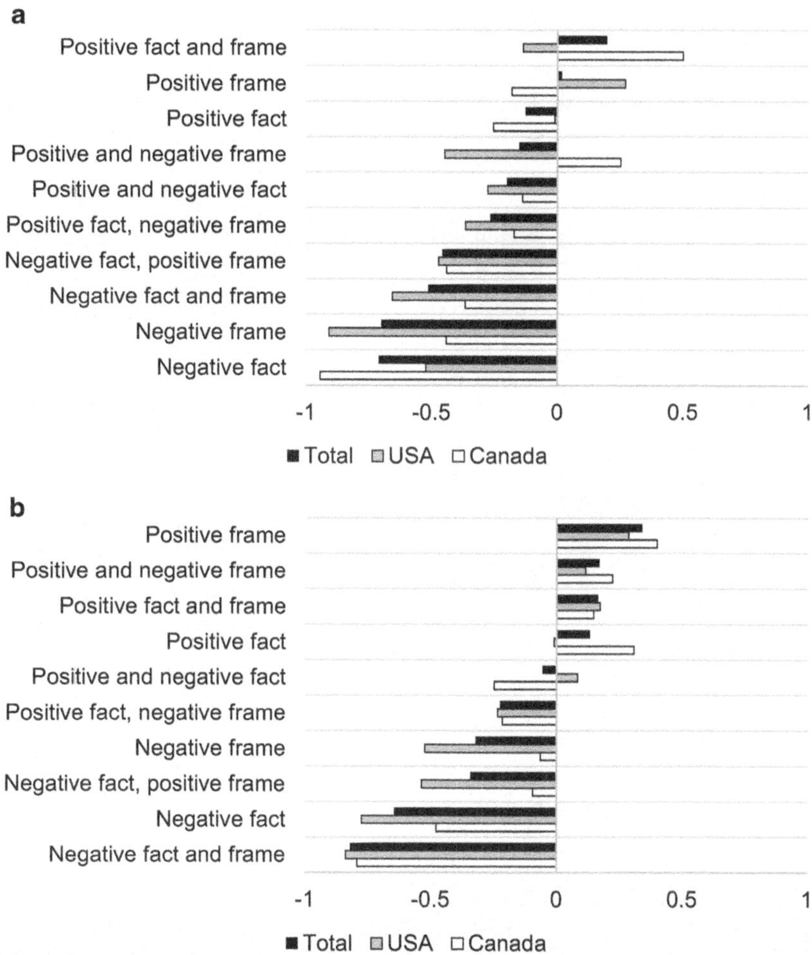

Fig. 12.6 (**a**) Lower trust in science; (**b**) Higher trust in science (*Source* Authors' own source)

to score lowest, but its score is statistically indistinguishable from that of the left. Those on the right, however, show greater support for fluoridation, and its score is statistically different than that of the left or right ($F = 28.124$, $p < 0.001$). Furthermore, the fact that those on the right

Table 12.4 Effect of narratives considering skepticism toward science

	Model 1: Lower trust				*Model 2: Higher trust*			
	B	*Std. Err*	*Beta*	*Sig*	*B*	*Std. Err*	*Beta*	*Sig*
Experimental group								
Positive fact	−0.122	0.224	−0.019	0.586	0.198	0.152	0.031	0.194
Positive frame	0.046	0.224	0.007	0.838	0.283	0.145	0.047	0.052
Negative fact	−0.775	0.225	−0.122	0.001	−0.592	0.153	−0.093	0.000
Negative frame	−0.730	0.227	−0.113	0.001	−0.308	0.154	−0.048	0.046
Positive fact and frame	0.232	0.220	0.037	0.293	0.211	0.151	0.034	0.163
Negative fact and frame	−0.595	0.237	−0.087	0.012	−0.752	0.150	−0.121	0.000
Positive and negative frame	−0.168	0.220	−0.027	0.445	0.225	0.152	0.036	0.139
Positive and negative fact	−0.248	0.239	−0.036	0.300	−0.039	0.145	−0.007	0.790

	Model 1: Lower trust				Model 2: Higher trust			
	B	*Std. Err*	*Beta*	*Sig*	*B*	*Std. Err*	*Beta*	*Sig*
Negative fact, positive frame	–0.391	0.222	–0.062	0.078	–0.299	0.152	–0.047	0.048
Positive fact, negative frame	–0.289	0.226	–0.045	0.201	–0.230	0.153	–0.036	0.133
Country	0.019	0.113	.005	0.867	–0.177	0.075	–0.051	0.018
Gender	–0.009	0.111	–0.003	0.935	–0.222	0.074	–0.064	0.003
Income	0.791	0.301	0.089	0.009	0.512	0.182	0.066	0.005
Education	0.702	0.244	0.094	0.004	0.914	0.155	0.129	0.000
Age	0.954	0.241	0.121	0.000	0.880	0.157	0.116	0.000
Intercept	3.318	0.226		0.000	3.891	0.151		0.000
Adj. R-sq	0.049				0.089			
S.E.E	1.729				1.653			
N	1052				2184			

Source Authors' own source

scored highest challenges some of the notions that anti-fluoridationists are conservatives.

The next step is to look at each ideological group to determine which narratives are most effective. Those on the left (see Fig. 12.7a) seem evenly split between scenarios that raise support for water fluoridation, and those that diminish it. But unlike expectations, those on the left are not necessarily more persuaded by communitarian narratives. Groups that contain a positive frame either show no significant effect, or they appear ineffective to counteract negative narrative. For instance, when the positive, communitarian, frame is paired up with a negative fact, support declines, which is significant at $p < 0.10$ (see Model 1 of Table 12.5). It should be noted that the sample size ($n = 460$) is a constraint on statistical power, and this may explain why what is visible in Fig. 12.6a may not yield significant results in Table 12.5.[12] Consequently, results here are not conclusive.

Those ideologically moderate (i.e., at the "centre") roughly show a similar pattern, but not nearly as split (see Fig. 12.7b), with the bulk of results resembling most previous results, whereby negative narratives seem more effective. But there are more instances when opinion improves over the control group. Respondents in Group 5 (combined positive fact and moral frame) show strengthening of support ($B = 0.363$, $p < 0.05$), while a similar boost appears for Group 2 ($B = 0.331$), although this is marginally significant at $p < 0.10$.

On the right (Fig. 12.6c), there is virtually no effect at all, except for negative prompts, regardless of whether they appear on their own or combined, even when combined with a positive prompt (see Model 3, Table 12.5). Results generally confirm expectations, where those on the right were expected to be moved by negative frames. Coefficients are highest for Groups 3 and 6.

Conclusion

If public health authorities wish to learn one thing from results reported here, it is that water fluoridation is one of those topics best managed

[12] The highly divergent pattern for Canadian and American respondents in Group 2 (positive frame) is not significant when tested in separate regression models for each country.

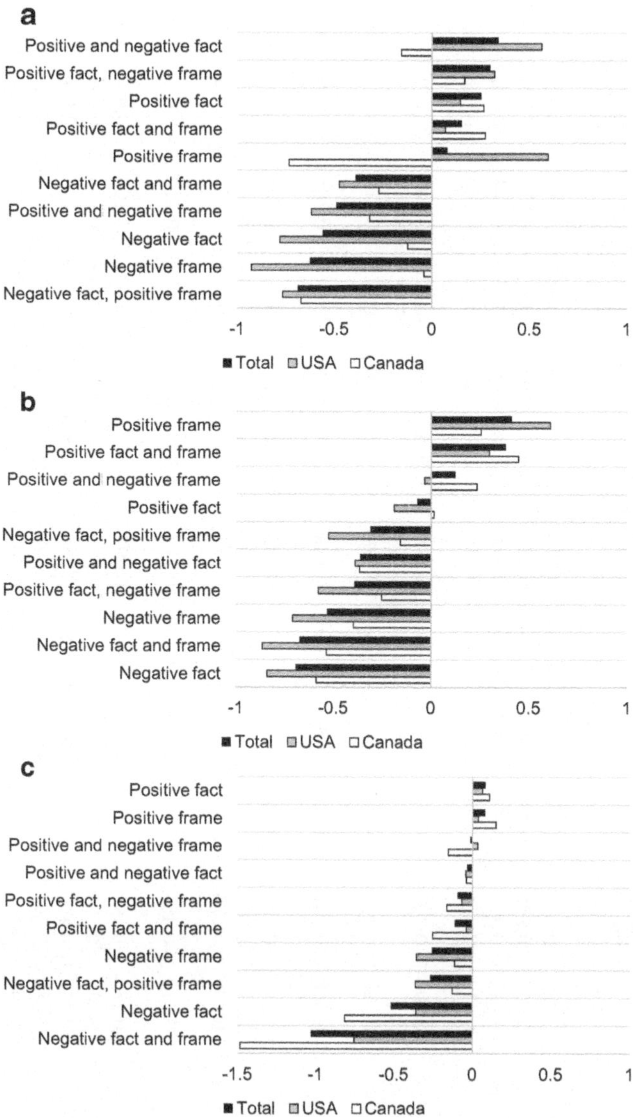

Fig. 12.7 (**a**) Ideology, Left; (**b**) Ideology, Centre; (**c**) Ideology, Right (*Source* Authors' own source)

Table 12.5 Effect of narratives considering ideology

	Model 1: Left				*Model 2: Centre*				*Model 3: Right*			
	B	*Std. Err*	*Beta*	*Sig*	*B*	*Std. Err*	*Beta*	*Sig*	*B*	*Std. Err*	*Beta*	*Sig*
Experimental group												
Positive fact	0.252	0.408	0.032	0.536	−0.033	0.178	−0.006	0.854	0.097	0.213	0.015	0.648
Positive frame	0.229	0.394	0.031	0.562	0.331	0.175	0.057	0.059	0.004	0.213	0.001	0.986
Negative fact	−0.685	0.399	−0.091	0.087	−0.679	0.174	−0.118	0.000	−0.556	0.224	−0.082	0.013
Negative frame	−0.613	0.374	−0.090	0.102	−0.521	0.186	−0.083	0.005	−0.249	0.220	−0.038	0.258
Positive fact and frame	0.290	0.398	0.039	0.466	0.363	0.178	0.061	0.041	−0.070	0.207	−0.011	0.737
Negative fact and frame	−0.365	0.357	−0.057	0.307	−0.761	0.194	−0.116	0.000	−0.975	0.211	−0.156	0.000
Positive and negative frame	−0.368	0.407	−0.047	0.366	0.166	0.183	0.027	0.364	−0.017	0.206	−0.003	0.935
Positive and negative fact	0.335	0.388	0.046	0.389	−0.407	0.183	−0.066	0.027	−0.029	0.207	−0.005	0.889
Negative fact, positive frame	−0.708	0.375	−0.103	0.060	−0.151	0.193	−0.023	0.434	−0.244	0.201	−0.042	0.226
Positive fact, negative frame	0.310	0.455	0.034	0.496	−0.428	0.177	−0.073	0.016	−0.093	0.210	−0.015	0.660
Country	−0.634	0.197	−0.143	0.001	0.037	0.092	0.011	0.687	−0.026	0.108	−0.007	0.812
Gender	−0.212	0.190	−0.050	0.266	−0.083	0.090	−0.025	0.357	−0.149	0.105	−0.044	0.157
Income	0.215	0.460	0.024	0.640	0.686	0.236	0.086	0.004	0.820	0.264	0.104	0.002
Education	2.040	0.399	0.247	0.000	0.680	0.198	0.097	0.001	0.617	0.225	0.086	0.006
Age	1.798	0.394	0.201	0.000	1.074	0.203	0.140	0.000	−0.029	0.213	−0.004	0.892
Intercept	2.859	0.374		0.000	3.448	0.187		0.000	4.342	0.217		0.000
Adj. R-sq	0.155				0.083				0.048			
S.E.E	1.907				1.592				1.669			

	Model 1: Left				*Model 2: Centre*				*Model 3: Right*			
	B	*Std. Err*	*Beta*	*Sig*	*B*	*Std. Err*	*Beta*	*Sig*	*B*	*Std. Err*	*Beta*	*Sig*
N	460				1363				1128			

Source Authors' own source

without too much engagement from the public. This is not an expression of doubt in the ability of democracy to deliver good governance, but the inability of democratic deliberations to manage discourse so that decisions are weighed on factual accounts. When deliberations are open to the public, they are open to a variety of interpretations and narratives. And the result, as we've seen both in actual practice and in the experimental survey design reported here, is for the public to grow wary. This is reflected by taking a big-picture view of all the results presented here. By counting all of the experimental groups that yield a significant effect at the 0.10 level in Tables 12.2 to 12.5, the 33 negative outcomes (i.e., coefficients) overwhelm the six that are positive. It is clear support for fluoridation can get knocked off balance, pressuring elected officials to follow the public will, whether that will is guided by science or not.

The problem is not just that deliberations are dominated by normative frames. As reported, narratives based on the negative fact that fluoride can cause bone disease are effective, far more than reassuring facts about its safety at low doses. Results also do show the negative factual narrative as more effective than the negative normative narrative. When paired, results show support for fluoridation weakens when negative facts are paired with a supportive factual or moral narrative. Of all the results presented in Tables 12.2 to 12.5, the most frequent significant results are those experimental groups that exposed respondents to negative facts or negative facts paired with negative moral frames. A close second is the group exposed to the libertarian negative frame, while third place (six times out of nine results) involves the group exposed to negative facts paired with a positive moral frame.

The potential for moral narratives to be as effective against factual accounts is apparent in some cases, but unexpectedly. For instance, it was expected for those with high levels of scientific knowledge to be more swayed by factual accounts. Instead, negative frames appear at least as effective. Even among those who show high levels of trust for science appear split on their susceptibility to factual versus normative narratives.

One could argue, then, that one way to challenge anti-fluoridationists is to mount an effective factually based counter narrative. Perhaps, but when the second round of experiments is considered, the more technically elaborate—and supposedly more benign—explanation of the chemical process involving hydrofluorosilicic acid does little to offset any fear. Once the public is even a bit fearful of something, the instinct of risk aversion predominates.

Arguably, one limitation here is the design of the experiment. While respondents were exposed to one of 10 different narratives (and combinations of narratives), the *dose* of each of these narratives was fixed. In reality, it is plausible that one side of a debate can assemble more resources and expend more energy to mount a stronger campaign. Such an asymmetry is likely in the context of fluoridation. The benefits of fluoridation, as important and impactful as they may be, are diffuse. Therefore, incentives for citizens to cooperate in its defense are small (Olson 1965). Instead, we are unlikely to see a groundswell of support behind a pro-fluoridation initiative, but are more likely to be left with a tyranny of an active and poorly informed minority.

References

Arnold, Jr., F. A. (1943). Role of fluorides in preventive dentistry. *Journal of the American Dental Association*, 30, 499–508.

Beauchamp, D. E. (1985). The neglected tradition of public health. *The Hastings Center Report*, 15(6), 28–36.

Block, L. E. (1986). Antifluoridationists Persist: The constitutional basis for fluoridation. *Journal of Public Health Dentistry*, 46(4), 188–198.

Boyd, E. M. (1959). The acute oral toxicity of caffeine. *Toxicology and Applied Pharmacology*, 1(3), 250–257.

Cacciatore, M. A., Scheufele, D. A., & Iyengar, S. (2016). The end of framing as we know it… and the future of media effects. *Mass Communication and Society*, 19, 7–23.

Catleugh, M., Delves, A., & Bellaby, P. (n.d.) A review of opinion polls concerning water fluoridation carried out in the United Kingdom between 1977 And 2007. Available at: https://bfsweb.org/wp-content/uploads/2020/09/NWFG-Opinion-poll-report-1997-to-2007-final-new0608-2.pdf

Camargo Jr., K., & Grant, R. (2015). Public health, science, and policy debate: being right is not enough. *American Journal of Public Health*, 105(2), 232–235.

Carstairs, C., & Elder, R. (2008). Expertise, health, and popular opinion: Debating water fluoridation, 1945–80. *Canadian Historical Review*, 89(3), 345–371.

Chari M., Ravaghi V., Sabbah W., Gomaa N., Singhal S., & Quiñonez C. (2022). Oral health inequality in Canada, the United States and United Kingdom. *PLoS ONE*, 17(5), e0268006.

Clemmesen, J. (1983). The alleged association between artificial fluoridation of water supplies and cancer: A review. *Bulletin of the World Health Organization*, 61(5), 871.

Collins, H., & Evans, R., (2008). *Rethinking expertise*. University of Chicago Press.

Collins H., & Evans. (2014). *Are we all scientific experts now?* John Wiley & Sons.

D'Antonio, M. (2019, Feb. 14). Anti-vaxxer rants are at home in Trump-world. *CNN*. https://www.cnn.com/2019/02/14/opinions/darla-shine-trump-vaccine-conspiracy-dantonio

Delli Carpini, M. X., & Keeter, S. (1996). *What Americans know about politics and why it matters*. Yale University Press.

Dillard, J.P., Tian, X., Cruz, S.M., Smith, R.A., & Shen, L. (2021): persuasive messages, social norms, and reactance: a study of masking behavior during a COVID-19 campus health campaign. *Health Communication*, https://doi.org/10.1080/10410236.2021.2007579

Druckman, J. N., & Bolsen, T. (2011). Framing, motivated reasoning, and opinions about emergent technologies. *Journal of Communication*, 61(4), 659–88.

Durant, J. R., Evans, G. A., & Thomas, G. P. (1989). The public understanding of science. *Nature*, 340(6228), 11–14.

Environmental Health & Safety. 2001. Materials safety data sheets: Sodium fluoride. MSDS # S3722.

Fournier, P. (2002). The uninformed Canadian voter. In J. Everitt & B. O'Neill (eds.), *Citizen politics research and theory in Canadian political behaviour* (pp. 92–109). Oxford University Press.

Gardner, D. 2009. *Risk: The science and politics of fear*. McLelland and Stewart.

Gupta, A.K., & Ayoob, S. (2016). *Fluoride in drinking water: Status, issues and solutions*. CRC Press.

Haneke, K. E., & Carson, B. L. (2001). *Sodium hexafluorosilicate [CASRN 16893–85–9] and fluorosilicic acid [CASRN 16961–83–4]: Review of toxicological literature. Prepared for Scott Masten*. Diss. Ph. D., National Institute of Environmental Health Sciences, Research Triangle Park, North Carolina. Contract.

Health Counts in Yates! n.d. https://www.health.ny.gov/statistics/chac/docs/yates2.pdfu

Hochschild, J.L., & Einstein, K.L. (2015). *Do facts matter?: Information and misinformation in American politics*. Norman, OK: University of Oklahoma Press.

Hochschild, J.L., & Sen, M. (2015). Technology optimism or pessimism about genomic science variation among experts and scholarly disciplines. *The Annals of the American Academy of Political and Social Science*, 658(1), 236–252.

Kahan, D. M., Braman, D., Cohen, G. L. Gastil, J., & Slovic, P. (2015). Who fears the HPV vaccine, who doesn't, and why? An experimental study of the mechanisms of cultural cognition. *Law and Human Behavior*, 34(6), 501–516.

Kahan, D.M, Jenkins-Smith, H., & Braman. D. 2011. Cultural cognition of scientific consensus. *Journal Of Risk Research,* 14(2), 147–174. https://doi.org/10.1080/13669877.2010.511246.

Kahan, D. M., Peters, E., Wittlin, M., Slovic, P., Ouellette, L. L., Braman, D., & Mandel, G. (2012). The polarizing impact of science literacy and numeracy on perceived climate change risks. *Nature Climate Change*, 2(10), 732–735.

Kanygin, J. 2021. Majority of Calgarians support re-introducing fluoride into water supply, new poll says. *CTV News*, Sept. 28. https://calgary.ctvnews.ca/majority-of-calgarians-support-re-introducing-fluoride-into-water-supply-new-poll-says-1.5604413

Kaul v. City of Chehalis. (1954). 45 Wash. 2d 616, 277 P. 2d 352, 356.

Kirst, M., Shankardass, K., Singhal, S., Lofters, A., Muntaner, C., & Quiñonez, C. (2017). Addressing health inequities in Ontario, Canada: What solutions do the public support? *BMC Public Health*, 17(1), 1–9.

Kiss, S. 2015. Where did all the baby bottles go? Risk perception, interest groups, media coverage and institutional imperatives in Canada's regulation of bisphenol A. *Canadian Journal of Political Science,* 47(4), 741–765. 1–25. https://doi.org/10.1017/S0008423914001127.

Kiss, S., Rath, K., & Perrella, A.M.L. 2018. Balanced journalism amplifies minority voices: A case study of a fluoridation plebiscite. *Canadian Journal of Communication*, 43(4), 633–645.

Kiss, S. J., Montpetit, É., & Lachapelle, E. (2020). Beyond regions and ideology: Using cultural theory to explain risk perception in Canada. *Canadian Journal of Political Science*, 53(2), 439–460.

Kraft, P.W., Lodge, M., & Taber, C.S. (2015). Why people "don't trust the evidence" motivated reasoning and scientific beliefs. *The Annals of the American Academy of Political and Social Science*, 658(1), 121–133.

Li, Y., Liang, C., Slemenda, C.W., Ji, R., Sun, S., Cao, J., Emsley, C.L., Ma, F., Wu, Y., Ying, P.O., & Zhang, Y. (2001). Effect of long-term exposure to fluoride in drinking water on risks of bone fractures. *Journal of Bone and Mineral Research*, 16(5), 932–939.

Marketing Resource Group. (2013, May 10). Poll: Three out of four Michigan voters want healthy kids dental program offered statewide. Press release.

Martin, B. (1991). *Scientific knowledge in controversy: The Social dynamics of the fluoridation debate.* State University of New York Press.

Mayer, B. (2014). How much nicotine kills a human? Tracing back the generally accepted lethal dose to dubious self-experiments in the Nineteenth Century. *Archives of Toxicology*, 88(1), 5–7.

McCright, A. M., Dentzman, K., Charters, M., & Dietz, T. (2013). The influence of political ideology on trust in science. *Environmental Research Letters*, 8(4), 1–9.

McKay, F. S. (1925). Mottled enamel: A fundamental problem in dentistry. *Dental Cosmos*, 67, 847–860.

Miller, J.M. (2020). Do COVID-19 conspiracy theory beliefs form a monological belief system?. *Canadian Journal of Political Science* 53(1), 319–326.

National Health Service, West Midlands. (2010). Public attitudes on dental health and water fluoridation in the West Midlands. Retrieved Sept. 7, 2019, from https://www.ipsos.com/sites/default/files/publication/1970-01/sri-attitudes-to-dental-health-in-west-midlands.pdf

Nelson, S. (2018). Safety data sheet. *Health* 2(3).

Nunn, C.Z., Crockett, H.J., & Williams, J.A. (1986). National study of tolerance and political ideology, 1973. Inter-university Consortium for Political and Social Research.

Olson, M. (1965). *The theory of collective action: public goods and the theory of groups*. Harvard University Press.

Perrella, A.M.L., & Kiss, S.J. (2015). Risk perception, psychological heuristics and the water fluoridation controversy. *Canadian Journal of Public Health*, 106(4), e197–e203.

Pratt, E., Rawson, R. D., & Rubin, M. (2002). Fluoridation at fifty: What have we learned? *Journal of Law, Medicine and Ethics*, 30, S117–S121.

Pratusha, N. G., Banji, O. J., Banji, D., Ragini, M., & Pavani, B. (2011). Fluoride toxicity–A harsh reality. *International Research Journal of Pharmacy*, 2, 79–85.

PHAC. (2017). *The State of Community Water Fluoridation across Canada*. Ottawa, ON. Retrieved August 28, 2019. https://www.canada.ca/en/services/health/publications/healthy-living/community-water-fluoridation-across-canada-2017.html

PHAC. (2018). *Public Health Agency of Canada. Key Health Inequalities in Canada: A National Portrait*. Ottawa: Public Health Agency of Canada. https://www.canada.ca/en/public-health/services/publications/science-research-data/inequalities-oral-health-canada-infographic.html

Quinonez, C. R., & Locker, D. (2009). Public opinions on community water fluoridation. *Canadian Journal of Public Health*, 100(2), 96–100.

Ran, T., Chattopadhyay, S. K., & Community Preventive Services Task Force. (2016). Economic evaluation of community water fluoridation: a community guide systematic review. *American Journal of Preventive Medicine*, 50(6), 790–96.

Shanahan, E. A., Jones, M. D., & McBeth, M. K. (2011). Policy narratives and policy processes. *Policy Studies Journal*, 39(3), 535–561.

Stadt, Z. M. (1954). Résumé of dental benefits of fluoride ingestion. In James A. Shaw (Ed.), *Fluoridation as a Public Health Measure* (pp. 1–35). American Association for the Advancement of Science.

Thompson, E. 2022. Liberals will fulfil promise of dental care for kids, Freeland says. *CBC News*. August 9. https://www.cbc.ca/news/politics/canada-dental-care-kids-1.6546224

Urbansky, E. T. (2002). Fate of fluorosilicate drinking water additives. *Chemical Reviews*, 102(8), 2837–2854.

WHO (2017). Guidelines for drinking-water quality, 4th edition incorporating the first addendum. Geneva, World Health Organization. Available at: https://apps.who.int/iris/bitstream/handle/10665/254637/9789241549950-eng.pdf.

Whyman, R. A., Mahoney, E. K., & Børsting, T. (2016). Community water fluoridation: attitudes and opinions from the New Zealand Oral Health Survey. *Australian and New Zealand Journal of Public Health*, 40(2), 186-192.

Wood, M.J., Douglas, K.M., & Sutton, R.M. (2012). Dead and alive: Beliefs in contradictory conspiracy theories. *Social Psychological and Personality Science* 3(6): 767–773.

PART V

COVID-19

CHAPTER 13

Exploring the Role of Information Sources in Vaccine Decision-Making Among Four Culturally and Linguistically Diverse Communities in the U.S.

Rukhsana Ahmed, Dina Refki, Jeanette Altarriba, Erting Sa, Mary C. Avery, and Sana Abdelkarim

Introduction

The fact that an individual's demographic characteristics can shape their health outcomes and predict their likelihood of accessing life-saving vaccines, services, or protection from exposure during a global pandemic has been well documented. The intersection of poverty, medical bias/racism, nativity, and immigration status along with a healthcare system that remains insufficiently responsive to the unique needs of

R. Ahmed (✉) · E. Sa
Department of Communication, University at Albany, State University of New York, Albany, NY, USA
e-mail: rahmed4@albany.edu

J. Altarriba · M. C. Avery
Department of Psychology, University at Albany, State University of New York, Albany, NY, USA

M. Gattinger (ed.), *Democratizing Risk Governance*,
https://doi.org/10.1007/978-3-031-24271-7_13

culturally and linguistically diverse communities and the social determinants of their health outcomes have proven deadly for many in these communities (Batalova, 2020; Ross et al., 2020).

During the COVID-19 pandemic, CALD communities in the U.S. had to shoulder a disproportionate burden of infections and fatalities due to (a) overrepresentation in the essential workers' sector including the healthcare industry; (b) lack of ability to maintain social distancing due to the nature of work and life in crowded quarters and neighborhoods; (c) lack of culturally and linguistically appropriate prevention materials; (d) inequitable access to healthcare services (Borjas, 2020); and (e) barriers to healthcare access and vaccine hesitancy (Abba-Ali et al., 2022). Limited English Proficient (LEP) individuals were 35% more likely than their English-speaking counterparts to die from COVID-19. Black, Latino, and Native American patients were more likely than their White counterparts to report higher hospitalization and death (Bebinger, 2021; Tai et al., 2021).

The miracle of modern medicine has provided a promise of an escape from the darkness of the pandemic. Vaccine rollout, however, has again exposed serious flaws in a system that has not yet integrated the needs of CALD individuals (i.e., those who speak a language other than English and subscribe to different cultural beliefs toward health care and disease prevention and treatment). As of July 2022, only 41 states and Washington D.C. reported vaccination data by race and ethnicity. Across these states, vaccination rates among Asians, Hispanics, and Whites were reported at 87, 67, and 64%, respectively. These groups were vaccinated at a higher rate than their Black counterparts, where only 59% of the total Blacks in these states were vaccinated. Furthermore, the share of Blacks and Latinos who received the first booster dose was lower compared to their white counterparts (Ndugga et al., 2021).

Refusal or delayed acceptance of COVID-19 vaccines can be influenced by certain contextual factors, including historical, cultural, socioeconomic, institutional, political, as well as vaccine safety or personal beliefs (Saied et al., 2021). Past research has shown that information source is an important factor in forming vaccine attitudes. For example, parents search

---

D. Refki · S. Abdelkarim
Department of Public Administration & Policy, University at Albany, State University of New York, Albany, NY, USA

the Internet to gather more information about vaccines, such as the side effects and negative consequences of vaccines (Harmsen et al., 2013). Other studies have shown some association between vaccination resistance and people's preferences for online information (Martin & Petrie, 2017).

While there is little evidence about how online information influences vaccination decisions (Meppelink et al., 2019), in their scoping review of parents' information-seeking related to vaccines from online sources and childhood vaccination decision-making, Ashfield and Donelle (2020) found evidence of significant childhood vaccine misinformation and risks online. The authors underscore the importance of digital health literacy, which is crucial in evaluating online vaccination information. Consequently, they put emphasis on further research of parents' "information-seeking practices, preferred resources, and ability to critically evaluate vaccination-related information" (p. 6). There is even less evidence on the information ecosystems (Susman-Peña et al., 2014) of CALD communities as they relate to vaccine decision-making. The current study inquires if members of CALD communities can obtain health information that they trust and the sources they rely on to make vaccination decisions. Understanding the contexts of the sources of health information they rely on can inform tailored packaging of vaccine-related information to ensure it appeals to and reaches targeted members of CALD communities so that they can make informed decisions about vaccination.

Literature Review

Vaccination Decision-Making

Widespread vaccination coverage is crucial to containing the COVID-19 pandemic. However, delayed acceptance or refusal to vaccinate, also referred to as vaccine hesitancy (World Health Organization, 2014), is widespread, working as a barrier to achieving the required vaccine coverage levels (Lin et al., 2021). Past research has identified concerns contributing to refusal or delayed acceptance of vaccination, including safety and effectiveness of vaccines (Kennedy et al., 2011; Freed et al., 2010), and low levels of confidence and trust in vaccine information from medical professionals, public health agencies, and the government (Gehlbach et al., 2022; Holroyd et al., 2021; Salmon et al., 2005). In terms of vaccine hesitancy, in a recent study, Marzo et al. (2022) found that almost half of the participants ($N = 10{,}447$) from 20 countries with cultural and linguistic differences, showed hesitancy toward COVID-19 vaccination. The level of their perceived COVID-19 vaccine acceptance, uptake decision-making, and hesitancy were significantly correlated with socio-demographic and economic characteristics, including country of residence, education, and employment. With regard to the unwillingness to receive the COVID-19 vaccine, available research has identified factors such as newness, safety, and potential side effects of the vaccine (Lu, 2022; Neumann-Böhme et al., 2020; Sherman et al., 2021).

Importantly, existing evidence has demonstrated that members of ethnic minority groups and with lower-income levels have more negative attitudes toward vaccines and are less willing to vaccinate against COVID-19 (Lee & Huang, 2022; Paul et al., 2021). Lower socioeconomic status was also associated with greater uncertainty and unwillingness to receive COVID-19 booster doses (Paul & Fancourt, 2022). In order to understand and address barriers to COVID-19 vaccine acceptance among CALD communities, it is important to examine determinants of vaccine decision-making in those communities to tailor public health communication programs accordingly.

There are some factors associated with vaccine decisions, including vaccine information sources. With regard to childhood immunization, parents who harbour varying vaccination attitudes regularly cited healthcare providers as the most trusted vaccine information source (Brunson 2013; Chung et al., 2017), similar to Marzo et al.'s (2022) research result that healthcare providers' advice was the top determinant for COVID-19 vaccine decision-making. Many also revealed that they relied on a personal network such as spouses/significant others/partners, friends (Chung et al., 2017), as well as online sources to find vaccine information (Brunson, 2013; Sobo, 2015). Although several studies have been conducted focusing on vaccination decisions among parents of young children and individuals' decision-making, there is a dearth of research on vaccination decision-making among CALD communities. Furthermore, little attention has been paid to factors such as health information sources influencing vaccine decision-making. This study aims to investigate the role of information sources in vaccine decision-making among four culturally and linguistically diverse communities—Arabic-speaking, Bengali, Chinese, and Spanish-speaking—in the U.S. Understanding health information source as a determinant of vaccine decision-making among communities with varying vaccination decisions can help inform the design of targeted and tailored interventions to increase vaccine uptake among those communities.

Social Determinants of Health and Vaccination Decisions

An individual's ability to actively pursue and acquire vaccination is a function of the presence of certain conditions. These conditions include (a) knowledge about the vaccine through the communication of information that is clear, understandable, and relatable; and (b) ability to obtain resources including time, sense of psychological safety and security, and geographic reach.

Availability of culturally and linguistically appropriate vaccine information. The research literature posits that Hispanic adults were more likely than their white counterparts to indicate that they did not have enough information about where or when they could get vaccinated (Ohlsen et al., 2022; Pradhan, 2021). This phenomenon can partly be explained by the inability to obtain vaccine information. Silva (2021) reported that there is limited vaccine information in languages other than English. Lack of translations of available information is exacerbated by the fact that these resources are developed with a monocultural lens and fail to provide messages that would resonate with culturally diverse individuals. Cultural barriers, as Silva (2021) contends, made it difficult for immigrants and non-English-speaking communities to get COVID-19 vaccines. Overcoming barriers to vaccination mandates the development of culturally adapted messages that respond to the different understanding and cultural beliefs of disease processes and progression (Thomas et al., 2021; Wu et al., 2022).

When information that is vital to obtaining life-saving vaccines is not provided in languages spoken, CALD communities often are forced to turn to non-credible information and become victims to false and inaccurate misinformation that has infested the global social media scene about the vaccine (Pradhan, 2021). Despite the vulnerabilities associated with culturally and linguistically diverse communities, information regarding vaccines in languages other than English was relatively delayed (e.g., NCDHHS, 2022). It is crucial to employ culturally and linguistically relevant methods to reach communities of colour and to tailor health information. Along these lines, Stadnick et al. (2022) identified key factors in increasing equitable COVID-19 vaccination uptake for communities of colour, that is, the development of culturally and linguistically appropriate COVID-19 programs. These programs included bilingual staff and trusted cultural and linguistic information with proper channels in the design of outreach and educational materials.

Digital literacy. Furthermore, vaccine-related materials that are disseminated through official online platforms require digital information literacy. Silva (2021) explains that on New York State's main COVID-19 page, users have to scroll through various graphics to get to the bottom of the page, where they can click on "language access" to find other languages. If people do not know that they need to scroll down or they cannot read English, their attempts to access information will be thwarted. On a national scale, Paz et al. (2022) reported that in the top 10 most populous cities in the U.S., the number of clicks made on each Department of Health website before reaching vaccine information, locations, and registration was significantly greater for Spanish-speakers. The researchers stressed the importance of making links for vaccine registration in Spanish more readily accessible (Paz et al., 2022). It is critical that digital access to vaccine information and resources be made a priority for CALD members, as opposed to an alternative hidden behind predominantly English webpages (Caldwell, 2022).

Additionally, when translations are provided, they tend to be limited to top languages. Hotlines set up to assist in scheduling vaccine appointments tend to offer information only in English and Spanish, ignoring other languages that are commonly spoken and those that are minority languages. Thirty-nine percent of adults who have received at least one dose of vaccine indicated that they needed someone to find or schedule an appointment for them (Pradhan, 2021).

Many immigrants live in lower socioeconomic communities and may not have digital capabilities. So, they face clear hurdles to learning about the vaccination sites and scheduling of vaccine appointments (Luu, 2021). Family members of non-English speakers had to spend a significant amount of time on English-based digital portals only to constantly receive the message of "No upcoming appointments available" (Woelfel, 2021). Moreover, many communities disproportionately affected by the pandemic rely on the traditional form of oral communication when seeking and acquiring information (AuYoung et al., 2022). The lack of digital capabilities may possibly isolate community members and hinder their access to comprehendible and trusted health information about COVID-19.

Vaccine roll out has largely depended on the Internet to schedule vaccine appointments. Jameel and Chen (2021) reported that for CALD communities that do not have access to the Internet, it was much more difficult to schedule vaccine appointments.

Material and psychological resources. Limited ability to obtain material resources almost always translates to limited availability of time. Poverty of time hinders the ability to vaccinate, especially when one has to balance meeting basic needs of survival with the need for a preventive healthcare measure. Hernandez (2021) attributed lower vaccination rates among Latinos to the fact that many Latinos work hourly jobs that make accommodating vaccine appointments during the day difficult. They also face language barriers and a difficult sign-up process. While for non-English speakers, language barriers can create fear and confusion, for poor residents, it is more difficult (and more expensive) to take a few hours or a day or two off work to access vaccination (Bloch et al., 2021).

Lack of ability to obtain material resources such as private transportation can also hinder vaccination uptake. Johnson (2021) reported that immigrants had been turned away from pharmacies and other places after being asked for driver's licenses, Social Security numbers, or health insurance cards. Even though these specific documents are not mandated by states or the federal government, they are often requested at vaccination sites across the country. These requests also are communicated in English, a language many of the vaccine-seekers do not fully understand. Only ten states and D.C., which have residency requirements, also allow undocumented immigrants to obtain driver's licenses or state identification cards. Persaud (2021) emphasized that the lack of paperwork and identification needed for vaccination erected a barrier to vaccination.

Geographic impediments can also create an additional barrier. Vaccination rates tend to be lower for those in rural areas and those without vehicles (Lee & Huang, 2022). Thomas et al. (2021) posit that due to storage requirements of the COVID-19 vaccines, it is often difficult to reach rural areas. Migrant populations who are mobile and work in rural areas are often unable to obtain vaccinations. Unless the vaccine is brought to people where they live, geographic impediments will remain a barrier (Thomas et al., 2021). In cases where access is limited by geography, a possible alternative is to utilize trusted community partners to assist with transportation and administration (Barrett, 2022; Malone et al., 2022), and to focus on outreach strategies that bring information into communities, as opposed to relying on English-only websites (AuYoung et al., 2022).

Moreover, obtaining COVID-19 vaccination requires confidence to engage with an unfamiliar and complex healthcare system without fear of fiscal implications, apprehension produced by possible negative experiences and encounters, or threats of deportation. There is a psychological empowerment that is critical to engaging in preventive behavior; empowerment that Abraido-Lanza et al. (2007) argue may have been depleted by the relentless assault of compounding factors. They suggest that the weight of social, political, and economic factors erodes empowerment and cultivates a learned behavior of helplessness and resignation. Such learned behavior may mandate significant efforts to unlearn and restore the psychological energy that drives investment in self.

Social, economic, and psychological disempowerment produce two devastating outcomes: lack of trust and motivation to obtain vaccinations. Recent studies of COVID-19 vaccination uptake found that trust predicts vaccine uptake, and demographic characteristics including socioeconomic status, race/ethnicity, and educational levels, are closely associated with trust (Latkin et al., 2021).

The literature shows that the main concerns of vaccine hesitators are the safety of the COVID-19 vaccine and its longer-term side effects (Alabdulla et al., 2021; Himmelstein et al., 2022; Robertson et al., 2021). A study by Park et al. (2021) reported that 76% of participants had concerns about the vaccines. They felt that vaccine trials were conducted too quickly and were skeptical about the efficacy, necessity, and safety of the vaccine. Distrust in vaccines is intensified when considering a distrust in government institutions and the healthcare system that is prevalent in many CALD communities (AuYoung et al., 2022; Thomas et al., 2021). Gonzalez et al. (2021) reported that 68% of adult immigrants in their study trusted state or local public health officials, 40% trusted elected officials in the community, 28% trusted religious leaders, and 18% trusted other sources of information. These figures are alarming considering the level of trust needed to engage in vaccine uptake.

Pre-migration experiences, medical racism, bias, and anti-immigration policies in the midst of the pandemic exacerbate distrust in CALD communities. The Trump administration's anti-immigrant policies have exacerbated distrust in CALD communities. Numerous executive orders were enacted that targeted border enforcement and internal deportations, exclusion of many immigrants from safety net policies when they were desperately needed, and restrictions on legal refugees, immigrants, and asylum seekers (Migration Policy Institute, 2020). Such policies had devastating impacts on many in CALD communities that will require intensive trust building efforts.

As discussed above, several factors intersect to dampen motivation for vaccine uptake among many CALD communities. Lack of motivation can be a symptom of a variety of causes, including English language barriers, lack of material and psychological resources, and lack of trust. Nevertheless, lack of motivation can also be an independent root cause that hinders vaccine uptake in many CALD communities. Khatib et al. (2014) suggest that intention barriers (e.g., low motivation, low self-efficacy) are associated with poor hypertension prevention and management.

Lack of motivation can be a function of a feeling of invincibility and an expression of a sense that "I am healthy, I will not get sick, my body is resilient" or it can be a deeply rooted religious belief that pursues reliance only on protection from the divine.

Notwithstanding religious prohibition against vaccination, in many traditional faith communities, resignation to the will of God is an element of worship, and a sense of fatalism is encouraged in several faith traditions. "Fatalism, the belief that health is predetermined by fate, relates to poorer adoption of risk-reducing health behaviors" (Gutierrez et al., 2017, p. 271). The belief that nothing can be done to prevent cardiovascular disease or cancer is more likely among Hispanics/Latinos than non-Hispanic Whites (Christian et al., 2007; Niederdeppe & Levy, 2007). Negative behaviours such as low utilization of cancer screening services, low utilization of protective measures against cardiovascular disease, and low adoption of behaviors such as avoidance of smoking, exercising, and healthy diets are associated with fatalistic beliefs (de Los Espinosa Monteros & Gallo, 2011; Mosca et al., 2006; Niederdeppe & Levy, 2007). Promoting an internal locus of control over health behaviours can contribute to addressing the harmful consequences of fatalistic behaviours. Faith-based interventions have been used to increase African American religious communities' engagement in health-promoting behaviours. They have been used to increase cancer knowledge, decrease cancer fatalism, and overall "cancer activism" (Morgan et al., 2008, p. 237).

Methods

This study forms part of a larger project investigating the efficacy of innovative health communication strategies focused on educating CALD individuals about public health protective measures to combat COVID-19. Employing a mixed-methods approach, online survey and one-on-one interview data were collected in two sequential phases with a pre- and post-test and post-test-only design. The current study analyzes survey data to explore the role of information sources in vaccine decision-making among four diverse communities in the U.S. Specifically, three research questions guided this study:

RQ1: Does the decision to vaccinate against COVID-19 differ across the members of the four CALD communities?
RQ2: Can the members of the four CALD communities find health information that they trust? Is there any difference between their ability to find health information that they trust and their vaccination status?
RQ3: What health information sources do COVID-19 vaccinated and intended-to-be vaccinated members of the four CALD communities use on a regular basis? How does this health information use compare across the members of these communities?

Sample and Procedures

In the larger project, CALD individuals from four communities in the U.S.—Arabic-speaking, Bengali, Chinese, and Spanish-speaking—were invited to complete an online survey that measured their knowledge, attitude, and practice (Zhong et al., 2020), information sources (Babalola et al., 2020), and vaccine decisions (Larson et al., 2015) pertaining to COVID-19. A total of 318 participants in the current study completed the vaccine decisions survey questions. Participants were recruited through a combination of strategies, including partnership through target community-based organizations; e-mail messages and letters of information; and social media postings. Eligible participants self-identified as 18 years of age or older, foreign-born and currently living in the U.S., and primarily speaking Arabic, Bengali, Chinese, or Spanish.

The survey links were emailed to community partners for distribution among interested CALD individuals in the target communities. The primary survey in English was translated into the four languages (i.e., Arabic, Bengali, Chinese, and Spanish) representing the four participating

CALD communities. Participants had the option to choose the language of the survey. From March to June, 2021, the survey was hosted on Qualtrics, a web-based survey tool. The survey was pilot tested with six individuals representing the four languages spoken by the sample to assess the readability and clarity of survey questions. A total of 318 participants in the current study completed the vaccine decisions survey questions. Each participant was offered a $20 Walmart eGift card as compensation for completing the survey. All procedures were approved by the Institutional Review Board at the University at Albany, State University of New York.

Measures

To determine the general characteristics of study participants, ten socio-demographic questions were used: gender, age, marital status, educational attainment, employment status, yearly household income, housing condition, length of residence in the U.S., citizenship status, and language spoken at home. Participants' decision to vaccinate against COVID-19 was determined by their vaccination status, which was measured by the question: "Have you received the COVID-19 vaccination?" Response options included: "yes," "intend to," or "no." Participants' ability to find health information that they trust was measured by the question: "Overall, do you think you can find information about health that you trust?" Response options included: "yes," "no," or "maybe." To determine health information sources used on a regular basis, participants were asked to check all that applied from nine options: "print media in my language," "digital media in my language," "ethnic TV and radio channels, "community leaders," "community organizations," "faith based organizations," "friends and family," "social media," and "other (please specify):________."

Data Analysis

Descriptive statistics were calculated for all demographic variables. To address RQ1, the number of participants for the total sample, and each CALD community, was calculated by their decision to vaccinate to identify differences across community groups (see Fig. 13.1). Vaccination status was assessed via the question "Have you received the COVID-19 vaccine?" To address RQ2, the number of participants for the total sample, and each CALD community, was calculated by their ability to find health information that they trust, to identify differences across community groups (see Table 13.2). To further explore participants' ability to find health information that they trust, subgroups of participants were calculated based on their decision to vaccinate (see Fig. 13.2). RQ3 focuses only on participants who reported that they had been vaccinated or intended to vaccinate as a subgroup. To address RQ3, the information sources these participants used were calculated for the total subgroup and in each CALD community. Sources of health information used were classified according to Street's (2003) ecological model, by mapping information source items to the communication contexts (e.g., "Media Context," "Organizational Context," and "Interpersonal Context") (see Table 13.3). All calculations (and charts) were performed in Excel and cross-checked by members of the research team.

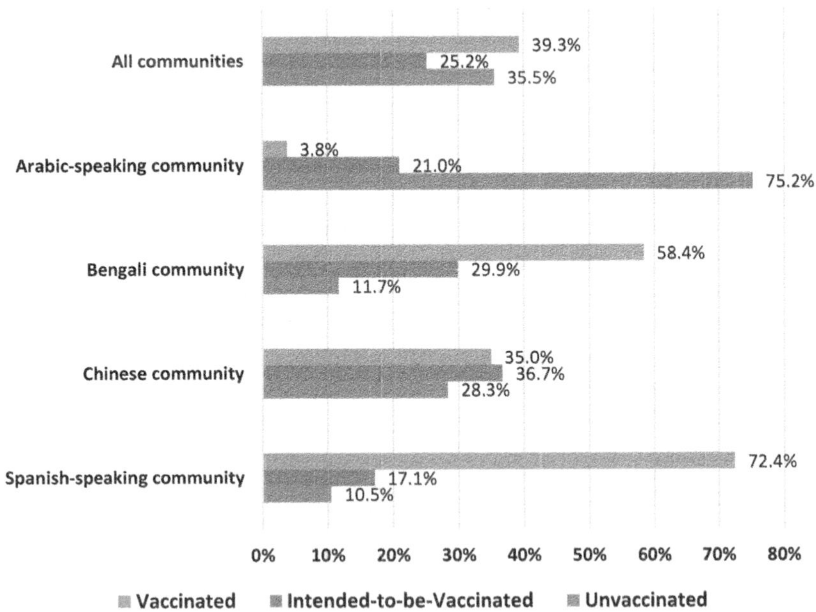

Fig. 13.1 COVID-19 vaccination status of participants (*Source* Authors' own source)

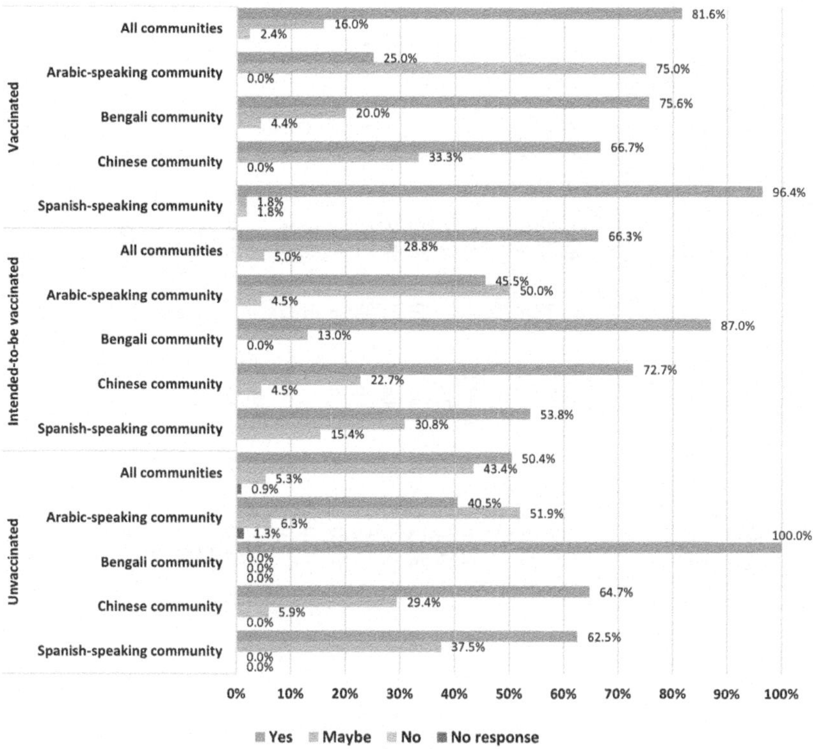

Fig. 13.2 Participant's ability to find health information that they trust by their vaccination status (*Source* Authors' own source)

Findings and Analysis

Selected Socio-Demographic Characteristics of Participants

Each of the four CALD communities is well-represented among the 318 survey participants (see Table 13.1). Members of the Arabic-speaking community, however, are slightly more represented (at 33.0%).

Gender. In the total sample, individuals identifying as females are more represented (at 65.7%). Females are over-represented in the Chinese community (66.7%), the Spanish-speaking community (59.2%), and especially in the Arabic-speaking community (90.5%). However, individuals

Table 13.1 Selected socio-demographic characteristics of participants (N = 318)

	All communities		*Arabic-Speaking community*		*Bengali community*		*Chinese community*		*Spanish-Speaking community*	
	(N = 318)		*(n_1 = 105)*		*(n_2 = 77)*		*(n_3 = 60)*		*(n_4 = 76)*	
Demographic characteristics	*n*	%	*n*	%	*n*	%	*n*	%	*n*	%
Gender										
Female	209	65.7	95	90.5	31	40.3	38	66.7	45	59.2
Male	101	31.8	5	4.8	46	59.7	19	33.3	31	30.7
Prefer not to answer	8	2.5	5	4.8	0	0.0	3	0.1	0	0.0
Age (years)										
18–24	33	10.4	24	22.9	6	7.8	1	1.7	2	2.6
25–29	49	15.4	21	20.0	19	24.7	1	1.7	8	10.5
30–39	128	40.3	37	35.2	27	35.1	20	33.3	44	57.9
40–49	74	23.3	16	21.6	12	15.6	29	48.3	17	22.4
50–59	18	5.7	7	6.7	5	6.5	1	1.7	5	6.6
60–69	10	3.1	0	0.0	4	5.2	6	0.1	0	0.0
70–79	6	1.9	0	0.0	4	5.2	2	3.3	0	0.0
80 and over	0	0.0	0	0.0	0	0.0	0	0.0	0	0.0
Marital status										
Married	269	84.6	84	80.0	66	85.7	54	90.0	65	85.5
Never married	29	9.1	13	12.4	7	9.1	2	3.3	7	9.2
Divorced or separated	12	3.8	2	1.9	2	2.6	4	6.7	4	5.3
Widowed	4	1.3	2	1.9	2	2.6	0	0.0	0	0.0
No response	4	1.3	4	3.8	0	0.0	0	0.0	0	0.0

	All communities		*Arabic-Speaking community*		*Bengali community*		*Chinese community*		*Spanish-Speaking community*	
	($N = 318$)		($n_1 = 105$)		($n_2 = 77$)		($n_3 = 60$)		($n_4 = 76$)	
Demographic characteristics	*n*	%	*n*	%	*n*	%	*n*	%	*n*	%
Educational attainment										
Less than high school	75	23.6	68	64.8	6	7.8	1	1.7	0	0.0
High school graduate (or equivalent)	66	20.8	29	27.6	15	19.5	5	8.3	17	22.4
Some college or Associate degree	51	16.0	1	1.0	22	28.6	11	18.3	17	22.4
Bachelor's degree	77	24.2	4	3.8	21	27.3	11	18.3	41	53.9
Graduate or professional degree	47	14.8	1	1.0	13	16.9	32	53.3	1	1.3
No response	2	0.6	2	1.9	0	0.0	0	0.0	0	0.0
Employment status										
Homemaker (stay at home dad/mom)	140	44.0	69	65.7	17	22.1	19	31.7	35	46.1
Student	17	5.3	13	12.4	2	2.6	2	3.3	0	0.0
Employed for salary	59	18.6	1	1.0	25	32.5	27	45.0	6	7.9
Self-employed	5	1.6	1	1.0	1	1.3	2	3.3	1	1.3
Out of work and looking for work	48	15.1	8	7.6	15	19.5	1	1.7	24	31.6
Out of work but not currently looking for work	21	6.6	4	3.8	9	11.7	2	3.3	6	7.9
Retired	18	5.7	2	1.9	8	10.4	6	10.0	2	2.6
Unable to work	4	1.3	4	3.8	0	0.0	0	0.0	0	0.0

(continued)

Table 13.1 (continued)

	All communities		*Arabic-Speaking community*		*Bengali community*		*Chinese community*		*Spanish-Speaking community*	
	($N = 318$)		($n_1 = 105$)		($n_2 = 77$)		($n_3 = 60$)		($n_4 = 76$)	
Demographic characteristics	*n*	%	*n*	%	*n*	%	*n*	%	*n*	%
Other	5	1.6	2	1.9	0	0.0	1	1.7	2	2.6
No response	1	0.3	1	1.0	0	0.0	0	0.0	0	0.0
Yearly household income										
Less than $10,000	31	9.7	14	13.3	12	15.6	1	1.7	4	5.3
$10,000–$19,999	33	10.4	14	13.3	7	9.1	5	8.3	7	9.2
$20,000–$29,999	38	11.9	11	10.5	11	14.3	5	8.3	11	14.5
$30,000–$39,999	19	6.0	0	0.0	10	13.0	5	8.3	4	5.3
$40,000–$49,999	12	3.8	1	1.0	9	11.7	1	1.7	1	1.3
$50,000–$59,999	9	2.8	0	0.0	7	9.1	1	1.7	1	1.3
$60,000–$69,999	18	5.7	2	1.9	2	2.6	3	5.0	11	14.5
$70,000–$79,999	24	7.5	0	0.0	2	2.6	1	1.7	21	27.6
$80,000 or more	50	15.7	1	1.0	11	14.3	27	45	11	14.5

Demographic characteristics	*All communities* ($N = 318$)		*Arabic-Speaking community* ($n_1 = 105$)		*Bengali community* ($n_2 = 77$)		*Chinese community* ($n_3 = 60$)		*Spanish-Speaking community* ($n_4 = 76$)	
	n	%	*n*	%	*n*	%	*n*	%	*n*	%
Prefer not to answer	82	25.8	60	57.1	6	7.8	11	18.3	5	6.6
No response	2	0.6	2	1.9	0	0.0	0	0.0	0	0.0
Housing condition										
Live in renter-occupied housing	138	43.4	85	81.0	24	31.2	10	16.7	19	25
Live in owner-occupied housing	173	54.4	16	15.2	53	68.8	50	83.3	54	71.1
No response	7	2.2	4	3.8	0	0.0	0	0.0	3	3.9
Length of residence in the U.S										
6 months or less	4	1.3	2	1.9	1	1.3	0	0.0	1	1.3
6 months–1 year	23	7.2	7	6.7	10	13.0	0	0.0	6	7.9
1 year–2 years	37	11.6	16	15.2	4	5.2	5	8.3	12	15.8
3–5 years	48	15.1	27	25.7	11	14.3	4	6.7	6	7.9
More than 5 years	203	63.8	51	48.6	51	66.2	50	83.3	51	67.1
No response	3	0.9	2	1.9	0	0.0	0	0.0	1	1.3
Naturalized U.S. citizen										
Yes	196	61.6	43	41.0	64	83.1	25	41.7	64	84.2
No	120	37.7	61	58.1	12	15.6	35	58.3	12	15.8
No response	2	0.6	1	1.0	1	1.3	0	0.0	0	0.0
Language spoken at home										
English only	61	19.2	1	1.0	37	48.1	0	0.0	23	30.3
English and their primary language	155	48.7	55	52.4	26	33.8	31	51.7	43	56.6
Primary language only	102	32.1	49	46.7	14	18.2	29	48.3	10	13.2

Note N = respondents from all communities; n_1, n_2, n_3, n_4 = respondents from individual communities; authors' own source

Table 13.2 Participants' ability to find health information that they trust

Response	*All communities (N = 318)*	*Arabic-speaking community (n_1 = 105)*	*Bengali community (n_2 = 77)*	*Chinese community (n_3 = 60)*	*Spanish-speaking community (n_4 = 76)*
Yes	212(66.7%)	43(41.0%)	63(81.8%)	41(68.3%)	65(85.5%)
No	13(4.1%)	6(5.7%)	2(2.6%)	2(3.3%)	3(3.9%)
Maybe	92(28.9%)	55(52.4%)	12(15.6%)	17(28.3%)	8(10.5%)
No response	1(0.3%)	1(1.0%)	0(0.0%)	0(0.0%)	0(0.0%)

Note N = respondents from all communities; n_1, n_2, n_3, n_4 = respondents from individual communities; authors' own source

identifying as females in the Bengali community are less represented (at 40.3%).

Age. Across the sample, those between the age of 30–39 are most represented (at 40.3%). This group was most represented in the Arabic-speaking community (35.2%), the Bengali community (35.1%), and the Spanish-speaking community (57.9%). In the Chinese community, individuals between the age of 40–49 were most represented (at 48.3%). In all communities, individuals over the age of 70 were underrepresented.

Marital status. Across the sample, most participants reported being married (at 84.6%). In all four CALD communities, too, married individuals are most represented.

Educational attainment. The sample varied in terms of educational attainment. Across the sample, slightly more than half of the participants had at least post-secondary education (at 55%). Most participants from the Arabic-speaking community had less than a high school degree (at 64.8%). Less than half of participants had at least a bachelor's degree. More than half of participants from the Spanish-speaking community had at least a bachelor's degree, and more than half from the Chinese community had a graduate or a professional degree.

Employment status. Across all four CALD communities, most participants identified as homemakers (at 44.0%). Relatively fewer participants were employed, most of whom were employed for salary (at 18.6%), and fewer were self-employed (at 1.6%). In contrast, 28.6% of the sample reported not working for various reasons (e.g., "out of work and looking," "unable to work," and so on).

Yearly household income. Across the sample, yearly household income was fairly distributed, though over a quarter of participants preferred not to answer (at 25.8%), many of whom were in the Arabic-speaking community (representing 73.2% of those who preferred not to answer). For the entire sample, 32.0% of participants reported a yearly household income of $29,999 or less. This pattern was consistent for all but the Chinese community (18.3%), with slightly fewer represented as lower-income. Most individuals in the Chinese community made a yearly household income of $80,000 or more.

Housing condition. Housing condition in the sample was somewhat split, though slightly more individuals reported living in owner-occupied housing (54.4%). Differences in housing conditions emerged between communities, with those in the Bengali (at 68.8%), Chinese (at 83.3%), and Spanish-speaking (at 71.1%) communities more represented in owner-occupied housing. Individuals in the Arabic-speaking community were more represented in rent-occupied housing (at 81.0%).

Length of residence in the U.S. For the majority of the sample, participants reported having resided in the U.S. for more than five years (63.8%). This pattern was consistent across all CALD community groups, though individuals in the Arabic-speaking community were less represented (at 48.6% for over five years of residence) and slightly more residing three to five years (at 25.7%). Cumulatively, over a third of participants in the sample reported residence for less than five years (at 35.2%).

Naturalized U.S. citizen. The total sample contained a good portion of naturalized U.S. citizens (61.6%) and a relatively small number of non-U.S. citizens (37.7%). The Spanish-speaking community (84.2%) and the Bengali community (83.1%) reported nearly twice as many U.S. citizens as the Chinese community (41.7%) and the Arabic-speaking community (41.0%). More than half of the individuals from the Chinese community (58.3%) and the Arabic-speaking community (58.1%) were non-U.S. citizens.

Language spoken at home. Most participants spoke both English and their primary language (48.7%) at home, as compared to those who spoke only their primary language (32.1%) or English (19.2%). Across all four CALD communities, individuals who spoke only their primary language at home were most represented by the Arabic-speaking (at 46.7%) and Chinese communities (at 48.3%), and individuals who spoke only English at home were most represented by the Bengali community (at 48.1%).

Vaccination Status Across CALD Communities

The first research question is concerned with differences across the members of the four CALD communities in their decision to vaccinate against COVID-19. The survey findings show that over a third (39.3%) of participants across the four CALD communities reported receiving a COVID-19 vaccine, with another quarter (25.2%) reported having an intention, while one third (35.5%) reported not receiving a vaccine (see Fig. 13.1). In other words, the largest proportion of the four participating CALD community members either decided to or planned to vaccinate against the risk of contracting COVID-19. Nevertheless, important variations exist when looking at the vaccination status in each community. For example, in the Spanish-speaking community, 58.4% reported receiving a COVID-19 vaccine, with another 29.9% reported having an intention, while only 9% reported not receiving a vaccine. In the Chinese community, 35% reported receiving a COVID-19 vaccine with another 36.7% reported having an intention, while 28.3% reported not receiving a vaccine. On the contrary, in the Arabic-speaking community, the largest proportion of survey participants (75.2%) reported not receiving a COVID-19 vaccine, with 21% reported having an intent and only 3.8% reported receiving a vaccine.

In summary, the findings indicate that there are differences across these four CALD communities in the rates at which they have been vaccinated or report the intention to be vaccinated. It may be the case that even though these groups may all represent culturally and linguistically diverse communities, particular language groups might find that messaging and communication are readily available in their native language (e.g., Spanish) as compared to those who predominantly speak other languages (e.g., Arabic). The rate at which individuals seek to obtain a vaccine or express the intention to do so might reside with their access to vaccines and exposure to information in their native or dominant language that would inform them on how to go about engaging in participation in vaccination sites and healthcare institutions. These findings would suggest that the expansion of materials and messaging related to acquiring vaccines need to appear in multiple languages and be culturally congruent with CALD communities, particularly those who stem from groups that represent linguistic minority speakers.

Ability to Find Trusted Health Information by Vaccination Status

The second research question is concerned with the ability of the members of the four CALD communities to find health information that they trust and any difference between their ability to find health information that they trust and their vaccination status.

The survey findings indicate that two-thirds (66.7%) of participants across the four CALD communities can find health information that they trust (see Table 13.2), with variation in the response rate among the individual communities: Spanish-speaking (85.5%), Bengali (81.8%), Chinese (68.3%), and Arabic-speaking (41.0%). It is important to note that only 4.1% of participants across the four communities reported that they could not find health information they trust, with little variation among the communities: Bengali (2.6%), Chinese (3.3%), Spanish-speaking (3.9%), and Arabic-speaking (5.7%). However, close to a third of participants (28.9%) across the four communities seem to be unsure about finding health information they trust, with variation among the communities: Arabic-speaking (52.4%), Chinese (28.3%), Bengali (15.6%), and Spanish-speaking (10.5%).

Relatedly, community members who are more likely to obtain a vaccine or show a high likelihood of their intention to do so also represent a community where their language may be more readily spoken (e.g., Spanish). As Spanish is one of the most highly spoken languages in the U.S., it is likely that Spanish-speaking communities have been exposed to messages and communication in their dominant language and trust the source and content of that information enough to promote action within these communities. Other cultural factors can also contribute to building a community's trust in vaccines, such as the status, gender, or perceived authority of the individuals communicating those messages to the members of the community (e.g., esteemed religious figures, medical personnel, and so on). With increased dissemination of knowledge in the dominant language of CALD communities, one might expect an increase in the trust or perceived value of the information being communicated, which can then promote an increase in actually obtaining a vaccine or having an intention to do so.

Ability of vaccinated participants to find health information that they trust. The survey findings provide important information regarding the association between participants' ability to find health information they trust and their vaccination status (see Fig. 13.2). For example, an overwhelming majority (81.6%) of participants who received a COVID-19 vaccine reported that they could find health information they trust, with more variation among the four communities: Spanish-speaking (96.4%), Bengali (75.6%), Chinese (66.7%), and Arabic-speaking (25.0%). It is worth noting that only a small proportion of vaccinated participants (2.4%) across the four communities reported that they could not find health information they trust. There was little variation among the four communities: Bengali (4.4%), Spanish-speaking (1.8%), Arabic-speaking (0.0%), and Chinese (0.0%). However, 16.0% of this subgroup of participants seem to be unsure about finding health information they trust, with many variations among the communities: Arabic-speaking (75.0%), Chinese (33.3%), Bengali (20.0%), and Spanish-speaking (1.8%).

These findings are consistent with other reported findings in the current chapter, namely, as information is more readily available in the native language of a CALD community member, the more likely that individual is to trust that information and engage in positive health-related behaviors. Across all communities, individuals report that they can locate trusted health information to make decisions about the vaccine at a rate that approaches 82%; however, those in the Spanish-speaking community report that they can find trusted information at a rate that is well above average, likely because information in Spanish is more readily available than in other languages. These findings imply that if messaging and communication are provided in other languages that are spoken by CALD communities in a culturally relevant manner, it is more likely that members of those communities will seek vaccination, trust the source of information regarding their health, and follow through when intending to seek health care. Thus, providing healthcare information in multiple languages is critical to increasing the overall rate of vaccination and in perpetuating trust in the source and content of healthcare information.

Ability of intended-to-be vaccinated to find health information that they trust. The survey findings suggest that two-thirds (66.3%) of participants who intended to receive a COVID-19 vaccine reported finding health information they trust, with important variation across the groups: Bengali (87.0%), Chinese (72.7%), Spanish-speaking (53.8%), and Arabic-speaking (45.5%). Similar to participants who received a COVID-19 vaccine, a limited number of participants with an intention to receive a COVID-19 vaccine (5.0%) reported that they could not find health information they trust, with some variation among the four communities: Spanish-speaking (15.4%), Arabic-speaking (4.5%), Chinese (4.5%), and Bengali (0.0%). Nevertheless, 28.8% of this subgroup of participants appear to be unsure about finding health information they trust, with some variation among the communities: Arabic-speaking (50.0%), Spanish-speaking (30.8%), Chinese (22.7%), and Bengali (13.0%).

Overall, it appears that a majority of participants who intended to receive a vaccine reported that they were able to find health information that they trusted. While this is a positive outcome of the current study, it is still the case that this value only represents roughly two-thirds of the overall set of participants. Thus, there is a likelihood that the remaining participants, though expressing an intention to be vaccinated, may not follow through on those intentions merely because they may not trust the information that they have received. It is possible that the information may have been provided in a language other than their native or dominant language, or, perhaps it was presented in a way that is not culturally sensitive in some manner, causing individuals to hesitate to follow through on their intentions. In effect, the fact that nearly 30% of the population that engaged in this study were uncertain as to whether or not they could find health information that they trusted to make a decision about the vaccine indicates that perhaps linguistic or cultural variables directly impacted the perceived validity of the information and thereby could have created vaccine hesitancy among those participants.

Ability of unvaccinated participants to find health information that they trust. Based on the survey findings, it is interesting to note that half (50.4%) of the participants who reported not receiving a COVID-19 vaccine can find health information they trust, with important variation among the four CALD communities: Bengali (100.0%), Chinese (64.7%), Spanish-speaking (62.5%), and Arabic-speaking (40.5%). It is also interesting to note that similar to participants who received and were intending to receive a COVID-19 vaccine, very few participants (5.3%) who did not receive a COVID-19 vaccine reported that they could not find health information they trust, with some variation among the four communities: Arabic-speaking (6.3%), Chinese (5.9%), Bengali (0.0%), and Spanish-speaking (0.0%). Notwithstanding this, it is important to note that a large number (43.4%) of this subgroup of participants happen to be unsure about finding health information they trust, with many variations among the communities: Arabic-speaking (51.9%), Spanish-speaking (37.5%), Chinese (29.4%), and Bengali (0.0%).

Overall, about half of the participants in the current study who did not receive a vaccine reported that they could find health information that they trusted to help decide about the COVID-19 vaccine. Even though this was the case, at the time of their participation, they still had not gotten a vaccine and did not report having an intention to do so. Thus, there is a likelihood that other factors above and beyond the perceived notion of trust might have moderated their hesitancy or decision not to engage in the vaccination process at all. With CALD communities, it is possible that cultural aspects that are not related to the trustworthiness of the information at all, or even the language in which it appears, could moderate vaccine-related behavior. For example, individuals may feel that a vaccine may not be warranted, as all of one's health or well-being is ultimately controlled by a higher power or that an outcome cannot be changed. Some beliefs or faith overall may encourage the notion that health outcomes are more readily determined by one's destiny, which is already preordained. All this said, it is important to point out that the other half of participants who noted they did not receive a vaccine indicated that they were unsure whether or not they were able to find the information they trusted or, indeed, were not able to find such information. Thus, the hesitancy or decision of this group to not engage in vaccination might stem from the notion of perceived trust in the information that could be derived by either the source of information, some aspect of its contents (e.g., language in which it appeared or cultural scheme within which it appeared) or indeed a variable such as an inability to comprehend information because it appeared in a language in which they had limited knowledge or experience. Detailed investigations of the role, nature, and scope of the sources of health-related information should be undertaken with CALD communities to understand more precisely how the nature, type, and delivery of information impact trust, which could indeed impact vaccine hesitancy or acceptance.

In summary, the current research findings suggest a few trends for the CALD communities under consideration and their responses concerning vaccination acceptance or hesitancy and trust in the source of health information. First, these data reveal that culturally and linguistically appropriate healthcare information and the existence and dissemination of that information may increase an individual's willingness to get vaccinated and may increase trust in the source of health information. The recommendation here is to expand existing materials across languages and in culturally appropriate ways to increase an understanding of the importance of vaccination, particularly during COVID-19. Second, it is clear that the Spanish language in particular, a predominant language in the U.S., is one in which health-related information can already be found, thus impacting the degree to which Spanish-speaking communities engage in vaccination and record lower vaccine hesitancy. Addressing the earlier recommendation of expanding information across different languages should serve to likewise increase vaccine acceptance in other CALD communities, leading to an overall decrease in vaccine hesitancy. Finally, vaccine hesitancy or the decision not to vaccinate at all may stem from a lack of trust in available health information, either due to the information's lack of cultural or linguistic match with the background of a particular community, or because the source was not considered credible. Future research should continue to examine the nature of the source of information and the variables that help to determine how they promote trust and adherence, so as to further increase the probability that individuals will seek vaccination and acceptance of public health guidance in general.

Health Information Source Use by Vaccinated and Intended-to-Be Vaccinated Participants Across CALD Communities

The third research question is concerned with the health information sources COVID-19 vaccinated and intended-to-be vaccinated members of the four CALD communities use regularly and how that compares across the members of these communities. Survey participants who reported receiving and intending to receive a COVID-19 vaccine were asked to identify the health information sources they use regularly. Based on their responses, these sources can be categorized into three communication contexts (Street's [2003] ecological model of communication [see Table 13.3]: media, organizational, and interpersonal).

Media context of health information sources. The survey findings demonstrate that out of the four types of media sources, digital media in native language have the highest rate (52.2%) of use across the four CALD communities with variation among the communities: Chinese (72.1%), Spanish-speaking (48.5%), Bengali (47%), and Arabic-speaking (42.3%). The use of ethnic TV and radio channels (42.9%) ranks second across the communities, with some variation: Spanish-speaking (54.4%), Bengali (48.5%), Arabic-speaking (34.6%), and Chinese (20.9%). The use of social media (42.9%) is tied for the second rank with some variation among the communities as well: Arabic-speaking (65.4%), Bengali (42.6%), Spanish-speaking (38.2%), and Chinese (37.2%). With a slightly

Table 13.3 Information sources used by vaccinated and intended-to-be vaccinated participants

Media context					
Health Information Source	*All communities (N = 205)*	*Arabic-speaking community (n_1 = 26)*	*Bengali community (n_2 = 68)*	*Chinese community (n_3 = 43)*	*Spanish-speaking community (n_4 = 68)*
Print media in native language	82(40.0%)	15(57.7%)	27(39.7%)	13(30.2%)	27(39.7%)
Digital media in native language	107(52.2%)	11(42.3%)	32(47.1%)	31(72.1%)	33(48.5%)
Ethnic TV and radio channels	88(42.9%)	9(34.6%)	33(48.5%)	9(20.9%)	37(54.4%)
Social media	88(42.9%)	17(65.4%)	29(42.6%)	16(37.2%)	26(38.2%)
Organizational context					
Health Information Source	All communities (N = 205)	Arabic-speaking community (n_1 = 26)	Bengali community (n_2 = 68)	Chinese community (n_3 = 43)	Spanish-speaking community (n_4 = 68)

Media context					
Health Information Source	*All communities (N = 205)*	*Arabic-speaking community (n_1 = 26)*	*Bengali community (n_2 = 68)*	*Chinese community (n_3 = 43)*	*Spanish-speaking community (n_4 = 68)*
Faith based organizations	51(24.9%)	7(26.9%)	18(26.5%)	1(2.3%)	25(36.8%)
Community leaders	54(26.3%)	10(38.5%)	15(22.1%)	3(0.1%)	26(38.2%)
Community organizations	77(37.6%)	11(42.3%)	31(45.6%)	7(16.3%)	28(41.2%)
Interpersonal context					
Health Information Source	All communities (N = 205)	Arabic-speaking community (n_1 = 26)	Bengali community (n_2 = 68)	Chinese community (n_3 = 43)	Spanish-speaking community (n_4 = 68)
Friends and family	86(42.0%)	11(42.3%)	34(50.0%)	17(39.5%)	24(35.3%)
Doctor	1(0.5%)	0(0.0%)	0(0.0%)	0(0.0%)	1(1.5%)
Health Information Source	All communities (N = 205)	Arabic-speaking community (n_1 = 26)	Bengali community (n_2 = 68)	Chinese community (n_3 = 43)	Spanish-speaking community (n_4 = 68)
Other	6(3.0%)	6(23.1%)	0(0.0%)	0(0.0%)	0(0.0%)

Note N = respondents from all communities; n_1, n_2, n_3, n_4 = respondents from individual communities; authors' own source lower rate (40.0%), the use of print media in native language varies among the Arabic-speaking (57.7%), Bengali (39.7%), Spanish-speaking (39.7%), and Chinese (30.2%).

In summary, these findings suggest that among the CALD communities included in the current study, digital media in the native language is the most typical source of health information, including health-related information. Given that individuals often use hand-held devices such as smartphones to communicate with others and gain knowledge, it is not unusual to expect that digital sources provide most individuals with news and information, particularly in their native language. The notion that ethnic TV and radio are other sources for health information indicates that again, for CALD communities, information that is available in their native language is often sought and considered more readily than information that is perhaps not as culturally or linguistically relevant, as in the case of information provided only in English. Thus, to reach these populations with health-related information that they are likely to consult and use on a regular basis, having that information appear in their native language is probably the best way to assure that the information is read and considered in decision-making, in the current context, vaccine decision-making. Interestingly, the notion that Arabic-speaking community members accrue to print media more readily than the other groups in the current study speaks to the notion that perhaps this is the more accessible format or source appearing in the Arabic language for the current communities. Thus, there is a call for health information to be presented across languages in various formats but assuring that digital and other related media sources make this information available in culturally and linguistically appropriate ways.

Organizational context of health information sources. With regard to the use of health information in the organizational context, survey participants across the four CALD communities reported relying on community organizations the most (37.6%), with many variations among the communities: Bengali (45.6%), Arabic-speaking (42.3%), Spanish-speaking (41.2%), and Chinese (16.3%). Community leaders (26.3%) are other sources of information across the four communities with substantial variation among the communities: Chinese (0.1%), Bengali (22.1%), Spanish-speaking (38.2%), and Arabic-speaking (38.5%). It should be noted that faith-based organizations (24.9%) are closely tied with community leaders as sources of health information across the four communities with noteworthy variation among the communities: Spanish-speaking (36.8%), Arabic-speaking (26.9%), Bengali (26.5%), and Chinese (2.3%).

The data on organizational contexts as sources of health information indicate that most members of these CALD communities rely on community organizations for finding health-related information. Clearly, the importance of networking and communicating within a group of individuals who are like-minded and share a set of values and beliefs cannot be underestimated. Moreover, these communities rely on faith-based organizations, particularly within the Spanish- and Arabic-speaking communities where religion plays a predominant role in shaping worldviews and influences the context in which members receive and comprehend information that impacts their decision-making. Trust in these sources plays a vital role in shaping the thoughts and beliefs of community members, particularly with respect to mental and physical health and well-being.

Interpersonal context of health information sources. As interpersonal sources of health information, friends and family are dominant (42.0%) across the four CALD communities with very little variation among the communities: Bengali (50.0%), Arabic-speaking (42.3%), Chinese (39.5%), and Spanish-speaking (35.3%).

The current survey findings suggest that these individuals, as one might expect, do rely heavily on interpersonal sources of health information such as that gathered from close family relations and friends. This finding appears pervasive across all of the four CALD communities surveyed, and indeed, with communities that represent collectivist cultural mores, it is not uncommon to note that individuals rely on interpersonal and intercultural communication as sources of information. Most importantly, these interactions can influence and guide healthcare decisions and the interpretation of data or information provided across media sources and help convert those notions into action. Moreover, it is often the case that individuals follow the information derived from their immediate cultural groups to be part of that "in-group" experience and demonstrate an affinity with the group, expressing the importance of their membership and allegiance to the culture at hand. These findings underscore the power of culturally and linguistically appropriate and valid information in informing and motivating behavior change and the selection of individuals to engage in certain healthcare options such as vaccination and disease prevention.

Discussion

The findings reported in this chapter provide comparative insights into the perceived trust of health information that participating members of four CALD communities in the U.S. obtain and the sources of health information they use that shape their COVID-19 vaccination decisions.

The picture that emerged is that Arabic-speaking participants have the lowest rates of vaccination (3.8%), followed by the Chinese participants (35%), Bengali (58.4%), and Spanish-speaking (72%). Arabic-speaking participants are more likely to indicate no intention to receive vaccination (75.2%), followed by the Chinese participants (28.3%), Bengali (11.7%) and Spanish-speaking (10.5%). Obviously, the critical mass of Spanish-speaking communities has the advantage of increasing the likelihood of finding information that is culturally and linguistically appropriate. In fact, most formal venues of vaccine communications in the U.S. are available in English and Spanish, including hotlines. Outreach to the Spanish-speaking communities through Spanish-speaking community-based organizations and through targeted culturally and linguistically sensitive resources can explain the high vaccination and intention to vaccinate rates in the Spanish-speaking community. However, what explains the high vaccination rates in the Bengali survey participants and the low rates in the Arabic and Chinese-speaking participants by comparison? Is it a lack of trust in the healthcare establishment? Or is it a lack of an ability to obtain information one trusts?

The diverse socio-demographic characteristics of survey participants can provide possible answers that are consistent with the literature on COVID-19 vaccination. On the one hand, the Arab and Chinese-speaking participants had a large proportion of individuals who spoke a language other than English (46.7 and 48.3% respectively), while the Bengali participants were mostly English speakers (48%). High rates of non-citizens were also prevalent in the Arabic and Chinese-speaking survey participants (58.1 and 58.3% respectively). In comparison, the Spanish and Bengali-speaking participants had high rates of naturalization (84.2 and 83.1% respectively). Approximately 65% of Arabic-speaking participants reported low educational attainment, 66% were homemakers, 90.5% were female, 49% indicated a length of stay less than 5 years, and 81% lived in rent-occupied housing. The Arabic-speaking participants and to a great extent their Chinese counterparts faced multiple layers of vulnerabilities when compared to their Spanish and Bengali-speaking counterparts, which most probably meant lower health literacy, and decreased access to health and social services, opportunity structures and other adverse social determinants of health.

The findings of this study are consistent with other research results on the impact of socioeconomic status on health information seeking, confidence, and trust. Richardson et al. (2012), for example, show that lower educational attainment and lower-income are associated with reduced information-seeking behavior and trust in doctors and other healthcare providers. Dimensions of the sociopolitical and cultural environment may also shape attitudes and decisions to engage with the healthcare system.

Implications

The number of confirmed and presumptive positive cases of COVID-19 disease recorded in the U.S. had surpassed 94 million as of September 4, 2022, with over one million deaths among these cases. As of the same date, more than 809.9 million doses of the COVID-19 vaccine were delivered, and more than 610 million doses were administered with 224.1 million individuals fully vaccinated (amounting to 67.5% of the population), 108.8 million received their first booster dose, and 22 million received their second booster dose in the U.S. (CDC, 2022). To manage and end the pandemic, higher vaccination coverage against COVID-19 will be required. It is of paramount importance to examine individuals' intention to be vaccinated, especially among priority groups, so that targeted health messages and strategies can be tailored to boost the public's confidence in COVID-19 vaccinations.

According to the CDC Internet panel survey (n = 3541) in 2020 to assess baseline perceptions of the COVID-19 vaccine and intentions to get vaccinated among priority groups, lack of COVID-19 vaccine confidence, side effects, and safety concerns about the vaccine were reported, and varied by demographic characteristics, including age group, sex, race/ethnicity, educational level, household income level, region, Metropolitan Statistical Area (MSA status), urbanicity, and health insurance status.

Within the backdrop of the increased health risks and challenges that CALD communities face during the COVID-19 pandemic, the findings of the current study bear important implications for policy and practice. Obtaining trusted health information contributed to COVID-19 vaccination decisions among members of the four participating CALD communities. Vaccine recipients in this study rely on multiple sources of information to protect themselves and their families. Healthcare providers and policymakers should target health information sources trusted by CALD communities for COVID-19 vaccine communication to these groups.

Adverse social determinants of health across different dimensions and levels of influence emerge as strong predictors of ability to obtain vaccination and decision to vaccinate. As emphasized in this chapter, socioeconomic status and educational attainment hinder the ability of individuals to obtain information that they trust and decrease the likelihood that they can obtain vaccination. Availability of culturally and linguistically appropriate resources and outreach to communities that are adversely affected by limited English proficiency and poverty is critical to the fight against COVID-19 and vaccine uptake goals. Finally, this study points to sources of information that are trusted by people who have chosen to vaccinate in the four participating CALD communities. These information sources can be more effectively leveraged to achieve increased diffusion of vaccine information and greater vaccine uptake, as well as mitigation of future health crises and effective dissemination of critical public health messages in general.

Limitations and Future Research Directions

The findings of this study should be considered in light of a number of limitations. First, the sample of four CALD communities is not representative of the entire population of CALD communities in the U.S. Hence, future research should strive to expand the sample size to recruit members of other CALD communities and account for any differences in the use of health information sources that may relate to the context and history of settlement. Second, there was a low representation of older individuals in the survey participant sample. This may be due to limited digital literacy and may mask a bleaker picture especially in communities where members have low English language proficiency, low documentation status, and higher vulnerability to COVID-19. Accordingly, future research should recruit a more diverse sample. Third, the study used descriptive statistics, hence, findings cannot extend beyond summarizing emerging patterns of participants' self-reported vaccination status, trusted health information obtained, and health information source use across the communities in shaping their vaccine decision-making. Future research should utilize complementary inferential statistics to make conclusions about relevant hypotheses, and thereby provide a more in-depth understanding of health information use practices of CALD communities by also examining their access to these sources and ability to understand and use the information to make informed vaccine and other preventive health decisions. It is also recommended that qualitative methods such as focus groups be employed with CALD communities to gain deeper insights into the influence of health information sources used by these communities for making vaccination decisions.

References

Abba-Aji, M., Stuckler, D., Galea, S., & McKee, M. (2022). Ethnic/racial minorities' and migrants' access to COVID-19 vaccines: A systematic review of barriers and facilitators. *Journal of Migration and Health*, 100086. https://doi.org/10.1016/j.jmh.2022.100086.

Abraido-Lanza, A. F., Viladrich, A., Florez, K. R., Cespedes, A., Aguirre, A. N., & De La Cruz, A. A. (2007). Fatalismo reconsidered: A cautionary note for health-related research and practice with Latino populations [Commentary]. *Ethnicity and Disease*, 17(1), 153–158. Retrieved from http://www.ncbi.nlm.nih.gov/pubmed/17274225.

Alabdulla, M., Reagu, S. M., Al-Khal, A., Elzain, M., & Jones, R. M. (2021). COVID-19 vaccine hesitancy and attitudes in Qatar: A national cross-sectional survey of a migrant-majority population. *Influenza and Other Respiratory Viruses*, 15(3), 361–370. https://doi.org/10.1111/irv.12847.

Ashfield, S., & Donelle, L. (2020). Parental online information access and childhood vaccination decisions in North America: Scoping review. *Journal of Medical Internet Research*, 22(10), Article e20002. https://doi.org/10.2196/20002.

AuYoung, M., Rodriguez Espinosa, P., Chen, W.-T., Juturu, P., Young, M.-E. D. T., Casillas, A., Adkins-Jackson, P., Hopfer, S., Kissam, E., Alo, A. K., Vargas, R. A., Brown, A. F., & And the STOP COVID-19 C. A. Communications Working Group. (2022). Addressing racial/ethnic inequities in vaccine hesitancy and uptake: Lessons learned from the California alliance against COVID-19. *Journal of Behavioral Medicine*, 1–14. https://doi.org/10.1007/s10865-022-00284-8.

Babalola, S., Krenn, S., Rimal, R., Serlemitsos, E., Shaivitz, M., Shattuck, D., & Storey, D. (2020). KAP COVID Dashboard.Johns Hopkins center for communication programs, Massachusetts Institute of Technology, Global outbreak alert and response network, Facebook Data for Good. Published

September 2020. Data Retrieved October 12, 2020, from https://ccp.jhu.edu/kap-covid/.

Barrett, C. (2022, March 3). *How a rural hospital broke language barriers to provide COVID vaccines to immigrants*. WFYI. Retrieved September 9, 2022, from https://www.wfyi.org/news/articles/seymour-rural-hospital-language-barriers-covid-vaccines-immigrants.

Batalova, J. (2020, May 14). *Immigrant health-care workers in the United States*. Migration Policy Institute. Retrieved November 30, 2020, from https://www.migrationpolicy.org/article/immigrant-health-care-workers-united-states-2018.

Bebinger, M. (2021, April 22). *The pandemic imperiled non-English speakers in a hospital*. NPR. Retrieved June 11, 2021, from https://www.npr.org/sections/health-shots/2021/04/23/989928262/the-pandemic-imperiled-non-english-speakers-in-a-hospital.

Bloch, M., Buchanan, L., Holder, J. (2021, March 26). See who has been vaccinated so far in New York City. *The New York Times*. https://www.nytimes.com/interactive/2021/03/26/nyregion/nyc-vaccination-rates-map.html.

Borjas, G. (2020, April). *Demographic determinants of testing incidence and COVID-19 infections in New York City neighbourhoods*. National Bureau of Economic Research. Retrieved June 11, 2021, from https://www.nber.org/papers/w26952 Article 26952. https://doi.org/10.3386/w26952.

Brunson, E. K. (2013). The impact of social networks on parents' vaccination decisions. *Pediatrics*, 131(5), e1397–e1404. https://doi.org/10.1542/peds.2012-2452.

Caldwell, A. (2022, February 9). *Spanish language vaccine resources harder to access, while Hispanic vaccination rates remain below overall average, study finds*. At the Forefront UChicagoMedicine. Retrieved September 9, 2022, from https://www.uchicagomedicine.org/forefront/research-and-discoveries-articles/spanish-language-vaccine-resources-harder-to-access.

Certified Languages International. (2021, January 25). *How trust can help combat vaccine hesitancy among non-English speakers*. Certified Languages International. Retrieved June 11, 2021, from https://certifiedlanguages.com/blog/combat-vaccine-hesitancy-among-non-english-speakers/.

CDC. (2022, September 9). COVID data tracker. Centers for Disease Control and Prevention. https://covid.cdc.gov/covid-data-tracker.

Christian, A. H., Rosamond, W., White, A. R., & Mosca, L. (2007). Nine-year trends and racial and ethnic disparities in women's awareness of heart disease and stroke: An American Heart Association national study. *Journal of Women's Health*, 16, 68–81. https://doi.org/10.1089/jwh.2006.M072.

Chung, Y., Schamel, J., Fisher, A., & Frew, P. M. (2017). Influences on immunization decision-making among US parents of young children. *Maternal and Child Health Journal*, 21(12), 2178–2187. https://doi.org/10.1007/s10995-017-2336-6.

de Los Espinosa Monteros, K., & Gallo, L. C. (2011). The relevance of fatalism in the study of Latinas' cancer screening behavior: A systematic review of the literature. *International Journal of Behavioral Medicine*, 18, 310–318. https://doi.org/10.1007/s12529-010-9119-4.

Freed, G. L., Clark, S. J., Butchart, A. T., Singer, D. C., & Davis, M. M. (2010). Parental vaccine safety concerns in 2009. *Pediatrics*, 125(4), 654–659. https://doi.org/10.1542/peds.2009-1962.

Gehlbach, D., Vázquez, E., Ortiz, G., Li, E., Sánchez, C. B., Rodríguez, S., Pozar, M., & Cheney, A. M. (2022). Perceptions of the Coronavirus and COVID-19 testing and vaccination in Latinx and Indigenous Mexican immigrant communities in the Eastern Coachella Valley. *BMC Public Health*, 22(1), 1019. https://doi.org/10.1186/s12889-022-13375-7.

Gonzalez, D., Karpman, M., & Bernstein, H. (2021, April). *COVID-19 vaccine attitudes among adults in immigrant families in California*. Urban Institute. Retrieved June 11, 2021, from https://www.urban.org/sites/default/files/publication/103973/covid-19-vaccine-attitudes-among-adults-in-immigrant-families-in-california_0.pdf.

Gutierrez, A. P., McCurley, J. L., Roesch, S. C., Gonzalez, P., Castañeda, S. F., Penedo, F. J., & Gallo, L. C. (2017). Fatalism and hypertension prevalence, awareness, treatment and control in US Hispanics/Latinos: Results from HCHS/SOL Sociocultural Ancillary Study. *Journal of Behavioral Medicine*, 40(2), 271–280.

Harmsen, I. A., Mollema, L., Ruiter, R. A., Paulussen, T. G., de Melker, H. E., & Kok, G. (2013). Why parents refuse childhood vaccination: A qualitative study using online focus groups. *BMC Public Health*, 13(1), 1–8. https://doi.org/10.1186/1471-2458-13-1183.

Hernandez, E. (2021, April 19). *Denver says it wants to help more Latinos get the COVID vaccine. Data shows that's not happening*. Denverite. Retrieved June 11, 2021, from https://denverite.com/2021/04/19/denver-says-it-wants-to-help-more-latinos-get-the-covid-vaccine-data-shows-thats-not-happening/.

Himmelstein, J., Himmelstein, D. U., Woolhandler, S., Dickman, S., Cai, C., & McCormick, D. (2022). COVID-19–related care for Hispanic elderly adults with limited English proficiency. *Annals of Internal Medicine*, 175(1), 143–145. https://doi.org/10.7326/M21-2900.

Holroyd, T. A., Howa, A. C., Delamater, P. L., Klein, N. P., Buttenheim, A. M., Limaye, R. J., Proveaux, T. M., Saad, B. O., & Salmon, D. A. (2021). Parental vaccine attitudes, beliefs, and practices: Initial evidence in California after a vaccine policy change. *Human Vaccines & Immunotherapeutics*, 17(6), 1675–1680. https://doi.org/10.1080/21645515.2020.1839293.

Jameel, M., & Chen, C. (2021, March 1). *How inequity gets built into America's vaccination system*. ProPublica. Retrieved June 11, 2021, from https://www.propublica.org/article/how-inequity-gets-built-into-americas-vaccination-system.

Johnson, A. (2021, April 10). For immigrants, I.D.s prove to be a barrier to a dose of protection. *The Washington Post*. https://www.washingtonpost.com/health/2021/04/10/covid-vaccine-immigrants-id/.

Kennedy, A., Basket, M., & Sheedy, K. (2011). Vaccine attitudes, concerns, and information sources reported by parents of young children: Results from the 2009 HealthStyles survey. *Pediatrics*, 127(Supplement 1), S92–S99. https://doi.org/10.1542/peds.2010-1722N.

Khatib, R., Schwalm, J. D., Yusuf, S., Haynes, R. B., McKee, M., Khan, M., & Nieuwlaat, R. (2014). Patient and healthcare provider barriers to hypertension awareness, treatment and follow up: A systematic review and meta-analysis of qualitative and quantitative studies. *PloS One*, 9(1), e84238.

Larson, H. J., Jarrett, C., Schulz, W. S., Chaudhuri, M., Zhou, Y., Dube, E., Schuster, M., MacDonald, N. E., Wilson, R., & SAGE Working Group on Vaccine Hesitancy, 2015 Larson, H. J., Jarrett, C., Schulz, W. S., Chaudhuri, M., Zhou, Y., Dube, E., Schuster, M., MacDonald, N. E., Wilson, R., & SAGE Working Group on Vaccine Hesitancy. (2015). Measuring vaccine hesitancy: The development of a survey tool. *Vaccine*, 33(34), 4165–4175. https://doi.org/10.1016/j.vaccine.2015.04.037.

Latkin, C. A., Dayton, L., Yi, G., Colon, B., & Kong, X. (2021). Mask usage, social distancing, racial, and gender correlates of COVID-19 vaccine intentions among adults in the US. *PLoS One*, 16(2), Article e0246970. https://doi.org/10.1371/journal.pone.0246970.

Lee, J., & Huang, Y. (2022). COVID-19 vaccine hesitancy: The role of socioeconomic factors and spatial effects. *Vaccines*, 10(3), 352. https://doi.org/10.3390/vaccines10030352.

Lin, C., Tu, P., & Beitsch, L. M. (2021). Confidence and receptivity for COVID-19 vaccines: A rapid systematic review. *Vaccines*, 9(1), 1–42. https://doi.org/10.3390/vaccines9010016.

Lu, J. G. (2022). Two large-scale global studies on COVID-19 vaccine hesitancy over time: Culture, uncertainty avoidance, and vaccine side-effect concerns. *Journal of Personality and Social Psychology*. https://doi.org/10.1037/pspa0000320.

Luu, L. (2021, March 4). *How pharmacists can assist patients with limited English proficiency*. GoodRX. https://www.goodrx.com/blog/pharmacist-resources-for-limited-english-proficient-populaitons/.

Malone, B., Kim, E., Jennings, R., Pacheco, R. A., & Kieu, A. (2022). COVID-19 vaccine distribution in a community with large numbers of immigrants and refugees. *American Journal of Public Health*, 112(3), 393–396. https://doi.org/10.2105/AJPH.2021.306608.

Martin, L. R., & Petrie, K. J. (2017). Understanding the dimensions of anti-vaccination attitudes: The vaccination attitudes examination (VAX) scale. *Annals of Behavioral Medicine*, 51(5), 652–660. https://doi.org/10.1007/s12160-017-9888-y.

Marzo, R. R., Ahmad, A., Islam, M. S., Essar, M. Y., Heidler, P., King, I., Thiyagarajan, A., Jermsittiparsert, K., Songwathana, K., Younus, D. A., El-Abasiri, R. A., Bicer, B. K., Pham, N. T., Respati, T., Fitriyana, S., Faller, E. M., Baldonado, A. M., Billah, M. A., Aung, Y., Hassan, S. M., ... Yi, S. (2022). Perceived COVID-19 vaccine effectiveness, acceptance, and drivers of vaccination decision-making among the general adult population: A global survey of 20 countries. *PLoS Neglected Tropical Diseases*, 16(1), e0010103. https://doi.org/10.1371/journal.pntd.0010103.

Meppelink, C. S., Smit, E. G., Fransen, M. L., & Diviani, N. (2019). "I was right about vaccination": Confirmation bias and health literacy in online health information seeking. *Journal of Health Communication*, 24(2), 129–140. https://doi.org/10.1080/10810730.2019.1583701.

Morgan, P. D., Tyler, I. D., & Fogel, J. (2008, November). Fatalism revisited. In *Seminars in oncology nursing* (Vol. 24, No. 4, pp. 237–245). WB Saunders.

Mosca, L., Mochari, H., Christian, A., Berra, K., Taubert, K., Mills, T., et al. (2006). National study of women's awareness, preventive action, and barriers to cardiovascular health. *Circulation*, 113, 525–534. https://doi.org/10.1161/CIRCULATIONAHA.105.588103.

NCDHHS. (2022, February 24). NCDHHS makes key COVID-19 vaccine information available in N.C.'s most used languages. Retrieved September 9, 2022, from https://www.ncdhhs.gov/news/press-releases/2022/02/24/ncdhhs-makes-key-covid-19-vaccine-information-available-ncs-most-used-languages.

Neumann-Böhme, S., Varghese, N. E., Sabat, I., Barros, P. P., Brouwer, W., van Exel, J., ... & Stargardt, T. (2020). Once we have it, will we use it? A European survey on willingness to be vaccinated against COVID-19. *The European Journal of Health Economic: HEPAC: Health Economics in Prevention and Care*, 21(7), 977–982. https://doi.org/10.1007/s10198-020-01208-6.

Niederdeppe, J., & Levy, A. G. (2007). Fatalistic beliefs about cancer prevention and three prevention behaviors. *Cancer Epidemiology, Biomarkers and Prevention*, 16(5), 998–1003. https://doi.org/10.1158/1055-9965.EPI-06-0608.

Ndugga, N., Pham, O., Hill, L., Artiga, S., Alam, R., & Parker, N. (2021). Latest data on COVID-19 vaccinations race/ethnicity. *Kais Family Found*. Accessed

September 4, 2022, https://www.kff.org/coronavirus-covid-19/issue-brief/latest-data-on-covid-19-vaccinations-by-race-ethnicity/.

Ohlsen, E. C., Yankey, D., Pezzi, C., Kriss, J. L., Lu, P. J., Hung, M. C., Dionicio Bernabe, M. I., Kumar, G. S., Jentes, E., Elam-Evans, L. D., Jackson, H., Black, C. L., Singleton, J. A., Ladva, C. N., Abad, N., & Lainz, A. R. (2022). COVID-19 vaccination coverage, intentions, attitudes and barriers by race/ethnicity, language of interview, and nativity, National Immunization Survey Adult COVID Module, April 22, 2021–January 29, 2022. *Clinical Infectious Diseases*, ciac508. https://doi.org/10.1093/cid/ciac508.

Park, T. V., Dougan, M., Meyer, O., Nam, B., Tzuang, M., Park, L., Vuong, Q., & Tsoh, J. (2021). Differences in COVID-19 vaccine concerns among Asian Americans and Pacific Islanders: The compass survey. *Journal of Racial and Ethnic Health Disparities*. https://doi.org/10.1007/s40615-021-01037-0.

Paul, E., & Fancourt, D. (2022). Predictors of uncertainty and unwillingness to receive the COVID-19 booster vaccine: An observational study of 22,139 fully vaccinated adults in the UK. *The Lancet Regional Health-Europe*, 14, 100317.https://doi.org/10.1016/j.lanepe.2022.100317.

Paul, E., Steptoe, A., & Fancourt, D. (2021). Attitudes towards vaccines and intention to vaccinate against COVID-19: Implications for public health communications. *The Lancet Regional Health-Europe*, 1, Article 100012. https://doi.org/10.1016/j.lanepe.2020.100012.

Paz, M. I., Marino-Nunez, D., Arora, V. M., & Baig, A. A. (2022). Spanish language access to COVID-19 vaccination information and registration in the 10 most populous cities in the USA. *Journal of General Internal Medicine*, 1–3. https://doi.org/10.1007/s11606-021-07325-z.

Persaud, C. (2021, April 5). Immigrant communities in PBC less likely to be vaccinated. *Palm Beach Post*. https://www.palmbeachpost.com/story/news/local/2021/04/05/covid-few-seniors-vaccinated-palm-beach-county-immigrant-areas/4822129001/.

Pradhan, R. (2021, March 23). *'Press 1 for English': Vaccination sign-ups prove daunting for speakers of other languages*. California Healthline. Retrieved June 11, 2021, from https://californiahealthline.org/news/article/press-1-for-english-vaccination-sign-ups-prove-daunting-for-speakers-of-other-languages/.

Richardson, A., Allen, J. A., Xiao, H., & Vallone, D. (2012). Effects of race/ethnicity and socioeconomic status on health information-seeking, confidence, and trust. *Journal of Health Care for the Poor and Underserved*, 23(4), 1477–1493. https://doi.org/10.1353/hpu.2012.0181.

Robertson, E., Reeve, K. S., Niedzwiedz, C. L., Moore, J., Blake, M., Green, M., Srinivasa, V. K., & Benzeval, M. J. (2021). Predictors of COVID-19 vaccine hesitancy in the U.K. household longitudinal study. *Brain, Behavior, and Immunity*, 94, 41–50. https://doi.org/10.1016/j.bbi.2021.03.008.

Ross, J., Diaz, C. M., & Starrels, J. L. (2020). The disproportionate burden of COVID-19 for immigrants in the Bronx, New York. *JAMA Internal Medicine*, 180(8), 1043–1044.

Saied, S. M., Saied, E. M., Kabbash, I. A., & Abdo, S. A. E. F. (2021). Vaccine hesitancy: Beliefs and barriers associated with COVID-19 vaccination among Egyptian medical students. *Journal of Medical Virology*, 93(7), 4280–4291. https://doi.org/10.1002/jmv.26910.

Salmon, D. A., Moulton, L. H., Omer, S. B., DeHart, M. P., Stokley, S., & Halsey, N. A. (2005). Factors associated with refusal of childhood vaccines among parents of school-aged children: A case-control study. *Archives of Pediatrics & Adolescent Medicine*, 159(5), 470–476. https://doi.org/10.1001/archpedi.159.5.470.

Sherman, S. M., Smith, L. E., Sim, J., Amlôt, R., Cutts, M., Dasch, H., Rubin, G. J., & Sevdalis, N. (2021). COVID-19 vaccination intention in the UK: Results from the COVID-19 vaccination acceptability study (CoVAccS), a nationally representative cross-sectional survey. *Human Vaccines & Immunotherapeutics*, 17(6), 1612–1621. https://doi.org/10.1080/21645515.2020.1846397.

Silva, C. (2021, February 22). *"It's life and death": Non-English speakers struggle to get COVID-19 vaccine across US*. USA TODAY. https://www.usatoday.com/story/news/nation/2021/02/22/covid-19-vaccine-registration-non-english-speakers-left-behind/4503655001/.

Sobo, E. J. (2015). Social cultivation of vaccine refusal and delay among Waldorf (Steiner) school parents. *Medical Anthropology Quarterly*, 29(3), 381–399. https://doi.org/10.1111/maq.12214.

Street, R. L. Jr. (2003).Communicating in medical encounters: An ecological perspective. In T. L. Thomson, Dorsely, T. L., A.M., Miller, K. I., & Parrott, R. (Eds.), *Handbook of health communication* (pp. 63–89). London: Lawrence Erlbaum Associates.

Stadnick, N. A., Cain, K. L., Oswald, W., Watson, P., Ibarra, M., Lagoc, R., Ayers, L. O., Salgin, L., Broyles, S. L., Laurent, L. C., Pezzoli, K., & Rabin, B. (2022). Co-creating a theory of change to advance COVID-19 testing and vaccine uptake in underserved communities. *Health Services Research*, 57(Suppl 1), 149–157. https://doi.org/10.1111/1475-6773.13910.

Susman-Peña, T., Audette, N., Funk, J., Mesich, A., Cook, T., Myers, M., Grad, I., Chang, N., Makino, A., & van Voorst, R. (2014). Why information matters: A foundation for resilience. Internews Center for Innovation & Learning. Retrieved June 13, 2021, from https://www.unhcr.org/innovation/wp-content/uploads/2017/10/150513-Internews_WhyInformationMatters.pdf.

Tai, D., Sia, I. G., Doubeni, C. A., & Wieland, M. L. (2021). Disproportionate impact of COVID-19 on Racial and Ethnic Minority Groups in the United

States: a 2021 Update. *Journal of Racial and Ethnic Health Disparities*, 1–6. Advance online publication. https://doi.org/10.1007/s40615-021-01170-w.

Thomas, C. M., Osterholm, M. T., & Stauffer, W. M. (2021). Critical considerations for COVID-19 vaccination of refugees, immigrants, and migrants. *The American Journal of Tropical Medicine and Hygiene*, 104(2), 433–435. https://doi.org/10.4269/ajtmh.20-1614.

Woelfel, M. (2021, March 2). *Language barriers are leaving non-English speakers behind in the vaccine rollout*. WBezChicago. Retrieved June 11, 2021, from https://www.wbez.org/stories/language-barriers-are-leaving-non-english-speakers-behind-in-the-vaccine-rollout/d6332d30-63d6-4fbc-b793-2c16679bd781.

World Health Organization. (2014). Report of the SAGE Working Group on vaccine hesitancy. Retrieved June 13, 2021, from https://www.who.int/immunization/sage/meetings/2014/october/1_Report_WORKING_GROUP_vaccine_hesitancy_final.pdf.

Wu, T. Y., Yang, X., Lally, S., Rainville, A. J., Ford, O., Bessire, R., & Donnelly, J. (2022). Using community engagement and geographic information systems to address COVID-19 vaccination disparities. *Tropical Medicine and Infectious Disease*, 7(8), 177. https://doi.org/10.3390/tropicalmed7080177.

Zhong, B. L., Luo, W., Li, H. M., Zhang, Q. Q., Liu, X. G., Li, W. T., & Li, Y. (2020). Knowledge, attitudes, and practices towards COVID-19 among Chinese residents during the rapid rise period of the COVID-19 outbreak: A quick online cross-sectional survey. *International Journal of Biological Sciences*, 16(10), 1745–1752. https://doi.org/10.7150/ijbs.45221.

CHAPTER 14

Opportunities and Perils of Public Consultation in the Creation of COVID-19 Vaccine Priority Groups

S. Michelle Driedger, Gabriela Capurro, Cindy Jardine, and Jordan Tustin

Introduction

In January 2020, the World Health Organization declared the spread of COVID-19 to be a Public Health Emergency of International Concern (PHEIC). A PHEIC is defined as: "an extraordinary event which is determined to constitute a public health risk to other States through the international spread of disease and to potentially require a coordinated international response" (World Health Organization, 2005, p. 9). A PHEIC declaration is the highest alert level by the WHO and carries with it legally binding requirements for countries in following International

S. M. Driedger (✉) · G. Capurro
Department of Community Health Sciences, University of Manitoba, Winnipeg, MB, Canada
e-mail: Michelle.driedger@umanitoba.ca

G. Capurro
e-mail: Gabriela.capurro@umabitoba.ca

M. Gattinger (ed.), *Democratizing Risk Governance*,
https://doi.org/10.1007/978-3-031-24271-7_14

Health Regulations (e.g., resource mobilization, data sharing). COVID-19 has caused over 600 million cases and over 6 million deaths worldwide, and more than 4 million cases and over 43,000 deaths in Canada, as of August 2022 (IPAC Canada, 2022; World Health Organization, 2022).

From a public audience perspective, the two most important factors in controlling a PHEIC event are: (1) good risk communication for actionable protective behaviors; and (2) once available, distribution of effective vaccine(s) to protect against severe illness or death. Since the WHO characterized COVID-19 as a PHEIC event, there has been considerable international scientific collaboration and state level funding to sequence the SARS-CoV-2 virus with the aim to develop effective vaccines. Canada, like many other governments, invested substantially through contractual agreements to purchase vaccines from pharmaceutical companies even without knowing which vaccines may successfully pass through clinical trials and regulatory authorizations. This type of investment was critical to reduce the length of time typically required for the development of a new vaccine (i.e., from 10–15 years to about 10–15 months). Canada also made commitments to COVAX (World Health Organization, 2021), a WHO initiative aimed at providing global equitable access to COVID-19 vaccines to low and middle income countries that might not otherwise have the resources to protect their citizens. It has been widely acknowledged that safe and effective vaccines as well as a global mass immunization program are necessary to end the COVID-19 pandemic (The Lancet Microbe, 2021). However, given anticipated scarcity in the availability of safe and effective vaccines, state actors were cognizant that prioritization decisions would need to be made, and that these decisions would need to be created in an ethical, equitable, and transparent manner.

C. Jardine
Faculty of Health Sciences, University of the Fraser Valley, Chilliwack, BC, Canada
e-mail: cindy.jardine@ufv.ca

J. Tustin
School of Occupational and Public Health, Toronto Metropolitan University, Toronto, ON, Canada
e-mail: jtustin@torontomu.ca

In this chapter, we explore aspects related to the risk governance of public health emergencies, including vaccines, in Canada, with a particular focus on addressing issues of equity and fairness under conditions of scarcity. We will describe how public health emergencies are regulated and managed, including the mechanisms used to establish preliminary recommendations for COVID-19 vaccine priority groups, and public and stakeholder engagement mechanisms. We will then document how some provinces made operational decisions on vaccine delivery within their jurisdiction. We will examine specific opportunities and challenges when the public is engaged in evaluating prioritization recommendations, particularly when those who are consulted may not necessarily reflect the population diversity underlying equity considerations. In order to explore these tensions, we share public comments about COVID-19 vaccine priority groups from age-stratified (18–34 years; 35–54 years; 55 years and older) mixed-gender focus groups in Vancouver, Winnipeg, Toronto, and Ottawa, conducted in December 2020.

Background: Regulating and Managing Public Health Emergencies

In Canada, both Health Canada and the Public Health Agency of Canada play important roles in protecting the health of Canadians. Health Canada is the regulator for any new health product (e.g., vaccine, drug, health technology/therapy) as well as being responsible for federal policy and financial contributions to the healthcare system, among other areas. The Public Health Agency of Canada (PHAC) was created following Canada's experiences with SARS (Severe Acute Respiratory Syndrome) in 2003, in order to have a stronger and more coordinated public health presence (Canada, 2003). It is through PHAC that some of Canada's obligations under the WHO International Health Regulations (e.g., surveillance) are met. In order to provide guidance on the use of vaccines approved in Canada, PHAC delegates this mandate to the National Advisory Committee on Immunization (NACI). NACI provides expert advice in vaccine recommendations by evaluating published literature, conducting reviews, and publishing statements and updates.

National immunization recommendations in Canada rely on a long-standing framework that focuses on traditional scientific (e.g., disease burden) and programmatic (e.g., feasibility, acceptability, ethics, costs) factors important for decision-makers (provinces/territories) (Erickson et al., 2005). Since the establishment of NACI in 1964, it has often focused on the scientific considerations in developing recommendations, and the consideration of programmatic factors was mainly done separately by a federal/provincial/territorial committee. This was changed in 2019 when NACI's mandate was officially expanded to evaluate programmatic factors (Ismail et al., 2020). Consequently, NACI sought to develop a systematic and transparent way of applying an Ethics, Equity, Feasibility, and Accessibility Framework (EEFA). This involved an extensive review of evidence on these aspects as they applied to vaccines, and surveys and consultations with experts and stakeholder groups. Consultations involved key informant interviews with representatives from organizations within Canada (e.g., Public Health Ethics Consultative Group, Canadian Task Force on Preventive Health Care), including representation from Indigenous Nations (e.g., Inuit Tapirit Kanatami, First Nations Information Governance Centre, Métis National Council), and outside Canada (e.g., Strategic Advisory Group of Experts World Health Organization). The full process and considerations are well documented by Ismail and colleagues (2020). Nonetheless, while NACI might make recommendations for immunization programs, it is the provincial and territorial governments that ultimately decide how to implement publicly funded immunization programs.

The COVID-19 pandemic has deepened health inequities in Canada, as some groups—particularly racial minorities and people living in poverty—have been more deeply affected (Public Health Agency of Canada, 2021). At the onset of COVID-19, NACI had the mandate within PHAC to create a set of priority vaccination groups in Canada even before Canada had any knowledge of when and which vaccines would become available, as will be described in the next section. This mandate is based in the need for fair and equitable vaccine distribution.

Creation of Priority Vaccination Groups in Canada

Canada approved the use of two COVID-19 vaccines in December 2020, Pfizer-BioNTech and Moderna, and in February and March 2021, respectively, AstraZeneca and Janssen COVID-19 vaccines were approved. With these emergency authorizations, and given the limited supply of vaccines, provinces and territories had to develop vaccine distribution plans and create priority groups who would have early access to the vaccine, making appropriate modifications as more vaccine supply became available. Based on the EEFA Framework (Ismail et al., 2020), NACI recommended the prioritization of four key populations for early access to the COVID-19 vaccine: (a) those at high risk of severe illness and death from COVID-19, including advanced age; (b) those most likely to transmit COVID-19 to high risk groups, and workers essential to maintaining the COVID-19 response, including healthcare workers, personal care workers, and caregivers providing care in long-term care facilities; (c) workers providing essential services for the functioning of society; and (d) people whose living or working conditions put them at high risk of infection and where infection could have disproportionate consequences, including Indigenous communities (ibid.).

A key component of the EEFA framework was public consultation (Ismail et al., 2020). NACI referred to an expert stakeholder survey conducted in Canada in July and August 2020 (Zhao et al., 2020), and time-series cross sectional surveys with the general population to gauge their attitudes toward the COVID-19 vaccine and group priority (Impact Canada, 2021). The surveys were conducted by Canada's COVID-19 Snapshot Monitoring Study (COSMO Canada) and are part of an ongoing study which has so far included 11 surveys between April 2020 and February 2021. NACI drew on the results of surveys 6 and 7, the most recent at the time (National Advisory Committee on Immunization, 2020).

The expert stakeholders consulted included members of clinical or public health expert groups involved with PHAC, patient or community advocacy representatives and/or experts from the CanCOVID network,[1] members of Canadian health professional associations, members of provincial and territorial committees and/or national Indigenous groups, and representatives of federal government departments (Zhao et al., 2020). Stakeholders generally ranked the strategies in the following order from most to least important:

1. Protect those who are most vulnerable to severe illness and death from COVID-19.
2. Protect healthcare capacity.

[1] CanCOVID Network is a rapid-response network of Canadian scientists and health and policy experts for facilitating COVID-19 research effort.

3. Minimize transmission of COVID-19.
4. Protect critical infrastructure.

The general population survey (Impact Canada, 2021), in which 2000 Canadians participated, revealed that the most commonly identified populations for priority immunization included: those with underlying medical conditions (57%); the elderly (53%); healthcare workers (22%); and frontline/essential workers (18%). However, when asked to rank a pre-determined list of groups to be prioritized to receive a COVID-19 vaccine before others in wave 7 of the survey, a different set of preferences emerged. The most commonly identified group for priority immunization was healthcare workers (40%), followed by individuals with high-risk medical conditions (19%), frontline workers (16%), seniors (12%), long-term care/nursing homes (10%), and children (2%) (ibid.).

Provincial Implementation of Priority Group Recommendations

All provinces and territories devised immunization plans in order to distribute the COVID-19 vaccine in phases, given that insufficient supply required vaccinating some people first while others had to wait. Phase 1 of the vaccine rollout began, for most provinces, between late December 2020 and early January 2021. In all cases, priority groups included frontline healthcare workers, staff and residents at long-term care facilities, and individuals over 80 years of age. Prioritizing these groups is in line with the recommendations developed by NACI explained above; however, definitions of frontline healthcare workers varied across provinces and territories. For example, in Alberta, Phase 1 included healthcare workers in intensive care units, respiratory therapists, and staff in long-term care and assisted living facilities (Alberta, 2021); meanwhile, in Saskatchewan, the category of healthcare workers was broader and included those working in ICUs, emergency departments, COVID-19 units and testing centres, respiratory therapists, code blue and trauma teams, and emergency medical services (Saskatchewan, 2021).

Indigenous populations (First Nations, Métis, and Inuit) were also included as a priority group in most provinces and territories, following NACI recommendations. Indigenous populations in Canada have been deeply affected by the COVID-19 pandemic, and they are particularly vulnerable to health risks due to "reduced access to adequate health care, healthy food and clean water, while also experiencing much greater levels of overcrowded housing, homelessness and incarceration" (Mosby & Swidrovich, 2021, p. E381).

Initial priority categories were expanded as vaccine availability increased, progressively broadening to include other "at risk" categories and the general population. Several provinces reserved the AstraZeneca vaccine for populations between 55 and 65 years of age after NACI revised its guidelines on March 29th due to reports of adverse reactions to the vaccine around the world. Reports emerged of people who had received the vaccine developing blood clots (Dangerfield, 2021; Reuters, 2021), an extremely rare side-effect (Rabson, 2021) that received much media attention. However, due to lack of demand for the AstraZeneca vaccine among the target population, NACI backtracked its decision and on April 16th once again recommended the vaccine for everyone over the age of 18 (Chung & Hogan, 2021). Below, we discuss the initial priority groups in three provinces (Ontario, Manitoba, and British Columbia) that represent the COVID-19 vaccine program context of our focus group participant perspectives reported later in the chapter.

Ontario

In Ontario, the COVID-19 vaccination program started in late December 2020. The province devised a three-phased plan, starting with those deemed at "greatest risk of severe illness and those who care for them" (Ontario, 2021b). The plan sought to prioritize the elderly, those at higher risk due to underlying health conditions, those who live in crowded conditions, those who live in "hot spots," defined as areas with higher rates of death, hospitalization, and transmission, and those who cannot work from home (ibid.). The province based its immunization plans on available scientific evidence that vaccinating the elderly and those in high-risk situations could prevent severe cases of the novel coronavirus and hospitalizations (Ontario, 2021a). Ontario also followed ethical frameworks and NACI's recommendations to ensure its vaccination plan was "equitable and fair" (ibid.).

Phase 1 began in late December 2020 and lasted until the end of March 2021. It targeted high-risk populations, including staff, essential caregivers, and residents in long-term care and retirement homes, First Nations elder care homes, healthcare workers identified as highest priority, followed by very high priority, based on the Ministry of Health's guidance, Indigenous[2] adults in northern remote and higher risk communities (on-reserve and urban), and adults ages 80 and older (Ontario, 2021b).

Phase 2 of the vaccination plan started in April, coinciding with the third wave of infection and amid strong criticism over the slow pace of the vaccination effort. This phase targeted, in 5-year decreasing intervals, adults aged 79 to 60, those in high-risk congregate settings (such as shelters, community living), individuals with high-risk chronic conditions and their caregivers, and those who cannot work from home (Ontario, 2021b). The third wave of COVID-19, however, quickly tested Ontario's healthcare system, as the number of new daily cases, new ICU admissions, and deaths climbed steadily. With pediatric hospitals making room for adult critical patients across the province, the vaccination plan was expanded ahead of schedule to include individuals over 50 years of age in "hot spots." Despite vaccine availability, many in the 55 and over age group rejected the AstraZeneca vaccine, thought to be due to media reports of clotting disorders (Laucius, 2021). Strong criticism and popular demand to offer the vaccine to younger populations led the government of Ontario to offer the vaccine to individuals over 40 years of age by April 19, months before populations under 60 were set to be vaccinated under phase 3 (July 2021 onward). The government was also criticized for offering the vaccine to the general population over 40 while younger educators and those with underlying health conditions remained unvaccinated (Taylor, 2021). The designation of "hot spots"[3] also generated controversy as some of those areas did not meet the requirements and some argued they had been included in the priority list due to political motivations (Payne, 2021).

Manitoba

Manitoba began vaccinating against COVID-19 in December 2020. The province announced that "every eligible person in Manitoba who wants to be immunized can get the COVID-19 vaccine"; however, "people who

[2] The Ontario plan only references Indigenous people generally.

[3] The Government of Ontario designated areas that have been disproportionately impacted by COVID-19 as "hot spots." These areas are created by postal code and include neighbourhoods and communities with historical and ongoing high rates of COVID-19 cases, deaths, and hospitalizations (Ontario, 2021b).

are most at risk will get the vaccine first" (Manitoba, 2021). The province announced that while it considered the guidelines provided by NACI regarding priority groups, it would adapt them to the particular context of the province and with the medical advice of local experts (ibid.).

The first phase of vaccination in Manitoba went from December 2020 to February 2021, and it was focused on individuals working in acute care facilities, COVID-19 immunization clinics and testing sites, COVID-19 isolation facilities, prisons, congregate living facilities, and personal care homes. Also included were paramedics, acute care workers, and home care workers born before 1975 (Manitoba, 2021). In March, vaccination was expanded to include other healthcare workers and essential workers, as well as Manitobans over the age of 80. This coverage progressively expanded and by April 21st Manitobans over the age of 50 and First Nations over the age of 30 were eligible to receive a COVID-19 vaccine.

Prioritizing Indigenous populations—defined as First Nation, Inuit, and Métis Citizens (Government of Canada, 2009)—was one of the recommendations outlined by NACI, and it has been adopted consistently across Canada, except for the province of Manitoba. In Manitoba, only First Nations were considered a priority group, while Inuit and Métis Citizens were considered "general population" Manitobans (Manitoba, 2021). This unequal treatment of Indigenous populations was heavily criticized by Métis and Inuit leaders (Robertson, 2021), particularly because Manitoba has the largest Métis population in the country and is considered the homeland of the Red River Métis Nation. The provincial government used the province's large Indigenous population as an argument to persuade the federal government to increase the province's vaccine quota (Robertson, 2020). The federal government agreed to increase the vaccine quota to account for First Nations on reserve, who are protected by federal law, but not for any First Nations, Inuit or Métis citizens living in cities. Nevertheless, the federal government asked provinces and territories to provide equitable access to those Indigenous Citizens in their jurisdiction (Robertson, 2021). Manitoba, however, did not amend at that time its priority groups to include Métis and Inuit populations (MacLean, 2021a).

The Manitoba Métis Federation (MMF) denounced the unequal treatment (Wong, 2021). In April 2021, the MMF demanded a public inquiry into the province's vaccination program and why the Métis People had not been prioritized (Stranger, 2021). To meet the needs of its Citizens, the MMF partnered with a private laboratory to conduct COVID-19 testing for Métis Citizens (Frew & Petz, 2021) and applied for access for vaccine delivery through its Métis-owned pharmacy in order to focus those doses on Métis Citizens (Frew, 2021). In addition to this, some First Nations Chiefs differentially supported Métis inclusion. Some called for the Métis Federation to demonstrate an evidence-based[4] need to be prioritized for vaccines (Sinclair, 2021), a position upheld by provincial medical leads (Stranger, 2021), whereas others donated surplus vaccine supplies that could not be used within their own reserve community (Frew, 2021) before vials expired. On May 3, 2021, Manitoba changed its prioritization with respect to Indigenous Peoples to include all Indigenous Peoples—First Nations, Inuit and Métis—ages 18 and up to receive a Pfizer or Moderna COVID-19 vaccine[5] in acknowledgment of the impacts of colonization on all Indigenous Peoples (MacLean, 2021b).

British Columbia

British Columbia also began vaccinating against COVID-19 in December 2020. The province's phased vaccination plan, including the creation of priority groups, followed NACI guidance as well as that of the B.C. Immunization Committee and the public health leadership committee (British Columbia, 2021). Phase 1 of the immunization plan was put into action between December 2020 and February 2021, focusing on residents and staff of long-term care facilities, individuals assessed for and awaiting long-term care, residents and staff of assisted living residences, essential visitors to long-term care facilities and assisted living residences, hospital healthcare workers who may provide care for COVID-19 patients in ICUs, emergency departments, paramedics, medical units, and surgical units. Remote and isolated Indigenous communities were also included in this first phase of vaccination (ibid.).

[4] The Manitoba government and First Nations leadership have a data sharing agreement in place to be able to identify and control how First Nations-specific COVID-19 case-based data will be collected and reported. One of the issues is that there is no such agreement in place with the Manitoba Métis Federation making it impossible to report Métis-specific COVID-19 case counts.

[5] The AstraZeneca age-based vaccine eligibility had not changed from adults 40 and older, or adults with certain underlying conditions between the ages of 30–39; although later in May 2021, Manitoba reduced its use of this vaccine substantially (Gray & Walsh, 2021).

In February 2021, B.C. opened public health immunization clinics to immunize senior citizens born in 1941 or earlier, as well as Indigenous Peoples (First Nations, Métis and Inuit) born in 1956 or earlier, Elders and additional Indigenous communities (British Columbia, 2021). Also included in this second phase were hospital staff, community general practitioners and medical specialists, vulnerable populations living and working in congregated settings, and staff in community homes.

Phase 3 began in April, initially opening up eligibility for the COVID-19 vaccine to British Columbians aged 69–79, Indigenous peoples aged 18 to 64, and people between 16 and 74 who are clinically extremely vulnerable. However, by mid-April, all British Columbians over the age of 40 were made eligible to be immunized (British Columbia, 2021).

Focus Groups

Our research team conducted 12 mixed-gender focus groups with 82 Canadians in Vancouver, Winnipeg, Ottawa, and Toronto between December 8–22, 2020. Participants were recruited using a market research firm (Prairie Research Associates, PRA) using a variety of methods (e.g., emails to individuals signed up as part of existing panels, random digit dialing, ads posted on Facebook). Participants were age-segregated into one of three groups (18–34 years, 35–54 years, 55+ years) for each city, where at least 2 people were recruited in smaller age groupings within each category. For example, in the 18–34 age group recruitment, the market research firm needed to identify individuals between 18–24, 25–30, and 31–34 to ensure a better cross-section of participants. Focus groups were fairly diverse; 49% of participants were White and 42% were People of Colour. In terms of income, 35% of participants had a total household income below $50,000 and 38% reported an income between $50,000 and below $100,000. Table 14.1 provides a detailed description of our sample population.

We used the videoconferencing platform Zoom to host online focus groups. One of two senior-level PRA staff moderated each group and at least one of the project leads attended every group, answering any participant questions at the end. Research team members could also unobtrusively send individual messages to the Group Moderator to explore particular aspects as the conversations occurred. Ethics approval was obtained from the University of Manitoba Health Research Ethics Board (H2020:510 linked to H2020:164) and the Research Ethics

Table 14.1 Socio-economic and demographic characteristics of participants, $N = 82$

Characteristic	*Count (%)*
Gender	
Male	40 (48.8)
Female	42 (51.2)
Ages	
18–24	10 (12.2)
25–30	11 (13.4)
31–34	8 (9.8)
35–40	9 (11.0)
41 to 48	10 (12.2)
49–54	8 (9.8)
55–60	6 (7.3)
61–68	11 (13.4)
69 or older	9 (11.0)
Marital Status	
Single (never married)	27 (32.9)
Married or common law	43 (52.4)
Divorced, separated or widowed	12 (14.6)
Households with Children	
None	57 (69.5)
1	10 (12.2)
2	10 (12.2)
3	5 (6.1)
Education	
High School	4 (4.9)
Some College/University	24 (29.3)
College/University Degree	54 (65.9)
Income	
Under $50,000	29 (35.4)
$50,000-$74,999	21 (25.6)
$75,000-$99,999	10 (12.2)
$100,000-$149,999	18 (22.0)
$150,000 or more	4 (4.9)
Race[6]	

(continued)

[6] Participants were asked to self-declare their ethnicity. To demonstrate the diversity of participants, individuals were grouped into broader racial categories of White, Black, Indigenous, and People of Colour. Participants grouped as White self-identified as White, Caucasian, Scottish, Irish, German, Danish, British, Ukrainian, Canadian, French-Canadian, Russian, and European/Caucasian. Participants grouped as Black indicated African-Canadian, North African. Participants grouped as Indigenous chose to self-identify as either Indigenous, First Nations, or Metis. The ethnicity for People of Colour included several self-identifications as South Asian, South East Asian, East Asian, Asian, Chinese, Filipino, Filipino-Canadian, Pakistan, East Indian, Indian, IndoCanadian, Latin American, Muslim, Middle Eastern, and Mixed (e.g., West Indian mixed black/Caucasian; Latino-Caucasian).

Table 14.1 (continued)

Characteristic	*Count (%)*
White	40 (48.8)
Black	2 (2.4)
Indigenous	6 (7.3)
People of Colour	34 (41.5)

Source Authors' own source

Board of Toronto Metropolitan University (formerly Ryerson University) (2020:445). Participants gave informed consent to participate in the focus groups and to have any data collected published, including comments made during the focus group discussions. Participants were given the option to choose a pseudonym to be used in publications. Focus groups were audio recorded and transcribed with voice attribution. Transcripts were analyzed using NVivo12, a qualitative research software package.

We developed an initial thematic guide for the focus group sessions in which the moderator asked participants' opinions regarding implementation of public health guidelines, compliance with infection prevention measures, information seeking behavior and trust, and attitudes toward immunization in general and then specifically the COVID-19 vaccines. In this chapter, we discuss participants' opinions on the latter theme, focusing on their views about immunization priority groups in Canada. As noted above, as part of the consent process, participants could indicate if they wanted their first name or a pseudonym used in reports and publications. Citations throughout the remainder of this section note in parentheses the city in which the focus group took place, along with the age group to which the participant belonged.

Participants' Priority Groups

In each focus group, participants were shown a list of population groups and asked to choose three top priority groups to access the COVID-19 vaccine. The focus group moderator then shared the results of the poll with the group and asked participants to explain their choices. The list of available population groups from which to choose included:

- Healthcare workers (e.g., doctors, nurses, healthcare aides, personal support workers, etc.).

- People with underlying medical conditions.
- Essential workers.
- People living in remote or isolated communities.
- Indigenous Peoples (First Nations, Inuit, Métis).
- Seniors (living independently).
- People living in long-term care facilities (e.g., nursing homes).
- Educators (teachers, ECE, etc.).
- Other.

Many priority group rankings aligned with how most jurisdictions were carrying out those early immunization plans: frontline health professionals who actively care for COVID-19 patients and people living and working in long-term care facilities. However, many participants felt that essential workers and people with serious underlying conditions should also be prioritized. When defining what they meant by essential workers, participants spoke in terms of people who have public-facing jobs, particularly those workers who were not in roles where there might be a certain assumed risk associated with that job. For example, someone working as a cashier or in the service sector would never have anticipated their job presented potentially greater risks to their safety prior to the arrival of COVID-19. These same participants equally favoured prioritizing essential workers like first responders and police, because of the public-facing nature of their work. The groups most frequently mentioned as the top three to access the vaccine were healthcare workers, essential workers, people living in long-term care facilities, and people with underlying conditions. Most participants justified their choice by arguing that by virtue of employment they are the most exposed and vulnerable to COVID-19:

Jay: When COVID cases started building up, the hospitals were getting over packed. [Health care workers] care for people that have COVID, but they care for patients that have surgeries, other infections and other diseases. [...] Because if they get COVID, the number of health care workers would be going down and there would be less help and other health workers would have to be pulled to work with COVID patients. So, health care workers should get the vaccine first (Vancouver, 18–34).

Ben: Healthcare aides or healthcare workers should be first. They would benefit the most from receiving the vaccine. They're the ones who are the most at risk (Ottawa, 18–34).

Two participants, however, noted the dual position of healthcare workers being both at risk of contracting COVID-19 and also posing a risk of spreading the infection into the community:

Riya: I think I have, oh gosh, six healthcare workers who live in my building of 16 units. And they go to work in a hospital, as well as a retirement facility [...] If they contract from somewhere else they're not going to only just [be] getting people in their retirement home sick [...] they'll bring it back to where they live (Ottawa, 18–34).

Adam: It is to protect them so that they do not pass it on to people that are vulnerable and people that they are working with who are at a high risk of suffering from COVID (Winnipeg, 35–54).

However, participants also noted that not all healthcare workers have the same importance during the pandemic, nor are they all facing the same level of risk. This is a nuance that is also reflected in the priority groups based on NACI recommendations. Participants identified some healthcare workers as more important to protect than others, such as doctors and nurses working in emergency departments, intensive care units, respiratory therapists, and healthcare aides. Participants also discussed that they would characterize Personal Support Workers (PSWs) working in long-term care facilities in this same category of patient-contact, even if the patient was not COVID-positive, because their work cannot really be done in a physically distant way. Therefore, participants prioritized those healthcare workers who are in closer contact with COVID-19 patients or populations at risk. For example:

Smir: I am saying emergency staff because they are the ones who just face the patient the first time. They don't know what they are dealing with. So they might be exposed. I would say that they should be given the vaccine as a priority (Ottawa, 35–54).

Raj: I would say the PSW because they're the ones that are going to be doing more the skin-to-skin contact. The people who come into my mom's house and take care of her (Ottawa, 35–54).

Steve: Healthcare workers like psychiatrists and optometrists that should not be on the priority list (Winnipeg, 55+).

Broadly speaking, the second priority group was often a tie between people with serious underlying medical conditions and essential workers. When discussing essential workers, participants distinguished between workers who were public-facing vs those who were not. Unlike health-care providers who assume a particular risk when entering the health profession, people who have been declared as essential workers (e.g., security/staff/cashiers in businesses selling essential goods like groceries, pharmacy products, hardware; transit workers, etc.), prior to COVID-19, never assumed such risks. Participants were also sensitized that many of these essential workers are minimum-waged staff who are not typically entitled to sick-leave benefits. Consequently, these essential workers not only carry a greater burden of physical risk, but they also face considerable financial risks if they need to take time from work to self-isolate and/or if they are infected with COVID-19.

Wendy: People with the most contacts should be given the vaccine first. A Manitoba hydro worker working on some line somewhere is not going to come in contact with a lot of people especially not in winter. A grocery store worker comes into contact with a lot of people. Some of those essential workers might not have sick leave provisions so they might still go to work even when they are not feeling well (Winnipeg, 55+).

Ali: First responders, ambulance drivers, police are people I would consider as essential workers. They need to go to work. Again, I am choosing my answers based on limiting the spread. To do their jobs effectively, they need to be protected as they protect the community (Winnipeg, 35–54).

Lael: I also feel like if you can prevent the spread and go from two different angles, those who are at risk and then also those who are the most in front of people then you have your best chance (Ottawa 35–54).

The category most likely to secure third place for priority groups was people living in long-term care facilities. However, people living in long-term care facilities were also some participants' first and second choice and the reasoning is well characterized by this conversational exchange in a Vancouver group:

Tammy: I put people in long-term care facilities because of their age and that they have been hit the hardest.

Murray: I feel the same way. I feel we have been dreadful of taking care of the long-term care facilities. A lot of people did not have COVID but died of starvation and everything else. COVID just brought a lot of those things to light. I think it would be wrong [to] overlook them again in terms of placing them at the back of the line (Vancouver, 55+).

Another participant, however, chose seniors living in long-term care facilities as a priority group because she was willing to sacrifice this group if the vaccines caused serious side-effects:

Rebecca: I picked people living in long-term care facilities. This is the cynic in me like I really want them to be protected, but if they're not and anybody has to go down at least they're not the ones that are out there taking care of everybody else that's out there working. You know? So, let's hope it really works (...) but if it doesn't, we're not going to devastate our population. Whereas, if we give it to all of our healthcare workers and it doesn't work, we're screwed (Ottawa, 35–54).

Some participants who did not choose the same categories as others in their 2nd and 3rd choices, outside of healthcare workers, had a different way of rationalizing their choices in terms of potential vulnerability to more serious effects from COVID-19 as in this exchange in the Toronto 18–34 group.

Calyx: My thinking was more about people's vulnerability and targeting the people who were the most vulnerable first. [So people in] long-term care facilities and second I [chose] underlying health conditions—the vulnerable was my thinking.

Kenneth: I chose essential workers because my criteria was the most potential of getting COVID or most exposure. So my thought was if people who are most exposed to it or had the potential to get exposed to it, get vaccines, maybe there's less transmission. And then after the first two I was kind of like, the criteria is more about the vulnerable, like who would have the most adverse effect of COVID.

Laura: I actually think I want to change my answer. I said people living in long-term care facilities but now that I'm looking at the list, I didn't really think too much about it, but I'm actually thinking now people, and not necessarily just Indigenous people, but people who are living on Indigenous reserves and in those communities have a really hard time getting access to healthcare. And if someone in one of those reserves or communities got sick it would probably spread very quickly and I think that would be in regards to any illness, and the further up north it's not easy to get to a hospital (Toronto, 18–34).

Contradictions and Cognitive Dissonance

Focus group participants expressed preferences over who should have priority access to COVID-19 vaccines based on who they perceived was facing the highest risk of exposure, serious outcomes, or death. While this reasoning coincides with NACI recommendations for priority groups, participants did not consider equitable access in the same way. For example, only a handful of participants referred to Indigenous Peoples and those living in remote areas, who have reduced access to health care and various resources, as priority groups. A participant, for example, referred to the logistical challenges of delivering the Pfizer vaccine to Indigenous Peoples living in remote communities as a reason to not prioritize this group. At the time of our focus groups, this vaccine was the only one approved in Canada and it needs to be stored at −70 degrees Celsius.

Steve: I personally don't think it's that feasible to get it to them yet. I think the manpower and money that needs to be spent to get them the vaccine rather than the people they know they can get it to right away outweighs it. The stuff has to be kept at minus 70 degrees. How can they do that when they're transporting it up north? You just can't so I think it would be a waste of resources (Winnipeg, 55+).

Those participants who considered Indigenous populations a high-risk group, however, emphasized poverty, inadequate housing and lack of health care as conditions putting Indigenous Peoples on and off reserves at greater risk of COVID-19. For example:

Margaret: I would say Indigenous people should be on the priority list. Indigenous people living in the north are living in poverty, substandard housing and have poor access to health care. They are extremely vulnerable and many have other health conditions as well. I believe those populations should be prioritized. For Indigenous people living in the city—we have a high population in Winnipeg—and unfortunately they also make a significant portion of our vulnerable population and might be homeless or living in poverty, so those also need to prioritized (Winnipeg, 55+).

Bryna: I chose [Indigenous Peoples] because they're a group that is more likely to acquire COVID and are very, very vulnerable. Many of them live in very cramped conditions. And I just think they deserve a break once in a while, they don't get many breaks (Ottawa, 55+).

However, some of our Indigenous participants in other groups, while acknowledging that they themselves would likely get the COVID-19 vaccine once available to them, highlighted that Indigenous Peoples more generally are hesitant about getting a COVID-19 vaccine as a priority group because of negative historical experiences. This was well described by one participant who self-identified as First Nations:

Leeann: I didn't choose them [First Nations, Inuit, Métis Peoples] because they're very hesitant and superstitious about getting the vaccination. I've spoken to members of the community. It's just based on history. They're very hesitant to take the

> vaccine if it's available. It's not that they won't, it's just that they're very hesitant. So instead, I picked people who are living in remote or isolated communities. It would be the same for Indigenous populations because a lot of our communities are remote. We have a whole First Nations that moved all their long-term care patients down to [a southern city] because they just don't have the medical facilities to deal with COVID. At the beginning of COVID a lot of the First Nations closed their doors to people coming in and out even if we were members of our communities, we could not go home. Because the First Nations only would allow people who resided in the First Nations 100% of the time into their communities. However, now we're seeing a mass outbreak in our First Nations communities which is devastating because they don't have the running water, they don't have the hospital or facilities, they don't have the medical personnel to deal with the outbreak that they're having now (Ottawa, 55+).

Contextualizing participant perspectives on prioritization for early access to COVID-19 vaccines is important. For example, despite clearly identifying priority groups as populations that should be protected and therefore should have early access to the vaccines, focus group participants also expressed doubts about the very vaccine they were willing to give to the most essential or most vulnerable. Many participants said that they themselves would not get vaccinated if eligible and instead they would rather wait for others to be immunized first to make sure COVID-19 vaccines are safe.

While a little over half of participants (56%) were strongly confident in decisions to receive a COVID-19 vaccine once it was their turn to be offered one, others had some open questions and wanted more information. Within this "wait and see" group, the time they felt comfortable in waiting ranged from 3 months to 1 or more years. However, their narratives suggested that several things were underlying their position: (1) feeling the vaccines were "rushed" and not tested enough, as well as some being genuinely unfamiliar with how agencies like Health Canada made assessments about vaccine safety and effectiveness prior to issuing authorizations; (2) wanting more safety data and general effectiveness data outside of clinical trial phases based on vaccine roll out to priority group recipients; (3) information about how effective COVID-19 vaccines were against emerging new variants; (4) more open and complete information about individual vaccines, particularly, as more vaccines would be approved over time; and (5) some generally felt their age or general health would allow them to weather any serious complications of contracting COVID-19 and therefore felt the uncertainty inherent in the risks of new vaccines was greater than the risk of the disease. Exploring participant perspectives of vaccine acceptance will be the subject of a future publication and will not be addressed in greater depth here.

Discussion

The COVID-19 pandemic has had a greater impact on racialized minorities in Canada (Public Health Agency of Canada, 2021), triggering calls to reduce health inequities, including vaccine access. Canada has shown a commitment to equitable distribution of vaccines through NACI guidelines, the way in which provinces developed priority groups, ensuring those facing a higher risk accessed the vaccine first, and also in the country's commitment to COVAX. Similarly, provincial and territorial governments sought to prioritize, first, those at higher risk of infection, followed by vaccine access to Indigenous Peoples and those who live in zones of higher transmission.

In our focus groups, however, participants' understanding of fairness and equity were variable and sometimes contradictory. Focus group participants identified some groups as facing higher risk of infection and death and therefore were willing to grant them priority access to a COVID-19 vaccine. The level of risk in the focus groups was determined mainly by age and contact with patients. However, other factors that cause health inequities were mostly disregarded, for example, poverty, racial discrimination, inadequate housing, or living in remote areas. In general, participants did not consider Indigenous Peoples a priority group.

With the initial production and approval of COVID-19 vaccines, developed nations around the world rushed to acquire as many doses as possible to protect their citizens, while developing countries, two years after the first vaccines were distributed, continue to face formidable challenges in their efforts to procure enough vaccine doses (World Health Organization, 2021). About 30% of the world population remains unvaccinated against COVID-19 (Holder, 2022). In this context, being eligible and having access to a COVID-19 vaccine is a privilege, particularly in Canada where the vaccines are funded through the public healthcare system. Many focus group participants, however, were not convinced about the safety of the vaccines despite the rigorous safety checks conducted by Health Canada prior to approval. These participants were unwilling to get vaccinated and adopted a “wait-and-see” approach, delaying their decision until more people had been vaccinated and had assumed the risk of potential side-effects of the vaccines. This vaccine hesitancy was not unique to our focus group participants but has been expressed by many Canadians, particularly those between the ages of 55 and 65, who rejected the AstraZeneca vaccine due to case reports of extremely rare clot disorders following vaccination (Cohen, 2021; Lofaro, 2021; Neustaeter, 2021).

Public engagement is a key aspect of health policymaking and risk communication (Hu & Qiu, 2020). Involving citizens in decision-making processes fosters trust and acceptance; however, there are some limitations to public engagement. First, engagement efforts tend to be self-selective, thus the same groups of people participate in these consultations, which amplifies their voices in detriment of others. Second, while fairness, equity, feasibility, and acceptability are all key components in health policymaking, in our focus groups we found that the general population rarely shares the same concerns and can often hold contradictory views simultaneously. In the focus groups, participants who were unwilling to be vaccinated due to vaccine safety concerns were eager to prioritize groups they considered especially vulnerable or as playing a crucial role in health care.

Conclusion

In times of vaccine scarcity, prioritization of vaccine access is a difficult task and the recommendations will not satisfy everyone. NACI released a set of recommendations based on application of its EEFA framework as well as expert and stakeholder engagement and public engagement. Nonetheless, it is entirely in the purview of provinces and territories to establish the implementation of immunization programs in their jurisdictions. The variability in the application of NACI recommendations was evident in the three provinces highlighted in this chapter. Assessing public preferences for prioritization through our focus groups equally highlighted the fundamental tensions that arise when engaging citizens who generally enjoy greater privilege in society (where privilege is defined as unearned benefits ascribed to an individual on the basis of their race, ability, gender, sexuality, and so forth).

Interestingly, the concept of "vulnerability" was the primary reasoning behind any prioritization recommendation regardless of source. Vulnerability was defined in terms of three factors: exposure, medical vulnerability, and disproportionate risk. Exposure vulnerability was based on a greater chance of exposure as a consequence of employment, which would capture healthcare workers who care for patients who have COVID-19 and essential workers (e.g., grocery workers, paramedics, etc.) who interact with the public at large, often for low pay and/or without sick-leave benefits. Medical vulnerability was defined as being at greater risk for more severe disease and death, which would capture people with serious underlying medical conditions and people of (often) advanced age living in long-term care facilities. Disproportionate risk vulnerability was defined as people living and working in conditions that put them at greater risk of infection with potentially disproportionate consequences, such as Indigenous Peoples. Yet, as Lemyre and colleagues (Lemyre et al., 2009) explain, there is a critical difference between vulnerability in terms of *susceptibility*, where people are more likely to become infected because of differential exposure, and *sensitivity,* where people are more likely to become seriously ill or die if they become infected.

NACI recommendations and stakeholder preferences focused first on sensitivity (underlying medical conditions, advanced age) before susceptibility (workers at greater risk of exposure due to environment). They also made efforts to address health inequities through the inclusion of Indigenous Peoples. Indigenous Peoples are both more susceptible to severe disease due to their living environments, often created through a shared colonial history of harm and trauma, as well as more sensitive due to a higher rate of multiple underlying chronic conditions relative to the general population. Subsequent waves of Canadian attitudes expressed through public opinion surveys, favoured susceptibility (through preferences for healthcare workers) over sensitivity (people of underlying conditions, advanced, age, etc.) (Impact Canada, 2021). The attitudes expressed in our public focus groups shared a similar pattern to that of Canadians overall. While acknowledged by NACI, considerations to address underlying inequities that are disproportionately borne by Indigenous Peoples and racialized groups, were rarely identified by focus group participants, and inconsistently reflected in the vaccine implementation plans of different provinces and territories. In fact, provinces and territories, unequally and selectively prioritized those who are at greater sensitivity (age, underlying medical conditions for those living in long-term care facilities, Indigenous populations) and focused mainly on susceptibility through risk of exposure (healthcare workforce).

There are several recommendations stemming from this work. First, it is important that independent bodies tasked with prioritization continue to go beyond systematic evaluations of evidence in terms of disease burden to include both expert/stakeholder groups and the public. This will aid in both informing and defining relevant values and perspectives. Second, when engaging broad audiences it may be important to discuss the relative weight each group should carry in informing final suggestions. The general public, as an entity, is challenged in its ability to represent all facets of diversity. Even with making efforts to ensure broader engagement, the capacity to engage those who are more marginalized in society is difficult, and almost impossible using typical public opinion methods (e.g., surveys and focus groups of reachable, willing and able participants). Without better ways to engage citizens living at the margins of society, public opinion tends to be fraught with contradiction and does not always consider all the key guiding principles of fairness and equity. Third, provinces and territories, as the jurisdictions ultimately tasked with making decisions after considering recommendations, need to be open and transparent in all their decisions for the inclusion/exclusion of different groups. In the three provinces focused on in this chapter, there was considerable variability in application of priority groups outside of general age considerations. With the explosion of new variant COVID-19 cases by May of 2021 some jurisdictions started prioritization of geographic "hot spots," which often included racialized populations. These decisions to increase vaccine access need to be accompanied by redirection of vaccine supply and other considerations, lest these moves appear to be more symbolic than genuine efforts to address inequities. As the SARS-CoV-2 virus continues to evolve and new variants emerge, new bivalent vaccines are being approved, targeting more than one strain (Canada, 2022). The initial implementation of the COVID-19 vaccine rollout provides important lessons that need to be documented and factored into plans for distribution of bivalent COVID-19 vaccines and for the next PHEIC event.

References

Alberta. (2021, January). *COVID-19 vaccine program*. https://www.alberta.ca/covid19-vaccine.aspx

British Columbia. (2021, April). *B.C.'s COVID-19 immunization plan*. Province of British Columbia. https://www2.gov.bc.ca/gov/content/covid-19/vaccine/plan

Canada (Ed.). (2003). *Learning from SARS: Renewal of public health in Canada—A report of the National Advisory Committee on SARS and Public Health*. National Advisory Committee on SARS and Public Health.

Canada, H. (2022, September 1). *Health Canada authorizes first bivalent COVID-19 booster for adults 18 years and older* [Statements]. https://www.canada.ca/en/health-canada/news/2022a/09/health-canada-authorizes-first-bivalent-covid-19-booster-for-adults-18-years-and-older.html

Chung, E., & Hogan, S. (2021, April 19). *How old do you have to be to get the AstraZeneca COVID-19 vaccine in each province?* CBC. https://www.cbc.ca/news/health/astrazeneca-vaccine-roundup-1.5993374

Cohen, B. (2021, March 12). People wrongfully denied AstraZeneca vaccine in Ontario rollout. *Toronto Star*. https://www.thestar.com/news/gta/2021/03/12/misinformation-people-being-turned-away-dog-ontarios-astrazeneca-vaccine-rollout.html

Dangerfield, K. (2021, April 13). *Canada reports 1st blood clot in AstraZeneca COVID-19 vaccine recipient*. Global News. https://globalnews.ca/news/7755376/astrazeneca-covid-19-vaccine-blood-clot/

Erickson, L. J., De Wals, P., & Farand, L. (2005). An analytical framework for immunization programs in Canada. *Vaccine*, *23*(19), 2470–2476. https://doi.org/10.1016/j.vaccine.2004.10.029

Frew, N. (2021, April 12). *Manitoba Metis Federation plans to give COVID-19 vaccines to 100 members by Saturday*. CBC. https://www.cbc.ca/news/canada/manitoba/manitoba-covid-19-vaccine-astrazeneca-mmf-1.5984533

Frew, N., & Petz, S. (2021, January 19). *Manitoba Metis Federation partners with local firm to create COVID-19 testing site for Métis people*. CBC. https://www.cbc.ca/news/canada/manitoba/manitoba-metis-federation-covid-19-testing-site-winnipeg-1.5878513

Government of Canada. (2009, January 12). *Indigenous peoples and communities* [Administrative page; fact sheet; resource list]. https://www.rcaanc-cirnac.gc.ca/eng/1100100013785/1529102490303

Gray, J., & Walsh, M. (2021, May 12). More provinces turn away from AstraZeneca COVID-19 vaccine. *The Globe and Mail*. https://www.theglobeandmail.com/canada/article-more-provinces-turn-away-from-astrazeneca-covid-19-vaccine/

Holder, J. (2022, August 15). Tracking coronavirus vaccinations around the world. *The New York Times*. https://www.nytimes.com/interactive/2021/world/covid-vaccinations-tracker.html

Hu, G., & Qiu, W. (2020). From guidance to practice: Promoting risk communication and community engagement for prevention and control of coronavirus disease (COVID-19) outbreak in China. *Journal of Evidence-Based Medicine*, *13*(2), 168–172. https://doi.org/10.1111/jebm.12387

Impact Canada. (2021, April 26). *COVID-19 Snapshot monitoring*. Impact Canada. https://impact.canada.ca/en/challenges/cosmo-canada/results

IPAC Canada. (2022, August 26). *Coronavirus (COVID-19)*. https://ipac-canada.org/coronavirus-resources

Ismail, S. J., Hardy, K., Tunis, M. C., Young, K., Sicard, N., & Quach, C. (2020). A framework for the systematic consideration of ethics, equity, feasibility, and acceptability in vaccine program recommendations. *Vaccine*, *38*(36), 5861–5876. https://doi.org/10.1016/j.vaccine.2020.05.051

Laucius, J. (2021, April 16). "It's demoralizing": Vaccine shoppers are declining AstraZeneca. *Ottawa Citizen*. https://ottawacitizen.com/news/local-news/its-demoralizing-vaccine-shoppers-are-declining-astrazeneca

Lemyre, L., Gibson, S., Zlepnig, J., Meyer-Macleod, R., & Boutette, P. (2009). Emergency preparedness for higher risk populations: Psychosocial considerations. *Radiation Protection Dosimetry*, *134*(3–4), 207–214. https://doi.org/10.1093/rpd/ncp084

Lofaro, J. (2021, March 15). *Some Montrealers refusing AstraZeneca vaccine; Canadian authorities assert it's safe*. CTV News. https://montreal.ctvnews.ca/some-montrealers-refusing-astrazeneca-vaccine-canadian-authorities-assert-it-s-safe-1.5347326

MacLean, C. (2021a, February 24). *Métis, Inuit in Manitoba want similar COVID-19 vaccine priority as First Nations | CBC News*. CBC. https://www.cbc.ca/news/canada/manitoba/indigenous-manitobans-vaccine-priority-1.5926939

MacLean, C. (2021b, May 3). *All Indigenous people in Manitoba now eligible for vaccine | CBC News*. CBC. https://www.cbc.ca/news/canada/manitoba/manitoba-covid19-vaccine-eligibility-1.6011557

Manitoba. (2021, February 25). *#ProtectMB*. https://protectmb.ca/eligible-groups/

Mosby, I., & Swidrovich, J. (2021). Medical experimentation and the roots of COVID-19 vaccine hesitancy among Indigenous Peoples in Canada. *CMAJ*, *193*(11), E381–E383. https://doi.org/10.1503/cmaj.210112

National Advisory Committee on Immunization. (2020, November 3). *COVID-19: Preliminary guidance on key populations for early immunization*.

Canada.Ca. https://www.canada.ca/en/public-health/services/immunization/national-advisory-committee-on-immunization-naci/guidance-key-populations-early-covid-19-immunization.html#a2

Neustaeter, B. (2021, April 15). *Ont. Doctor admits 3 COVID-19 patients to ICU after they turned down vaccines*. CTV News. https://www.ctvnews.ca/health/coronavirus/ont-doctor-admits-3-covid-19-patients-to-icu-after-they-turned-down-vaccines-1.5389713

Ontario. (2021a, January 11). *Ethical framework for COVID-19 vaccine distribution*. Ontario.Ca. https://www.ontario.ca/page/ethical-framework-covid-19-vaccine-distribution

Ontario. (2021b, April 19). *Ontario's COVID-19 vaccination plan*. COVID-19 (Coronavirus) in Ontario. https://covid-19.ontario.ca/ontarios-covid-19-vaccination-plan

Payne, E. (2021, April 15). COVID-19 science advisory table says it did not select hot spot postal codes. *Ottawa Citizen*. https://ottawacitizen.com/news/local-news/covid-19-science-advisory-table-says-it-did-not-select-hot-spot-postal-codes

Public Health Agency of Canada. (2021, February 21). *The impact of COVID-19 on racialized communities* [Statements]. CPHO Sunday Edition. https://www.canada.ca/en/public-health/news/2021/02/cpho-sunday-edition-the-impact-of-covid-19-on-racialized-communities.html

Rabson, M. (2021, April 14). *AstraZeneca very safe, effective, with extremely rare risk of clots: Health Canada*. CTV News. https://www.ctvnews.ca/health/coronavirus/astrazeneca-very-safe-effective-with-extremely-rare-risk-of-clots-health-canada-1.5387083

Reuters. (2021, April 16). *Australia reports first blood clot death "likely" linked to AstraZeneca vaccine*. Reuters. https://www.reuters.com/business/healthcare-pharmaceuticals/australia-reports-first-death-blood-clots-likely-linked-astrazeneca-vaccine-2021-04-16/

Robertson, D. (2020, December 12). Ottawa tabs extra vaccine doses for on-reserve First Nations use. Winnipeg Free Press. https://www.winnipegfreepress.com/special/coronavirus/ottawa-tabs-extra-vaccine-doses-for-on-reserve-first-nations-use-573376751.html

Robertson, D. (2021, March 9). *MMF, MKO air frustrations on remote priority plan*. Winnipeg Free Press. https://www.winnipegfreepress.com/special/coronavirus/mmf-mko-air-frustrations-on-remote-priority-plan-573951072.html

Saskatchewan. (2021, April). *COVID-19 Vaccine delivery phases*. Government of Saskatchewan. https://www.saskatchewan.ca/government/health-care-administration-and-provider-resources/treatment-procedures-and-guidelines/emerging-public-health-issues/2019-novel-coronavirus/covid-19-vaccine/vaccine-delivery-phases

Sinclair, N. (2021, February 27). *Include Métis, Inuit in indigenous vaccination plans*. Winnipeg Free Press, A4.

Stranger, D. (2021, April 1). *Manitoba Métis call for public inquiry into vaccine rollout campaign*. APTN News. https://www.aptnnews.ca/national-news/manitoba-metis-call-for-public-inquiry-for-not-being-prioritized-in-covid-19-vaccine-campaign/

Taylor, B. (2021, April 7). *Ont. Teachers push for COVID-19 vaccines over spring break*. CTV News. https://www.ctvnews.ca/health/coronavirus/ont-teachers-push-for-covid-19-vaccines-over-spring-break-1.5377979

The Lancet Microbe. (2021). COVID-19 vaccines: The pandemic will not end overnight. *The Lancet Microbe*, *2*(1), e1. https://doi.org/10.1016/S2666-5247(20)30226-3

Wong, N. (2021, January 16). Manitoba Metis Federation slams provincial government over lack of input into COVID vaccination plan. *Toronto Star*. https://www.thestar.com/news/canada/2021/01/16/manitoba-metis-federation-slams-provincial-government-over-lack-of-input-into-covid-vaccination-plan.html

World Health Organization. (2005). *Annex 2 of the International Health Regulations*. WHO; World Health Organization. http://www.who.int/ihr/annex_2/en/

World Health Organization. (2021, April 20). *COVAX*. COVAX: Working for global equitable access to COVID-19 vaccines. https://www.who.int/initiatives/act-accelerator/covax

World Health Organization. (2022, September 2). *WHO Coronavirus (COVID-19) dashboard*. https://covid19.who.int

Zhao, L., Ismail, S. J., & Tunis, M. C. (2020). Ranking the relative importance of COVID-19 immunisation strategies: A survey of expert stakeholders in Canada. *MedRxiv*, 2020.09.16.20196295. https://doi.org/10.1101/2020.09.16.20196295

Index

M. Gattinger (ed.), *Democratizing Risk Governance*,
https://doi.org/10.1007/978-3-031-24271-7

V

W

The manufacturer's authorised representative in the EU is Springer Nature Customer Service Centre GmbH, Europaplatz 3, 69115 Heidelberg, Germany. If you have any concerns regarding our products, please contact ProductSafety@springernature.com

Printed and bound by CPI Group (UK) Ltd, Croydon, CR0 4YY
28/09/2026
02224060-0007